Understanding Complex Systems

For further volumes:
http://www.springer.com/series/5394

Understanding Complex Systems

Future scientific and technological developments in many fields will necessarily depend upon coming to grips with complex systems. Such systems are complex in both their composition—typically many different kinds of components interacting simultaneously and nonlinearly with each other and their environments on multiple levels—and in the rich diversity of behavior of which they are capable.

The Springer Series in Understanding Complex Systems series (UCS) promotes new strategies and paradigms for understanding and realizing applications of complex systems research in a wide variety of fields and endeavors. UCS is explicitly transdisciplinary. It has three main goals: First, to elaborate the concepts, methods, and tools of complex systems at all levels of description and in all scientific fields, especially newly emerging areas within the life, social, behavioral, economic, neuro, and cognitive sciences (and derivatives thereof); second, to encourage novel applications of these ideas in various fields of engineering and computation such as robotics, nano-technology, and informatics; third, to provide a single forum within which commonalities and differences in the workings of complex systems may be discerned, hence leading to deeper insight and understanding.

UCS will publish monographs, lecture notes, and selected edited contributions aimed at communicating new findings to a large multidisciplinary audience.

Springer Complexity

Springer Complexity is an interdisciplinary program publishing the best research and academic-level teaching on both fundamental and applied aspects of complex systems—cutting across all traditional disciplines of the natural and life sciences, engineering, economics, medicine, neuroscience, social, and computer science.

Complex Systems are systems that comprise many interacting parts with the ability to generate a new quality of macroscopic collective behavior the manifestations of which are the spontaneous formation of distinctive temporal, spatial, or functional structures. Models of such systems can be successfully mapped onto quite diverse "real-life" situations like the climate, the coherent emission of light from lasers, chemical reaction-diffusion systems, biological cellular networks, the dynamics of stock markets and of the internet, earthquake statistics and prediction, freeway traffic, the human brain, or the formation of opinions in social systems, to name just some of the popular applications.

Although their scope and methodologies overlap somewhat, one can distinguish the following main concepts and tools: self-organization, nonlinear dynamics, synergetics, turbulence, dynamical systems, catastrophes, instabilities, stochastic processes, chaos, graphs and networks, cellular automata, adaptive systems, genetic algorithms, and computational intelligence.

The two major book publication platforms of the Springer Complexity program are the monograph series "Understanding Complex Systems" focusing on the various applications of complexity, and the "Springer Series in Synergetics", which is devoted to the quantitative theoretical and methodological foundations. In addition to the books in these two core series, the program also incorporates individual titles ranging from textbooks to major reference works.

Hassan Qudrat-Ullah
Editor

Energy Policy Modeling in the 21st Century

Springer

Editor
Hassan Qudrat-Ullah
School of Administrative Studies
York University
Toronto
Canada

ISSN 1860-0832 ISSN 1860-0840 (electronic)
ISBN 978-1-4614-8605-3 ISBN 978-1-4614-8606-0 (eBook)
DOI 10.1007/978-1-4614-8606-0
Springer New York Heidelberg Dordrecht London

Library of Congress Control Number: 2013950223

Printed on acid-free paper

Springer is part of Springer Science+Business Media (www.springer.com)

This book is dedicated to all the prophets, from *Adam* (*peace be upon him*) to *Muhammad* (*peace be upon him*), and to all those human beings who did and are doing their best in bringing humanity from the deep darkness to the beckoning path of and for life in this world.

Preface and Acknowledgements

Decision making in energy systems is a complex task at best. Complexity stems primarily from the nonlinear nature of energy–economy–environment interactions. Given the global trend of focusing on low carbon economies, energy policy design and assessment has become a dynamic problem. Uncertainties such as the dynamics of energy supply and demand, the price of fossil fuels, various regulatory regimes, the advances and challenges of energy production and consumption-related technologies, and impact of energy related emissions abound. Therefore, the use of model-based analysis and scenarios in energy policy design and assessment has seen phenomenal growth during the past several decades.

The primary aim of this book is to disseminate the roles and applications of various modeling approaches aimed at improving the usefulness of energy policy models in public decision making. The key focus is on the development, validation, and applications of system dynamics and agent-based models in service of energy policy design and assessment in the twenty-first century. Invitations were sent all around the globe. Several renowned authors were also specially invited to contribute. Each prospective contributor was initially asked to prepare a two to three page proposal. These proposals were reviewed by the editor and suggestions were made to prepare the full papers. The submitted papers were then reviewed by independent reviewer panels. Each panel consisted of three members—the editor and two independent experts in the field. The final acceptance/rejection decisions were made by the editor based on the revised papers submitted by the contributors.

The book contains three parts. Part I, "Energy Policy Modeling for the 21st Century: An Introduction" has one chapter. It introduces key aspects of major modeling approaches and presents an overview of all the chapters of this book. Part II of the book, "Modeling Approaches and Energy Policy Decisions", consists of six chapters and deals with the range of tools, methods, and technologies that support decision making in complex, dynamic energy systems including Thinking about the Future: System Dynamics and the Process of Electricity Deregulation, Fuzzy System Dynamics: A Framework for Modeling Renewable Energy Policies, The Diffusion of Eco-Technologies: A Model-Based Theory, Managing the Energy Basket in the Face of Limits, Power Plant Relocation Policy versus Investments in Transmission Network Infrastructure: a Study on the Italian Energy Market, and Simulation of Greenhouse Gas Cap-and-Trade Systems with ENERGY 2020. Part III of the book, "System Dynamics and Agent-Based Models in Actions", has six chapters and provides the empirical evidence to the application of both system dynamics and agent-based modeling approaches to energy policy design and assessment issues including Energy Policy Planning for Climate Resilient Low-Carbon Development, Understanding the Dynamics of Electricity Supply and Demand in Ontario, Adoption of Renewable Energy Technologies: A Fuzzy System Dynamics Perspective, Resurrecting a Forgotten Model: Updating Mashayekhi's Model of Iranian Economic Development, Making Progress Towards Emissions Mitigation: Modeling Low-carbon Power Generation Policy, and Exploring Energy and Economic Futures using Agent-based Modeling and Scenario Discovery.

We are grateful to the authors of the various chapters for their contributions. It had been a bit long process from the initial outlines to developing the full chapters and then revising them in the light of reviewers' comments. We sincerely acknowledge the authors' willingness to go through this process. We also acknowledge the work and knowledge of the members of our review panels, many of which had to be done at short notice.

Thanks to all the people at Springer, USA especially Christopher, HoYing, and Brian with whom we corresponded for their advice and facilitation in the production of this book.

Mrs. Puja Kumari, Crest Premedia Solutions (P) Ltd. prepared a camera-ready copy of the manuscript with her usual professionalism and cooperation and we wish to record our thanks to her.

Finally, I am grateful to my family, Tahira (for her incredible and selfless support all the time), Anam (for her professional proofreading and spiritual perspective on things around us), Ali (for sparing time from his kingdom), Umer (for his occasional smiles on my work and his work too), Umael (for bearing with me on Writing Assignments, and rollerblades stuff), and my father, Safdar Khan and my mother, Fazeelat Begum (for their sacrifice, support, and prayers all along)—all the very source of my inspiration and desire to embark on this journey. It would be unfair not to acknowledge the constant and consistent prayers and cares she extends to me, Saira Bano, my mother-in-law.

Toronto, Canada Hassan Qudrat-Ullah
July, 2013

Contents

Contributors

Jeffrey S. Amlin Xenia, OH, USA

Santiago Arango Faculty of Mines, Universidad Nacional de Colombia, Medellin, Colombia

Andrea M. Bassi Founder and CEO, KnowlEdge Srl (KE) and Director, Centre for Systemic Planning (CSP) Castellanza (VA) Italy

Laura M. Cardenas Facultad de Minas, Universidad Nacional de Colombia, Medellin, Colombia

Silvano Cincotti DIME-CINEF, University of Genoa, Genova, Italy

Pål Davidsen System Dynamics Group, University of Bergen, Bergen, Norway

Prakash (Sanju) Deenapanray Director, Ecological Living in Action (ELIA) and Director, Centre for Systemic Planning (CSP) La Gaulette, Mauritius

Isaac Dyner CIIEN, CEIBA, Facultad de Minas, Universidad Nacional de Colombia, sede Medellín, Colombia

Carlos J. Franco CIIEN, CEIBA, Facultad de Minas, Universidad Nacional de Colombia, Medellin, Colombia

Giulia Gallo DIME-CINEF, University of Genoa, Genova, Italy

M. D. Gerst Dartmouth College, Thayer School of Engineering, Hanover, NH, USA

Ruth Kaufmann-Hayoz Interdisciplinary Centre for General Ecology (IKAÖ), University of Bern, Bern, Switzerland

Saeed P. Langarudi Department of Social Science & Policy Studies, Worcester Polytechnic Institute, Worcester MA, USA

Erik R. Larsen Institute of Management, University of Lugano, Lugano, Switzerland

Matthias Otto Müller Greenwood Strategic Advisors AG, Unterägeri, Switzerland

Charles Mbohwa School of Mechanical and Industrial Engineering, University of Johannesburg, Bunting Road Campus, Johannesburg, South Africa

Michael Mutingi School of Mechanical and Industrial Engineering, University of Johannesburg, Bunting Road Campus, Johannesburg, South Africa

Hassan Qudrat-Ullah School of Administrative Studies, York University, Toronto, ON, Canada

Michael J. Radzicki Department of Social Science & Policy Studies, Worcester Polytechnic Institute, Worcester MA, USA

Khalid Saeed Social Science and Policy Studies Department, Worcester Polytechnic Institute, Worcester, MA, USA

Markus Schwaninger Institute of Management (IfB), University of St. Gallen, St. Gallen, Switzerland

Silvia Ulli-Beer Dynamics of Innovative Systems (DIS), Paul Scherrer Institut, Villigen, Switzerland

P. Wang Dartmouth College, Thayer School of Engineering, Hanover, NH, USA

Editorial Panel Members

Part I
Energy Policy Modeling in Twenty-First Century: An Introduction

Energy Policy Modeling in the 21st Century: An Introduction

Hassan Qudrat-Ullah

Introduction

Decision making in energy systems is a complex task at best. Complexity stems primarily from the nonlinear nature of energy–economy–environment interactions. Given the global trend of focusing on low-carbon economies, energy policy design and assessment has become a dynamic problem. Uncertainties such as dynamics of energy supply and demand, price of fossil fuels, various regulatory regimes, advances and challenges of energy production and consumption, related technologies, and the impact of energy-related emissions abound. Therefore, the use of model-based analysis and scenarios in energy policy design and assessment has seen phenomenal growth during the past several decades.

For the most part, large-scale optimization and econometric methods including the input–output analysis approach and computable general equilibrium (CGE) approaches have been used to serve the planning and strategic decision-making needs of energy policy decision makers (Bun and Larsen 1992; Qudrat-Ullah and Karakul 2007). However, given the emerging twenty-first century with concerns about climate change and energy security that pose unique modeling challenges, a critical look at these traditional energy modeling techniques reveals a number of methodological limitations, shortcomings, and constraints including (Qudrat-Ullah and Karakul 2007):

- Intertemporal interactions and feedback among energy–economy–environment related variables are not modeled explicitly in and across various sectors.
- The framework for transient behavior (e.g., the necessary adjustments for new energy production technologies in response to resource price dynamics typically create transient behavior) is missing.

H. Qudrat-Ullah (✉)
School of Administrative Studies, York University, 4700 Keele Street, M3J 1P3, Toronto, ON, Canada
e-mail: hassanq@yorku.ca

H. Qudrat-Ullah (ed.), *Energy Policy Modeling in the 21st Century*, Understanding Complex Systems, DOI 10.1007/978-1-4614-8606-0_1,

- The variability in the elasticity of substitution among competing energy production technologies is rarely reflected in these models.
- Time delays and other distortions in perceiving the true value of the energy–economy–environment related variables are not modeled explicitly.
- Desired and actual variable magnitudes (e.g., energy production capital and power plant operational efficiency) are rarely distinguished in the model.
- Nonlinear responses to actions (e.g., rate of return on capital investments) are not explicitly represented.
- Soft variables (e.g., operator morale, a key determinant of successful operations of power plants) are indexed to some probability numbers.
- Often demand and GDP are assumed exogenous (Bun and Larsen 1992).

In order to provide useful energy policy design and assessment, therefore, the required modeling approach needs to overcome the above-mentioned limitations. Researchers, especially from system dynamics and agent-based modeling approaches, have found these models capable of addressing these unique challenges. For instance, system dynamics models have been successfully used to study national energy policy evaluation (Qudrat-Ullah and BaekSeo 2010; Bun and Larsen 1992; Davidsen et al. 1990), energy investments and uncertainty (Bun and Larsen 1992), conservation policy analysis (Ford and Bull 1989), interfuel substitution (Moxnes 1990), privatization of the electricity industry (Qudrat-Ullah and Davidsen 2001; Bun and Larsen 1992; Assili et al. 2008), energy efficiency and electricity substitution (Assili et al. 2008; Adelino and João 2011), energy consumption analysis (Ansari and Seifi 2012), and electricity-related emission assessments (Anand et al. 2005).

Thus, in an attempt to provide some viable solutions for energy policy modeling in the twenty-first century, we issued the call for contributions to this volume. Specifically, we sought help from the system dynamics and agent-based modeling community. To provide credibility and enhance the appeal of these models, we also sought special emphasis on the validation of models used in each of these contributions. Consequently, several different examples of modeling approaches and models are provided in this volume.

Methodology

In our call for contributions to *Energy Policy Modeling in the 21st Century* we went through various email lists of professional bodies. We also posted the call for chapters on the message boards of a few international conferences on related topics. Personal invitations were sent to specific authors as well. We received 23 "one-page" abstracts as the expressions of interest. Based on the initial screening by our review panel, the authors of 17 chapters were invited to submit the complete chapter. All 17 chapters received from the contributors went through a double-blind process. The reports from the independent reviewers were sent to the authors to address the issues and incorporate the suggestions made by the reviewers. Only 12 chapters made it to the final stage of acceptance. The final versions of the chapters have been edited and included in this volume.

Research Categories

The chapters thus compiled are classified into three categories following the structure of the book. The first category, the current one, presents the introduction and preview of *Energy Policy Modeling in the 21st Century*. The second category examines the modeling approaches and energy policy decision making including Thinking About the Future: System Dynamics and the Process of Electricity Deregulation, Fuzzy System Dynamics: A Framework for Modeling Renewable Energy Policies, The Diffusion of Eco-Technologies: A Model-Based Theory, Managing the Energy Basket in the Face of Limits, Power Plant Relocation Policy Versus Investments in Transmission Network Infrastructure: A Study of the Italian Energy Market, and Simulation of Greenhouse Gas Cap-and-Trade Systems with ENERGY 2020.

Finally, the third category provides the empirical evidence for the application of both system dynamics and agent-based modeling approaches to energy policy designs and assessment issues including Energy Policy Planning for Climate-Resilient Low-Carbon Development, Understanding the Dynamics of Electricity Supply and Demand in Ontario, Adoption of Renewable Energy Technologies: A Fuzzy System Dynamics Perspective, Resurrecting a Forgotten Model: Updating Mashayekhi's Model of Iranian Economic Development, Making Progress Towards Emissions Mitigation: Modeling Low-Carbon Power Generation Policy, and Exploring Energy and Economic Futures Using Agent-Based Modeling and Scenario Discovery.

Modeling Approaches and Energy Policy Decisions

Energy Policy Decisions in Deregulated Markets

When it comes to energy policy decision making, deregulated energy markets have taken center stage. In Chapter 2, "Thinking about the Future: System Dynamics and the Process of Electricity Deregulation," by Erik Larsen and Santiago Arango, the risks and challenges faced by the companies operating in the deregulated energy markets are discussed. They mention two main challenges: internal (due to major internal changes in the organization), and external (demand-side issues and financial resource availability due to outside competition). They also identify three areas of risk that a deregulated company faces: regulatory, market, and organizational risks. They consider "time lags and feedback" among the variables of a deregulated energy system regime that constitute decision making as a difficult task. In search of a better solution to this issue, they posit that system dynamics, as a modeling approach, is well suited to help managers make sense of their new industry.

Making Renewable Energy Policy Decisions

The search of "a suitable modeling approach" as a theme continues in Chapter 3 with "Fuzzy System Dynamics: A Framework for Modeling Renewable Energy Policies" by Michael Mutingi and Charles Mbohwa, but, for contrast, quite a different modeling framework is described here. Here the use of fuzzy system dynamics rather than system dynamics provides the underlying modeling approach. The context also changes to "renewable energy policy formulation and evaluation." With the complex dynamics prevalent in energy systems, neither the objective to achieve a sustainable low-carbon energy economy nor the development of a robust long-term renewable energy policy is a simple task. Instead, to meet future energy demand while keeping CO_2 emissions at a sustainable level, effective renewable energy policies have to be put into place, the authors assert. In this chapter, they present a framework for evaluating renewable energy policies based on a fuzzy system dynamics paradigm. First, they describe the renewable energy policy problem, with a case study example. Second, they present a framework for fuzzy system dynamics modeling. Third, they propose a high-level causal loop analysis to capture the complex dynamic interactions among various energy demand and supply factors. Finally, they propose a fuzzy system dynamics model for renewable energy policy modeling and evaluation.

Decision Making in the Context of Eco-Technologies

In Chapter 4, "The Diffusion of Eco-Technologies: A Model-Based Theory," by Matthias Otto Müller, Ruth Kaufmann-Hayoz, Markus Schwaninger, and Silvia Ulli-Beer, we have a further modeling variation applied to the issue of the diffusion of eco-technologies. They propose a generic theory of the diffusion of eco-technologies that integrates properties of the market, technology, policy change, and public policy interventions, equally applicable to energy policy decisions. Methodologically, they rely on system dynamics modeling and simulation to arrive at a dynamic, causally explicit, and endogenous explanation of the key feedback loops driving (or inhibiting) such diffusion processes. In addition to a description of their theory, they provide an extensive discussion of how the system dynamics methodology can be used to conduct research and support policy making in the context of eco-technologies.

Integration of Policy Options Pertaining to Energy and Environment

Closely related to the assertive conclusion of Chapter 4 is the modeling solution presented in Chapter 5, "Managing the Energy Basket in the Face of Limits," by Khalid Saeed. Given the intertwined nature of energy use and its environmental repercussions, he addresses the challenge of integrating policy options pertaining to energy and the environment. He builds on a simple model suggested in Saeed (1985), which deals not with the human activity modeled in the Limits project, but with the ecosystem affected by human activity. Thus, it incorporates the policy space needed for managing the ecosystem rather than the demand for resources, which helps to delineate the operational means to avoid the impending catastrophe predicted in the Limits study, he asserts. Then he also explores the operational means for managing the environmental impact of energy use, the principles of which are outlined in Saeed (2004). These principles call for integrating environmental restoration into the market activity, he claims. Then he deals with the policy issues pertaining to energy use and environmental restoration in separate models following the problem-partitioning principles outlined in Saeed (1992).

Infrastructural Decision Making in Energy Markets

The development and application of optimization modeling continue as a methodological theme in Chapter 6 with a presentation of "Power Plant Relocation Policy Versus Investments in Transmission Network Infrastructure: A Study of the Italian Energy Market," by Silvano Cincotti and Giulia Gallo. First they present two important assertions: (i) in a zonal pricing mechanism network, congestion arises when the transmission network is not able to serve zones with the necessary electricity, contributing to higher zonal prices, and (ii) investing in transmission network infrastructure and establishing a uniform price is considered an efficient solution, especially in Europe. Then they analyze the Italian Power Exchange (IPEX) and propose a framework for evaluating the effects of policy mechanisms. In particular, they discuss a comparison between investments in transmission networks and generation capacity relocation policy under a zonal pricing mechanism. Their results point out that a proper localization of a reduced set of power plants is able to increase consumers' social welfare by taking advantage of the zonal splitting mechanism and its transmission capacity constraints.

Energy Policy in the Context of Climate Change

The final chapter of this section, Chapter 6, "Simulation of Greenhouse Gas Cap-and-Trade Systems with ENERGY 2020," by Jeffrey Amlin, presents greenhouse gas cap-and-trade systems as a partial solution to climate change due to greenhouse gas emissions. His simulation model, ENERGY 2020, has been used to simulate cap-and-trade systems for 15 years. He provides an overview of the ENERGY 2020 simulation model, describes using ENERGY 2020 to simulate various greenhouse gas cap-and-trade systems, and reflects on the lessons learned in the modeling process. ENERGY 2020 is an integrated, multiregion, energy model that has been actively used by state, provincial, and national governments as well as private energy companies since the early 1980s to conduct energy- and emission-related policy analysis and forecasting, he claims. The utility of his model, ENERGY 2020, is demonstrated by its application in 1998 to analyze several different cap-and-trade systems in the United States and Canada.

System Dynamics and Agent-Based Modeling in Actions

In this section we present six state-of-the-art applications of systems dynamics and agent-based simulation models. The focus of these modeling and simulation applications is to design and assess energy policies with the objective of achieving low-carbon economies.

Energy Policy Planning for Climate-Resilient Low-Carbon Development

In the background of the use of new market mechanisms (NMMs) for achieving global reductions in GHG emissions which was adopted at COP16 in Cancun (2010), and further referenced at COP17 in Durban (2011), Andrea Bassi, Prakash (Sanju) Deenapanray, and Pål Davidsen in Chapter 7, "Energy Policy Planning for Climate-Resilient Low-Carbon Development," present a possible solution model, based on system dynamics. They demonstrate the practical use of system dynamics modeling (SDM) for policy planning to achieve climate-resilient, low-carbon development pathways in the context of national development planning. In particular, they use examples from developing (Mauritius and Kenya) and developed (United States) countries to make the case for the use of SDM for climate-proofing the energy sector and to develop nationally appropriate mitigation actions (NAMAs) as one type of NMM.

In all cases, they discuss the impact of policy interventions on selected indicators disaggregated into three categories: for agenda setting (problem identification), policy formulation, and policy evaluation. In this respect, they show that SDM can be an effective tool to study cross-sectorial impacts of energy policies, in the context of green economy strategies, for instance (encompassing climate mitigation and adaptation as well).

Dynamics of Electricity Supply and Demand: The Case of Canada

The electricity supply and demand system of Canada is a highly dynamic one, consisting of a nearly immeasurable amount of variables. To better understand the dynamics of the electricity supply and demand system, Hassan Qudrat-Ullah presents in Chapter 8, "Understanding the Dynamics of Electricity Supply and Demand: The Case of Canada," a dynamic simulation model as a decision-making aid. Again drawing on system dynamics methodology, he describes the development, validation, and application of a system dynamics model. Founded on a model-based scenario analysis, he finds that, in addition to the traditional investments to account for the retired capacity, substantial new investments in electricity generation capacity, and in the system's efficiency enhancement mechanisms are needed to achieve a sustainable and balanced supply-and-demand system in Canada.

Adoption of Renewable Energy Technologies

In Chapter 9, "Adoption of Renewable Energy Technologies: A Fuzzy System Dynamics Perspective," Michael Mutingi presents the application of a fuzzy system model as a decisional aid for the adoption of renewable energy technologies (RETs). The adoption of RETs has been facing a number of barriers and constraints due to the dynamic interaction between potential technology adopters, adapters, imitators, inhibitors, and the technology policies in place, he asserts. However, the major challenge in modeling RET adoption is the existence of linguistic or fuzzy variables that often confront the decision maker. Linguistic and time-dependent variables lead to uncertainties in the impact of decisions taken. In this context, he develops a fuzzy system dynamics approach to improve the usefulness of energy policy system models characterized with linguistic variables. Sensitivity experiments and further "what-if" experiments are conducted in this study. He then draws managerial insights from the simulation results, relevant for policy makers concerned with renewable energy technologies. Fuzzy logic and system dynamics methodologies are integrated from a systems perspective to model typical RET scenarios, he explains. He anticipates that his developed methodology, a fuzzy system dynamic model, will be vital for real-world energy policy design and assessment in the twenty-first century.

Understanding the Dynamics of Oil Dependencies: The Case of Iran

Continuing with the theme of the application of system dynamics models to better understand the complex energy systems of regional and national economies, Saeed Langarud and Michael Radzicki in Chapter 10, "Resurrecting a Forgotten Model: Updating Mashayekhi's Model of Iranian Economic Development," resurrect, update, and validate a classic system dynamics model. The goal of the model is to investigate the issue of Iranian oil dependency. The original model had predicted that Iran would face a harsh economic recession during the 1980s due to a steep fall in oil revenue caused by natural resource depletion. Thirty-five years later, however, Iran's oil reserves remain intact and the country hasn't encountered the sort of severe depression that was predicted, they claim.

By examination of the original M-model, they showed that it did not contain the structure necessary to capture the dynamics of the Islamic revolution or the war with Iraq that occurred during the 1980s. Updating the M-model's exogenous variables, modifying some of its assumptions, and recalibrating some of its parameters significantly improved its ability to reproduce Iranian economic history, they posit.

Revalidation of the M-model has shown that it is fairly robust and generally reliable. Because its boundary was drawn somewhat narrowly, however, although it is an excellent tool for analyzing questions directly related to the issue of Iranian oil dependency, it is an inadequate platform for analyzing many contemporary Iranian macroeconomic policies. Broadening the boundary of the M-model by adding sectors such as a financial market, a foreign exchange market, a labor market, and an energy market would greatly enhance its versatility. They argue that the M-model can serve as a foundational platform for future Iranian macroeconomic modeling efforts.

Finally, they claim that this chapter can serve as a starting point and archetype for those who wish to develop a system dynamics macroeconomic model of a resource-dependent developing nation. They identified and suggested that future research involving the use of the M-model should address these issues:

1. As previously mentioned, the boundary of the M-model should be broadened to include a financial market, foreign exchange market, labor market, and an energy market.
2. The energy sector of the M-model should be revised to address energy–economy interactions. For example, the original M-model and its current modified version contain only one source of energy, oil. The boundary of the energy sector needs to be broadened to include alternative sources of energy and the economics of their substitutability.
3. The importation of energy is impossible in both the original M-model and its current modified version. This is not acceptable, particularly when the purpose of the model is to analyze energy–economy interactions.
4. The production functions in both the original M-model and its current modified version are very sensitive to their elasticity parameters. The formulation of these functions should be modified to eliminate this fragility.
5. The modified M-model should be recalibrated to see if it can reproduce the behavior of other capital-deficient oil-exporting nations that have large populations and significant agricultural sectors such as Nigeria, Algeria, Indonesia, Venezuela, Ecuador, or Mexico.

Modeling Low-Carbon Power Generation Policy

As the focus of this section is on "low-carbon" energy policies, Isaac Dyner, Carlos Franco, and Laura Cardenas in Chapter 11, "Making Progress Towards Emissions Mitigation: Modeling Low-Carbon Power Generation Policy," deal with the issue of "emissions mitigation." They focus on environment-related issues, including emissions, renewables-based technology, and change in consumer use patterns. In this context, policy aims at preserving and maintaining security of supply as well as a competitive environment within both power generation and energy-intensive industries, they claim. There are enormous uncertainties regarding the effect of GHGs on climate change in Latin America and on the structure of the electricity sector in the future. In spite of the obvious threats, these conditions also provide opportunities not yet explored. A low-carbon policy aims at changes regarding regulation, demand, supply, market structure, management, and, in general, the competitiveness of the power generation industry, they assert. In this context, they assess the effect of GHG policy on the Colombian electricity sector, based on system dynamics simulation; they also indicate how emission costs and incentives in the electricity sector induce technology changes leading towards a low carbon economy.

An Exploration into Energy and Economic Futures

Finally, in Chapter 12, "Exploring Energy and Economic Futures using Agent-based Modeling and Scenario Discovery," P. Wang, M. D. Gerst, and M. E. Borsuk present their state-of-the art application of agent-based modeling and scenario discovery approach. They demonstrate how the process of scenario discovery as applied to the results of ENGAGE, a stochastic, dynamic, agent-based model, might be used to generate socioeconomic scenarios relevant to a given emissions target, or RCP.

Although in their current contribution they have overcome some of the key limitations of earlier versions of ENGAGE by allowing for a growing population and uncertain fuel price, there are still a number of simplifying assumptions that they believe are too great to allow direct application of their current results to real-world policy questions. For example, the current simplicity of the energy sector may overlook opportunities for technology innovation and adoption. In particular, they only represent one energy production firm, and it is assumed to utilize the full lifetime of its energy technologies. Thus, it will not prematurely scrap any of its existing stock when improved carbon-light or carbon-free technology becomes available. Also, cost is currently the only factor in the model determining new technology adoption, precluding early adoption to meet moral obligation or public relations objectives. These factors add a significant lag to the achievement of carbon emission reductions in the model.

Finally, the decision rules of households and firms in their current model are homogeneous and simplified, they say. For example, firms cannot focus R&D efforts towards specific machine attributes or make decisions to hedge against anticipated energy price increases. Similarly, households have homogeneous preferences for needs that do not represent the true diversity of personal values and beliefs. They are currently working to alleviate these limitations by defining a suite of decision rules that households use to select goods that meet both their individual and social needs.

The authors recognize that further progress is necessary for ENGAGE to provide useful support for climate policy evaluation and formulation. Nevertheless, they believe that their proposed combination of stochastic, agent-based modeling and multidimensional scenario discovery can contribute to the ongoing climate scenario development effort by complementing traditional approaches.

Concluding Remarks

We started our journey in search of modeling approaches and models for energy policy with a particular focus on low-carbon economic development of regions and states in the world. In this quest, we have been successful in presenting 12 unique contributions. With regard to the theme of "modeling approaches" for energy policy in the twenty-first century, we have six leading contributions on "system dynamics methodology," "model-based theory," "fuzzy system dynamics framework," and an "optimization modeling approach." Consistent with the objective of this volume, "the development, validation, and applications of system dynamics and agent-based models in service of energy policy design and assessment in the twenty-first century," we have six state-of-the-art applications of system dynamics and agent-based models.

It is worth noting that most of the model-based contributions in this volume have addressed "model validation" explicitly. By utilizing the validation techniques and procedures that are effectively demonstrated in these contributions, researchers and practitioners in the energy systems domain can increase the appeal and wider acceptance of their policy models. Likewise, several chapters in this book present future research opportunities for the energy policy modeling community.

References

Adelino, P., João, S.: Generation expansion planning (GEP) -A long-term approach using system dynamics and genetic algorithms (GAs). Energy. **36**, 5180–5199 (2011).

Anand, S., Vrat, P., Dahiya, R.P.: Application of a system dynamics approach for assessment and mitigation of CO2 emissions from the cement industry. J. Environ. Manage. **79,** 83–398 (2005).

Ansari, N., Seifi, A.: A system dynamics analysis of energy consumption and corrective policies in Iranian iron and steel industry. Energy. **43**, 334–343 (2012).

Assili, M., Javidi, H., Ghazi, R.: An improved mechanism for capacity payment based on system dynamics modeling for investment planning in competitive electricity environment. Energy Policy. **36**(10), 3703–3713 (2008).

Bun, D., Larsen, E.: Sensitivity reserve margin to factors influencing investment behavior in the electricity market of England and Wales. Energy Policy. **29**, 420–429 (1992).

Davidsen, P., Sterman, J., Richardson, G.: A Petroleum life cycle model for the U.S. with endogenous technology, exploration, recovery, and demand. Syst. Dyn. Rev. **6**(1), 66–93 (1990).

Ford, A., Bull, M.: Using system dynamics for conservation policy analysis in the Pacific Northwest. Syst. Dyn. Rev. **15**(1),1–16 (1989).

Moxnes, E.: Interfuel substitution in OECD–European electricity production. Syst. Dyn. Rev. **6**(1), 44–65 (1990).

Naill, R.: A system dynamics model for natural energy policy planning. Syst. Dyn. Rev. **8**(1), 1–19 (1992).

Qudrat-Ullah, H., BaekSeo, S.: How to do structural validity of a system dynamics type simulation model: The case of an energy policy model. Energy Policy. **38**(5), 2216–224 (2010).

Qudrat-Ullah, H., Davidsen, P.: Understanding the dynamics of electricity supply, resources and pollution: Pakistan's case. Energy. **26**(6), 595–606 (2001).

Qudrat-Ullah, H., Karakul, M.: Modeling for policy assessment in the electricity supply sector of Pakistan. Int. J. Energy Sec. Manage, 3(1), 240–256 (2007).

Saeed, K.: An attempt to determine criteria for sensible rates of use of material resources. Technol. Forecast. Soc. Chang. **28**, 311–323 (1985).

Saeed, K.: Slicing a complex problem for system dynamics modeling. Syst. Dyn. Rev. **8**(3), 251–261 (1992).

Saeed, K.: Designing an environmental mitigation banking institution for linking the size of economic activity to environmental capacity. J. Econ. Issues. **38**(4), 909–937 (2004).

Part II
Modeling Approaches and Energy Policy Decisions

Thinking About the Future

System Dynamics and the Process of Electricity Deregulation

Erik R. Larsen and Santiago Arango

Introduction

The deregulation of energy markets in general, and the electricity and gas markets in particular, over the last two decades has created a large number of new challenges and opportunities for companies in this sector, as well as for companies who might diversify into the sector. Thirty years back, there were few who believed or even discussed the possibility of a large-scale deregulation of the electricity sector. Most of the issues discussed at that time concerned how to improve various aspects of the existing monopolistic structure of the industry, not how the whole sector could be transformed, as has happened in many countries over the last 10–20 years (Munasinghe and Meier 1993; Helm 2003). As deregulation spread across the world, companies in the sector have found that competitive complexity and intensity are increasing significantly as deregulated companies find themselves competing in newly created industries, with new rules, often with new owners, against unfamiliar competitors, and with rules and regulations that are often poorly understood by all the stakeholders. A wide range of competence-destroying innovations are making the links between the past and the future increasingly tenuous for electricity and gas companies, and a new wave of disruptive changes may be just around the corner as the industry is beginning to consolidate in some regions (Lomi and Larsen 1999). What kind of competencies should electricity companies build to prosper in an institutional and competitive environment in which the past seems to contain so little information about the future (Dyner and Larsen 2001)?

S. Arango (✉)
Faculty of Mines, Universidad Nacional de Colombia, Carrera 80 # 65–223,
Bloque M8a, Medellin, Colombia
e-mail: saarango@unal.edu.co

E. R. Larsen
Institute of Management, University of Lugano, Via Buffi 13,
6900 Lugano, Switzerland
e-mail: erik.larsen@lu.unisi.ch

H. Qudrat-Ullah (ed.), *Energy Policy Modeling in the 21st Century*,
Understanding Complex Systems, DOI 10.1007/978-1-4614-8606-0_2,

While we have accumulated more than 20 years of experience of deregulation, we might have learned only relatively few general principles. There are a number of reasons why this may be so, which we shall discuss later in this chapter. However, we can also see this by looking at the various problems that deregulated markets have had over this period of time. If we had learned from the "deregulation experience" across the world, we should have expected to "move down the learning curve" and there should have been fewer and fewer problems with deregulation—i.e. we should have learned to manage the process. However, there is currently no indication that this has happened (Sioshansi and Pfaffenberger 2006). Recent deregulations do not seem to be taking place more smoothly than deregulations that took place 15 years ago. This should not be misunderstood as saying that deregulation has not been a success in many of the countries where it has taken place—it definitely has been a success, and it has solved many problems that might otherwise have created major disruptions; however, there seems to be little learning taking place at the "process" level (Sioshansi and Pfaffenberger 2006). Why is this so?

One element that makes this question difficult to answer is that learning is required at many different levels, e.g. both at the institutional level, the actual design of the deregulation framework, and also at the company level. At the institutional level, it turns out that apparently small national differences and history might make large differences in the future performance of the system, e.g. generation technology (Dyner and Larsen 2001). At the company level, thriving in the new competitive world of electricity seems to imply a paradoxical combination of organizational learning and lack of experience (Lomi and Larsen 1999). One direct implication of this argument is that companies should reconsider the value of their typically large investments in backward-looking information systems, or at least see them as complementary to new *kinds* of forward-looking decision technologies (Dyner and Larsen 2001; Lomi and Larsen 1999).

It is also clear that traditional economics is not adequate for a full understanding of the new deregulated markets. Economics provides only a partial answer to many of the issues at the macro level (i.e. design of the deregulated market) and offers much less guidance as to how companies can make sufficient sense of these new markets to make the early investment required to "keep the lights on", and eventually exploit the markets to grow profitably. In this chapter, we try to classify the types of problem that many companies face when they are part of a deregulation, problems that seem to be quite similar across countries although the exact manifestation might differ from country to country.

The chapter is organized in the following way: We start with a short summary of the changes in the electricity sector that motivate our classification of problems, followed by a brief review of System Dynamics, the one methodology we will draw on in this chapter. We then discuss the three types of problem we have identified, as well as how System Dynamics can help in this process. Finally, we conclude the chapter with some observations on how companies should approach deregulation.

Types of Risk Associated with the Transition to Deregulated Markets

Many types of risk have long been acknowledged as critical in relation to energy markets; chief among them the financial and technical risk. Financial risk relates, among others, to the trading of energy, e.g. the loss of more than US$ 5 billion on natural gas by Amaranth Advisors, a Connecticut-based hedge fund, after making an estimated US$ 1 billion on similar trades the previous year. Another aspect of financial risk is the investment risk associated with investing in newly created markets, e.g. political risk, as seen in Argentina where interventions in the gas sector have created a number of market imperfections (Ponzo et al. 2011). Technical risk also has large-scale consequences, such as the blackouts in New England and Italy in 2003, as the transmission systems were unable to cope with the sudden changes in load. Other types of risk, in particular those associated with the transformation from monopolies to deregulated markets, are much less explored and less well understood (Larsen and Bunn 1999). In many cases, these risks are at least as serious as, if not, in many cases, even larger than the risk associated with the technical and trading arrangements in the market, as we shall discuss below. Furthermore, from a systemic point of view, these other risk factors might significantly increase the financial and technical risk in the deregulated market. In this chapter, we discuss some of these non-financial and technical risks associated with deregulated electricity markets, and methods that might be used to mitigate such risks. The way we classify the risk here is based on the experience of a large number of companies that have gone through this transformation, with more or less success. We conceptualize these issues as "risks" rather than "problems" because—if appropriately managed—they may provide unique opportunities for a company to establish the foundations of a sustainable competitive advantage (Lomi and Larsen 1999; Larsen and Bunn 1999; Emmons 2000). The three types of risk that we focus on here are:

- *Regulatory Risk*: The risk that is inherent in all markets where a regulatory institution has significant influence, as is the case with most deregulated electricity markets. The short- and long-term consequences, and the frequency of interventions in the market, are often poorly understood.
- *Market Risk*: The risk arising from having to learn to operate in a competitive market, where, before, the company was a monopoly and as such was in control of most aspects of the industry in the region or country. This is made more difficult due to the structure of most electricity markets and the limited understanding of the long-term consequences of the rules and regulation that are governing the industry after deregulation.
- *Organizational Risk*: The risk associated with the internal transformation most incumbent companies need to go through to adjust their structure, routines, practices, and understanding of the newly deregulated industry.

One might argue that there is a fourth type of risk, institutional risk, related to the design and implementation of the deregulation. While this risk is at least as important, we focus in this chapter on the three types listed above as they relate more or less directly to participants in the deregulated market. It should be clear that institutional risk is an indirect part of these company risks, and can increase these risks considerably. For more discussion about the institutional setting see Amobi (2004). In the following sections, we will outline one method that can help in limiting some of these risks by creating a better understanding of the issues in newly deregulated markets; we will then in turn discuss how System Dynamics can help to mitigate some of these risks.

Simulation in the Deregulated Electricity Sector

The traditional planning methods used within most electricity companies have been operational research methods, such as optimization using integer, linear, and dynamic programming. The models developed and used over a period of 30 years have proved to be extremely successful. It is probably fair to say that the electricity industry is one of the least disputable success stories of operational research (Dyner and Larsen 2001). However, as the industry changes, planning methods also need to change significantly. System Dynamics has many of the characteristics that make it a desirable addition to the toolbox of a deregulated electricity company.

When an industry undergoes disruptive change, incumbent companies face a difficult and dangerous transition. While managers may realize that their companies need to undergo massive transformation, they have no managerial experience or cognitive models that can meaningfully bridge the gap that they face. In this chapter, we argue that this challenge can be successfully addressed through a System Dynamics-based simulation approach that facilitates organizational learning about post-disruption behaviours and their consequences. We argue that System Dynamics can be used to create a rich learning experience that helps managers to more accurately understand the risks they face and the concrete steps they need to consider in order to avoid them. Our observations suggest that the use of this technique not only provides better and more informed decisions, but also produces higher levels of decision-making commitment.

Although deregulation typically is introduced stepwise, investment in, e.g. the electricity industry, has typically a lifetime of at least 30 years, and if the decision made by the companies in the (often very long) transition period is wrong, it has major consequences. This is true if the companies invest too much (low prices and possible bankruptcy) or if they do not invest enough (shortage and possible blackouts—with the accusations that follow from that). System Dynamics, as a method of feedback modelling, offers one of the only ways in which a management team can think through the consequences of these types of major disruption.

There is a long tradition in System Dynamics of using modelling for learning (e.g. Morecroft and Sterman 1994), which is needed in this case where we do not have the data or understanding of how the industry might evolve. Even when we are well into the deregulation period, new challenges arise that might change the competitive environment significantly, e.g. a number of countries, such as Switzerland and Germany, have decided to close their nuclear power plants in a relatively short time span (as a consequence of the Fukushima nuclear accident in Japan in 2011). Deregulation of the electricity sector has many of the characteristics that make the use of the modelling-for-learning framework applicable, including:

Lags: The building time, including planning permission, etc., for a small CCGT power station is 3–4 years, while for a large hydro plant or nuclear plant it can be up to 10 years. The economic lifetime is typically of the order of 25–30 years for conventional plants, and significantly longer for large hydro projects.

Unclear rules: Nobody knows what the long-term implications of a set of "new" liberalized rules are. In the UK, 10 years after deregulation, the market rules and the price setting were changed fundamentally.

Interdependence: Most decisions in this industry are highly interlinked: The regulatory framework influences the behaviour of the players in the industry, which will influence the investment, pricing, type of technology, fuel choices, etc., so that it is difficult to get an overview of the causal chain due to the lags in the reactions of the different segments.

You cannot just do one thing: There will always be a tendency to try to solve problems as they arise, a tendency that has become stronger as "evolving regulation" has emerged as the preferred way of controlling the industry. However, this way of setting up and regulating the industry will increasingly lead to unanticipated consequences, which then in turn will require even more selective changes, etc.

Many "stakeholders": Where before a deregulation the stakeholders have more or less aligned interests, not only are there new stakeholders added by deregulation, e.g. financial institutions, new competitors, electricity traders, etc., but also the nature of the interactions will change in many cases, making for a more hostile and confrontational environment.

System Dynamics can deal with these aspects of uncertainty. When change is rapid and past experience is irrelevant for navigating the future, we argue that simulated experience is a useful and necessary substitute. By "simulated experience" we mean allowing executives to play-out the future of their company with computer simulations. This way of rehearsing change through simulation is already widely accepted in many professions, but not in management where, arguably, it has the most to offer. Consider how architects, urban planners, engineers, and military strategists regularly use simulators to help them imagine and design new-yet-feasible buildings, highways, aircrafts, and battle plans. These days, even children design and build imaginary cities and homes using Sim-City and other Sim products

System Dynamics has a long history of being used in the energy sector, including Nail (1977), Ford (2001, 2002), Bunn and Larsen (1992) and Bunn et al. (1997). For a review, see Ford (1997), and for models used in the last decade see Arango and Larsen (2011).

Understanding Risk

We now return to a more detailed discussion of the different kinds of risks utility companies face, and how some of these risks can be mitigated by the use of System Dynamics. It should be made clear that there are many other methods that should also be considered and used in the deregulated company, though previously they were not seen as essential, such as Financial Risk Modeling (Jorion 1997; Humphreys and McClain 1998), Game Theory (Ferrero et al. 1998; Day 1999), Competitive Analysis (Grant 1998; Dyner and Larsen 2001), Real Options (Alleman and Noam 2000; Brennan and Trigeorgis 1999) and Scenarios (Schwartz 1991; Smith et al. 2005). However, we can only focus on one method, to keep the chapter within reasonable length.

Organizational Risks

Organizational risks are associated with the transformation of the company from a traditional monopoly to a commercial or market-oriented organization. A traditional monopoly is more like a governmental agency than a commercial company, particularly with respect to the amount of uncertainty it faces and the organizational structure. Most of the employees, including the middle and senior management, tend to have strong technical competencies or be political appointees. However, deregulation changes this significantly: the company will need to achieve a much greater focus on commercial aspects of the business, as the environment will become increasingly volatile, e.g. price and the need for change in the organizational structure to become more responsive. In many monopolies there are too many employees, as cost is not one of the major concerns; this fact led, in England and Wales over a period of 5 years, to a 60 % reduction of the workforce in the generation sector (Bunn 1994), and in Colombia too (Cavaliere et al. 2007). These changes are by no means trivial; the organizational changes that we have observed in the electricity industry are probably among the largest restructuring events recorded in any industry over the last 50 years. In fact, in discussions with managers of utilities companies, we observe that such companies (and people working in them) still behave as monopolies nowadays in Colombia, even after around two decades of deregulation. These transformations create organizational risks, as new capabilities have to be introduced, while at the same time the workforce has to be reduced without losing the technical competencies, which is a major challenge.

A number of other changes are summarized in Table 1. The result of all these changes and the corresponding adjustments within the company is a higher level of uncertainty, which for many former monopolies is a very uncomfortable situation. There are no easy ways in which this overall transition can take place. However, it is also clear that the faster the company can adjust to the new situation, the better off it will be. As pointed out earlier, in the formulation of strategy and the raising of finance, these companies were behaving as agents of government policy, and were

Table 1 Examples of changes taking place in the organization in the transition from monopoly to competitive market. (Larsen and Bunn 1999)

Attribute	Monopolistic market	Competitive market
Company focus	Best technical solution	Best cost-efficient solution
Management focus	Technical	Commercial
Customer focus	The customer has no choice	Retail competition forces a customer focus
Stakeholders	Relatively few, mainly government, and regulator	Many, including shareholders, customers, regulators, financial markets, NGOs
Planning methods	Classic operational research (OR) planning methods used successfully	New methods linking strategic thinking, uncertainty, and limited information
Level of uncertainty	Relatively low	High (price, demand, investments, etc.)
Outsourcing	Little or none	Increasing interest
Business rational	Social optimum	Shareholder value

relatively inexperienced in risk taking. Such companies have often been accustomed to receiving government subsidies, which together with monopoly power, encouraged them to increase assets and manpower instead of becoming leaner and more productive, as has been the experience in, e.g. the UK, as described above. Similar experiences can be found in many other countries such as, e.g. Colombia and Spain, and show that a successful transformation can take place, however painful it might be.

How can simulation and System Dynamics help in this situation? The use of simulation here is mainly in two areas: communication and management development/training. In both cases the simulation models are normally combined with a user-friendly interface to create what is known as a microworld, which is a sort of computer game developed and designed for use with teams (Dyner et al. 2009; Sterman 2000; Graham et al. 1992). Microworlds can be used both as a tool for communicating why certain actions needs to be taken, and also as a general tool for management development. For example, the Colombian market operator, ISA, was concerned about the missing depth and volume in electricity trading. After investigating, their conclusion was that this was in part because many potential participants felt that they did not understand the market, and in particular the risk involved in trading (Dyner et al. 2009). To help solve this problem, the EnerBiz microworld was developed, which has since been used to teach both trading and risk management in the Colombian market (Dyner et al. 2009).

Market Risks

The second type of risk, market risk, represents another major set of problems in the deregulation process. In a monopolistic electricity company, price formation is well understood, customers are captive, tariffs are negotiated with—or imposed by—the government, a relatively large amount of information about the industry is public, and as there is only one or a limited number of suppliers, expansion is based on a centralized planning process. Competition is not something that the management is focused on or even considers relevant, as exemplified by the quote below from a mid-level manager in a state-owned monopoly, when he was asked about competition:

> Competition may be good and well in many sectors, but how can we possibly do better than we are doing now? After all, if we have been doing this for a long time without major changes it is because we must be doing something right. (Lomi and Larsen 1999)

This type of statement shows the typical "mindset" of managers in monopolies (while this quote came from an electricity company, it is much the same in most other monopolies). This way of thinking makes the transition to a market-based organization even harder, as competition creates—among other things—consumer choice, price volatility, asymmetric information, new and possibly aggressive entrants, financial uncertainty, and the loss of "cost-plus" pricing leading to variable rates of return (Dyner and Larsen 2001). Much research has shown that it is very difficult to change the way in which managers think about their organization. This ultimately leads to an increasing level of stress, both for the individual managers but also for the organization as a whole. If the managers are not able to adapt to these changes they will eventually be replaced, as has been observed in many electricity companies, where the incumbent management has been replaced by managers with experience of competition from outside the electricity industry.

This problem with the necessary shift in management thinking comes together with a large number of other "problems" at the industry level that the company faces at the moment it becomes deregulated; to some extent, these are problems that are also shared with the regulator of the industry. A number of these problems are collected in Table 2. Such problems are related to the way in which the industry functioned under monopoly, as compared with operation under a deregulated regime.

Among the issues in Table 2, market power and investment decisions might have received the most attention. For long periods, market power was the focus of the regulator in England and Wales, companies in the England and Wales market were constantly under scrutiny from the regulator; every move they made was looked at through the lens of market power, forcing them to justify commercial strategies in greater detail than would have been expected, and limiting their options. Similar concerns have been raised in many other deregulated markets, such as California and Colombia. Nowadays, investment decisions have been increasingly important due to the concerns about long-term security of supply (Arango and Larsen 2011); this issue has been raised from the very beginning of deregulation in England and Wales (Bunn and Larsen 1992).

Another equally important, but not yet as-much-discussed issue, is the issue of energy savings: where initiatives such as demand-side management made perfect sense in a monopoly market, the logic breaks down when the industry becomes deregulated. From a rational or economic point of view, the companies in the industry have lost all incentives for contributing to energy saving and can only justify this in terms of corporate social responsibility. The first step towards a market-based

Table 2 Changes taking place at industry level when an electricity sector is deregulated. (Larsen and Bunn 1999)

Attribute	Monopolistic market	Competitive market
Business environment	Stable with only gradual adjustment, technically driven changes. Uncertainties in demand and costs	Unstable, volatile prices, new stakeholders, with diverse objectives. Market, corporate and regulatory environment
Information	Open and public domain information. Planned future	Information becomes secret. Future signals misleading
Investment decisions	Central long-term planning, based on optimization (minimize total system cost)	Agent-based decisions, based on firm's strategy to maximize profits
Regulatory environment	Concerned with social welfare	Awkward balance between interests of customers and new entrants
Market power	Not an issue as there was a regulated monopoly	Now crucial for regulators and companies
Conservation and environment	Easily incorporated into energy policy	Adds one more layer to regulatory risk
Public research and development (R&D)	Public R&D was seen as an important part of long-term obligation	Companies cannot justify public domain R&D

environmental policy has been taken by emissions trading, but there is still a long way to go before all the policies that were rational in a monopoly have been replaced with truly market-based policies.

We can describe the initial period after deregulation (although it is measured in years) as a state in which companies have to function suspended in time, without any relevant history that can guide decision making (Lomi and Larsen 1999). This situation, where there is no relevant history to learn from, creates major problems for most of the companies that become deregulated. An agreed upon past provides the basis for an understanding of how both competitors and customers behave and react to changes, and how prices might move given certain demand and supply conditions. The newly deregulated industry has not evolved over time as most industries, where a co-evolution between the companies and the industry has created a mutual adjustment and understanding. The deregulated industry has been "designed" by the regulator and the government, and there is no history that can guide the decisions of the company, as the industry did not exist as a competitive market place "yesterday". This uncertainty is not only affecting the companies, but also the regulator and the political institution that has been involved in the organization of the industry. The challenge for the company is thus to understand how the industry works and the nature of its weaknesses and strengths, enabling it to develop strategies either for competitive exploration or for political lobbying to influence future change.

Simulation can provide understanding of market risk in two ways: first, by making up for the lack of history or future plausible market evolutions, and second, in evaluating strategies. There is a need for companies to understand the possibilities and threats that they face in a deregulated industry, and to create long-term strategies and visions of where the company is heading. However, to be able to do this, a structured way of understanding the future is needed, without having access to a past (that does not exist). There is also a need to capture the dynamic elements and unintended consequences in the artificial market, i.e. a market made up of partly free and partly regulated market elements. Deregulated industries can be seen as complex systems, with many unanticipated consequences that the conventional economic and financial analysis will have difficulties in anticipating or discovering. An approach based on feedback, with explicit recognition of delays and representation of decision rules, as well as soft variables, has the necessary ingredients to be useful in an analysis of a situation such as this. Furthermore, simulation models at an early state in a deregulation cannot be validated empirically (as no data exists), but they can be developed to represent how the system is designed to operate and therefore, from such a prototypical basis, generate insights into the strategic opportunities created by the market's potential instability to shocks, parameter uncertainties, and market imperfections. Such models can thereby identify the sorts of business risks that might follow from a variety of scenarios for market structure and behaviour.

There are a number of examples of the use of System Dynamics (SD) for this purpose, including modelling of the England and Wales market (Bunn and Larsen 1992). The SD model of the England and Wales market highlighted at an early stage the potential problems that might arise as investment in generation capacity would become cyclical, following the pattern of capacity in similar capital-intensive industries (Larsen and Bunn 1999; Arango and Larsen 2011). Other examples include Colombia (Arango 2007), and California (Ford 2001, 2002), and many others listed in Arango and Larsen (2011).

Regulatory Risk

The final risk in our typology is regulatory risk. As electricity is a critical resource in all countries, after deregulation the power system maintains a regulator in some form (normally as an independent or semi-independent body), watching to see that the deregulation is carried out in the way it was intended. Typically, the function includes monitoring the market for anti-competitive practices, making adjustments to the regulation as the market evolves, etc. The regulator must choose how to balance controls on prices, investment, divestment, anti-competitive behaviour, security of supply, and protect possible remaining captive customers as well as moving the market forward. These duties have to be performed in the same uncertain and poorly understood markets the companies operate in; furthermore, the regulator is likely to have even less information than the companies operating in the industry, as the companies will tend to disclose only the absolute minimum amount of information required. From the point of view of individual companies, regulators become less predictable, and in many countries, the regulatory institution has the power to change, at least within some boundaries, the market and its competitive and organizational context within which companies operate (Cross 1996). Given these potential, and to some degree unpredictable changes, it is important for the companies to understand, as far as possible, how the regulator may react to any future incidents and to start thinking about regulation in strategic terms.

The use of simulation in this area has many similarities with the use in the market risk area. Again, the reason for using simulation in this case is that it might alert thinking and understanding to various unintended consequences that might trigger the regulator, or government, into reaction. An example of a simulation model used to explore the regulatory problems is Bunn et al. (1997). Here, a simulation model was used to explore the consequences of arbitrage across the short-term electricity and gas markets. The model showed how a dominant generator could influence prices in both markets and how the regulators in gas and electricity will have difficulties in dealing with it as long as they are separate institutions. The dominant generator can gain by creating increasing volatility in the electricity pool, thereby increasing the quantity of contracts that the customers are willing to sign at a premium to the otherwise "fair" price. If the generator owns any retail business, they will not suffer so much by this and will be in a better competitive position. Other case studies can be found in Ford (2001), Arango (2007), and Ponzo et al. (2011), among many others.

Discussion and Conclusion

In this chapter, we have outlined some of the common problems that most incumbent companies face when a deregulation takes place. While the exact manifestation of these risks might vary from country to country, the types of problem discussed above can almost always be found. Furthermore, in most deregulated systems these problems exist long after deregulation initially took place (Dyner and Larsen 2001). This is consistent with the view expressed earlier, that the transition period from monopoly to fully competitive industry is a very long one, in most markets.

The transition is even more complicated due to the interaction of these types of risk, i.e. the organizational transformation has to take place at the same time as the newly deregulated market is evolving and the regulatory institution is trying to understand its role, powers, and responsibility. In fact, the co-evolution of companies, markets, and regulation is a delicate balancing act, to which all the stakeholders in the power system need to pay careful attention. At a more theoretical level, it might be possible to argue that the problems, which we have observed in Chile and California, have resulted from this co-evolutionary process getting out of balance, as one part of the market developed faster than the other parts. Deregulation is a process rather than an event, i.e. the day on which deregulation takes place is just the beginning of a journey towards a well-functioning electricity market. As the market develops, companies get reorganized and begin learning to act in the new scheme, regulators understand the problems and opportunities to be found within the regulatory framework, and customers and other stakeholders start to explore the possibilities open to them. However, during these processes, there will be a number of unintended consequences resulting from the way in which the deregulation was implemented and the regulatory framework was composed, so that the market's functioning will have to be adjusted and, in some cases, will require significant changes. Sometimes, the lack of such adjustment would lead to the emergence of major problems at a later stage. For example, this is the case in England and Wales, where there is now a widespread consensus that there were too few companies created when the industry was deregulated in 1990 (Helm 2003). The regulator had to solve this problem 6 years later by providing incentives to the incumbent companies to sell off some of their generation capacity in return for being allowed to own distribution companies (Helm 2003).

The other two types of risk are much harder to understand via comparisons with other countries that have gone through the deregulation process. Colombia adapted the regulatory framework used in England and Wales (Arango et al. 2006), and it would be sensible to believe that Colombian companies could have learned from the experience of the English companies as England deregulated 5 years prior to Colombia. However, there is little in common between the evolution of the electricity price in England and Colombia, even though they have had the same market system. The main reason for this is the very different proportions of hydroelectric generation: only 4 % in England, as against 70 % in Colombia, which produces completely different price dynamics. For a comparison of the (very different) evolution of countries in Latin America, see (Arango et al. 2006).

While we related the risks and the possible mitigation of them to the use of System Dynamics, there are other simulation frameworks that provide insights into the working of new markets. As we pointed out earlier, we do not go through all the possible ways in which new markets can be modelled, but nor do we want to leave the reader with the impression that System Dynamics is the only way. Like all methods, System Dynamics has advantages and limitations. It is particularly strong when dealing with complex problems influenced by lags and feedback, where the rationality of decision is explicitly modelled. However, it is a method that builds on aggregated entities and structural relationships that need to be more or less constant during the period of study (Dyner et al. 2003). Recently, agent-based models have also been used in utility markets, and there are many other types of simulation that can be used.

While deregulation continues around the world, we need to improve our understanding of the long-term consequences. We have pointed out the areas that seem to us to lead to the main problems when electricity sectors are deregulated. Liberalized markets are significantly different from country to country, based on natural resources, generation technology, industry structure, network topology, etc. (Larsen and Bunn 1999). This means that there will be a need for each country to adapt or combine existing models, or invent a model that is suited to itself, and for each electricity company in each country to understand, learn, and develop efficient strategies tailor-made to that country. Simulation models should play a major role in this development; in particular, this development can benefit from the behavioural, high-level, and feedback characteristics of System Dynamics to deal with the special modelling challenges of restructured industries.

References

Alleman, J., Noam, E. (eds.): The New Investment Theory of Real Options and Its Implications. Kluwer, Boston (2000)

Amobi, M.C.: Electricity deregulation: evidence, analysis and public policy. Unpublished Ph.D. thesis, Cass Business School, London (2004)

Arango, S.: Simulation of alternative regulations in the Colombian electricity market. Socio. Econ. Plan. Sci. **41**(4), 305–319 (2007)

Arango, S., Larsen, E.R.: Cycles in deregulated electricity markets: empirical evidence from two decades. Energ. Policy **39**, 2457–2466 (2011)

Arango, S., Dyner, I., Larsen, E.R.: Lessons from deregulation: understanding electricity markets in South America. Util. Policy **14**, 197–207 (2006)

Brennan, M.J., Trigeorgis, L. (eds.): Project Flexibility, Agency, and Product Market Competition: New Developments in the Theory and Application of Real Options Analysis. Oxford University Press, Oxford (1999)

Bunn, D.W.: Evaluating the effects of privatisation of electricity. J. Oper. Res. Soc. **45**, 367–375 (1994)

Bunn, D.W., Larsen E.R.: Sensitivity of reserve margin to factors influencing investment in the UK electricity market. Energ. Policy **May**, 490–502 (1992)

Bunn, D.W., Dyner, I., Larsen E.R.: Modelling latent market power across gas and electricity markets. Syst. Dynam. Rev. **13**(4), 271–288 (1997)

Cavaliere, C., Larsen, E., Dyner, I.: The privatization of EEB: from cash drain to major contributor. Energ. Policy **35**(3), 1884–1895 (2007)

Cross, E.D.: Electric Utility Regulation in the European Union. Wiley, London (1996)

Day, C.J.: A computational approach to modelling strategic behaviour in electric pools. Unpublished Ph.D. thesis, London Business School, London (1999)

de Geus, A.: Planning as learning. Harvard Bus. Rev. **March–April**, 70–74 (1988)

Dyner, I.: Energy modelling platforms for policy and strategy support. J. Oper. Res. Soc. **51**(2), 136–144 (2000)

Dyner, I., Larsen, E.R.: From planning to strategy in the electricity industry. Energ. Policy **29**(13), 1145–1154 (2001)

Dyner, I., Larsen, E.R., Lomi, A.: Helping companies to learn without experience in competitive electricity markets: the role and experience of using systems dynamics. In: Ku, A. (ed.) Risk and Flexibility in Energy Markets, pp. 205–234. Risk Books (with I. Dyner and A. Lomi) (2003)

Dyner, I., Larsen, E.R., Franco, C.J.: Computer games for electricity traders: understanding risk in a deregulated industry. Energ. Policy **37**(2), 456–471 (2009)

Emmons, W.: The Evolving Bargain. Harvard Business School Press, Cambridge (2000)

Ferrero, R.W., Rivera, J.F., Shahidehpour, S.M.: Application of games with incomplete information for pricing electricity in deregulated power pools. IEEE T. Power Syst. **13**(1), 184–189 (1998)

Ford, A.: System dynamics and the electric power industry. Syst. Dynam. Rev. **13**(1), 57–85 (1997)

Ford, A.: Waiting for the boom: a simulation study of power plant construction in California. Energ. Policy **29**, 847–869 (2001)

Ford, A.: Boom and bust in power plant construction: lessons from the California electricity crisis. J. Ind. Compet. Trade **2**, 59–74 (2002)

Forrester, J.W.: Industrial Dynamics. MIT-Press, Cambridge (1961)

Forrester, J.W.: Policies, decisions, and information sources for modeling. Eur. J. Oper. Res. **59**(1), 42–63 (1992)

Graham, A.K., Morecroft, J.D.W., Senge, P.M., Sterman, J.D.: Model-supported case studies for management education. Eur. J. Oper. Res. **59**(1), 151–166 (1992)

Grant, R.: Contemporary Strategy Analysis. Blackwell, Malden (1998)

Helm, D.: Energy, the State and the Market: British Energy Policy Since 1979. Oxford University Press, Oxford (2003)

Humphreys, B., McClain, K.: Reducing the impacts of energy price volatility through dynamic portfolio selection. Energ. J. **19**(3), 107–131 (1998)
Jorion, P.: Value at Risk. Irwin, New York (1997)
Larsen, E.R., Bunn, D.W.: Deregulation in the electricity sector: understanding strategic and regulatory risk. J. Oper. Res. Soc. **50**, 337–344 (1999)
Lomi, A., Larsen, E.R. Learning without experience: understanding the strategic implications of deregulation in the electricity industry. Eur. Manage. J. **17**(2), 151–163 (1999)
Morecroft, J.D.W.: Executive knowledge, models and learning. Eur. J. Oper. Res. **59**(1), 9–27 (1992)
Morecroft, J.D.W., Sterman, J.D.: Modeling for Learning Organizations. Productivity Press, Portland (1994)
Munasinghe, M., Meier, P.: Energy Policy Analysis and Modelling. Cambridge University Press, Cambridge (1993)
Nail, R.: Managing the Energy Transition. Ballinger, Cambridge (1977)
Ponzo, R., Dyner, I., Arango, S., Larsen, E. Regulation and development of the Argentinean gas market. Energ. Policy **39**, 1070–1079 (2011)
Schwartz, P.: The Art of the Long View. Currency, New York (1991)
Sioshansi, F.P., Pfaffenberger, W. (eds.): Electricity Market Reform. Elsevier, Amsterdam (2006)
Smith, R., Cadena, A., Boman, U., Vesga, D., Larsen, E.R., Dyner, I.: Energy scenarios for Colombia: process and content. Futures **37**(1), 1–17 (2005)
Sterman, J.D.: Business Dynamics. McGraw Hill, New York (2000)
Sweeney, J.L.: The California Electricity Crisis. Hoover, Stanford (2002)
Vennix, J.A.M.: Group model building. Wiley, Chichester (1996)

Fuzzy System Dynamics: A Framework for Modeling Renewable Energy Policies

Michael Mutingi and Charles Mbohwa

Introduction

Renewable energy policy formulation and evaluation is an important subject matter at island, country, regional, and global levels. Industrial development has increased the demand for fossil fuels such as coal, petroleum, and natural gas. Due to high-potential social and environmental repercussions of global warming and the consequential climate change, the international community has emphasized the need to conserve energy and to mitigate carbon emissions. The Intergovernmental Panel on Climate Change (IPCC) estimated that about 90 % of global temperature rise is likely to be caused by greenhouse gas emissions. In addition, IPCC reported that the earth's average temperature is likely to rise by about 1.1–6.4 °C by the end of the twenty-first century (IPCC 2007), which is broadly consistent with earlier estimates (IPCC 2001). The Kyoto Protocol of United Nations Framework Convention on Climate Change called for a determined reduction of the emissions of greenhouse gases so as to mitigate climate change (United Nations 1998). Several countries have since participated in the global actions targeted at reducing carbon dioxide (CO_2) emissions by putting in place a set of greenhouse gas control strategies (Peters 2008; Chang et al. 2010). Consequently, the concepts of low-carbon economies, low-carbon islands, low-carbon regions, and low-carbon cities and societies have increasingly become central issues, aimed at building economies that consider the 3Es dimensions, that is, energy, economic development, and the environment (Qudrat-Ullah 2005; Trappey et al. 2012a).

Several countries have engaged themselves into developing low-carbon island policies by establishing renewable energy sources in an attempt to reduce CO_2 emissions to an acceptable level (Trappey et al. 2011; Chen et al. 2007). For instance, a number of interesting low-carbon island projects exist in the literature, including empirical studies in Gökçeada in Turkey (Demiroren and Yilmaz 2010),

M. Mutingi (✉) · C. Mbohwa
School of Mechanical and Industrial Engineering, University of Johannesburg,
Bunting Road Campus, PO Box 524, Auckland Park 2006, Johannesburg, South Africa
e-mail: mmutingi@gmail.com

H. Qudrat-Ullah (ed.), *Energy Policy Modeling in the 21st Century*,
Understanding Complex Systems, DOI 10.1007/978-1-4614-8606-0_3,

Kinmen Island in Taiwan (Liu and Wu 2010), Taiwan (Trappey et al. 2012a), Yakushima Island in Japan (Uemura et al. 2003), Dodecanese Islands in Greece (Oikonomou et al. 2009), Penghu Island administrative region in Taiwan (Trappey et al. 2012a), and in other countries such as Pakistan (Qudrat-Ullah and Davidsen 2001), the United States (Vicki and Tomas 2008; Ernest and Matthew 2009; GPO 2009), China (John et al. 1998; Han and Hayashi 2008; Li et al. 2009; Huang 2009), India (Huang 2009; Peter 2010; Chandrasekara and Tara 2007), Columbia (Dyner et al. 1995), among others (Krushna and Leif 2008). Among the several empirical studies, the central conclusion is that governments and stakeholders need to actively increase renewable energy adoption and promote effective policy incentives and policy controls so as to reduce the CO_2 emissions prevalent in their countries and regions. Possible policies in this regard include promoting solar energy industry (photovoltaic and solar thermal sectors), promoting solar energy adoption, promoting wind energy adoption, as well as promoting the adoption of other renewable energy sources such as tides, waves, and geothermal heat. Subsidies, price cuts, campaigns, promotions, and other control policies have a potential to contribute significantly to the popularity and adoption of renewable energy technologies (RETs). It is anticipated that this endeavor will ultimately reduce CO_2 emissions in the medium to long term.

Modeling renewable energy policies is a crucial undertaking that calls for system-wide analysis capabilities so as to obtain an in-depth understanding of the complex renewable energy systems. Understanding the complex interactions between the variables, the possible alternative decisions, and the likely consequences of the actions taken is highly imperative. A number of factors related to environment, economy, and the community have to be considered from a systems engineering point of view. This implies that the population, ground forest, industrial activities, commercial activities, transportation, daily domestic energy usage, and CO_2 generation are among the several factors that need to be taken into consideration when designing and evaluating renewable energy policies. All these and other factors form a complex dynamic system with complex causal relationships as far as energy consumption and carbon emissions are concerned.

Systems dynamics (SD) has been applied to a number of problems concerned with formulation of energy policies (Naill 1992; Qudrat-Ullah and Davidsen 2001; Raja et al. 2006; Qudrat-Ullah and Seong 2010; Trappey et al. 2012a) and assessment of environmental impact (Ford 1997; Jan and Hsiao 2004; Trappey et al. 2011; Mutingi and Matope 2013). Developing robust long-term policies is nontrivial due to complex dynamics prevalent in those energy systems. However, no attempts have been made to consider capturing the fuzzy imprecise variables in renewable energy policy design. It is known that low-carbon energy economies are human systems characterized with linguistic variables that are difficult to interpret and model using conventional systems simulation models. Clearly, the presence of fuzzy variables makes policy design and evaluation a complex responsibility for the policy maker who has to base his decisions on imprecise variables by observing the trends in the renewable energy marketplace. For instance, the policy maker may need to cautiously formulate investment decisions aimed at positively impacting renewable energy adoption which eventually leads to a low-carbon economy. The task is to utilize the fuzzy information at hand to formulate effective renewable energy policies in anticipation of long-term improvements in the economy–environment–energy system. Therefore, it is important to develop a systems simulation methodology that can address the complex dynamic features and the fuzzy characteristics inherent in renewable energy systems.

Motivated by the above energy and environmental issues, the purpose of this chapter is to present a framework for evaluating renewable energy policies based on a fuzzy system dynamics (FSD) paradigm. In this connection, the objectives of this chapter are as follows:

1. To present a causal loop analysis for the complex dynamic interactions between various energy-related factors
2. To present the proposed FSD framework incorporating system dynamics and fuzzy logic concepts
3. To present an application of the FSD framework to a case example in renewable energy policy evaluation

The rest of the chapter is organized as follows: The next section provides a background to FSD. The section "Fuzzy System Dynamics Framework" gives a description of the proposed FSD framework for renewable energy policy design and evaluation. The section "Case Example: South Africa" presents policy scenarios for a simulation study, based on a case study example, together with relevant discussions. Finally, we provide conclusions and further research prospects in the section "Concluding Remarks and Further Research".

Fuzzy System Dynamics: A Background

System dynamics (SD) (Forester 1961; Morecroft 2007) and fuzzy logic (Zadeh 1965, 1978) are powerful and viable tools for modeling complex systems. Tessem and Davidsen (1994) emphasized the need to include a qualitative approach to simulation and analysis of complex dynamics systems, based on the theory of fuzzy sets and fuzzy numbers. FSD is a systems simulation tool that incorporates fuzzy variables into system dynamics models so as to cater for systems whose structures, state, or behavior cannot be described with exact numerical precision (Levary 1990; Tessem and Davidsen 1994; Mutingi and Mbohwa 2012). The FSD paradigm utilizes the strengths of the widely applied system dynamics and fuzzy logic methodologies.

System Dynamics Applications

The strengths of SD can be seen from its wide application in related studies. SD has been utilized to assess environmental issues and CO_2 emissions (Vizayakumar and Mohapatra 1993; Anand et al. 2005; Qudrat-Ullah and Davidsen 2001). Jin et al. (2009) proposed a dynamic ecological footprint forecasting model for policy modeling of urban sustainability. In the same vein, Han and Hayashi (2008) investigated inter-city passenger transport in China using an SD model to assess CO_2 mitigation policy. Furthermore, Trappey et al. (2011) used SD to model life cycle dynamics to control mass customization carbon footprints. Related applications also exist in the literature (Trappey et al. 2012b, c). However, none of these SD applications considered the presence of fuzzy variables. Though the SD paradigm can be used effectively in system modeling of complex dynamic systems, there is need to add to the approach a method of capturing fuzzy linguistic variables that often exist in real world systems. Fuzzy variables that take linguistic values can be captured effectively by the use of fuzzy set theoretic applications such as fuzzy logic. Formal fuzzy logic tools can provide a useful way of accommodating linguistic values into policy design and evaluation models.

Fuzzy Logic System

A fuzzy logic system is a logical system that utilizes the theory of fuzzy sets. Fuzzy set theory relates to classes of objects that have non-crisp boundaries to which membership is a matter of degree (Zadeh 1965, 1978). The most important component of every fuzzy logic system is a set of rules that converts inputs to outputs based on the theory of fuzzy sets (Kosko 1992a, 1994, 1995). In practice, it is implemented using the fuzzy approximation theorem (FAT) (Kosko 1992b). Usually, the inputs to a fuzzy logic system are the information that relates to the state of the system, and the output is a specification of the action to be taken. As such, fuzzy logic incorporates a rule base that contains a set of "if-then" rules of the form:

$$\text{IF} x \text{is} A \text{THEN} y \text{ is } B \tag{1}$$

where, A and B are linguistic values defined by fuzzy sets on the ranges (universes of discourse) X and Y, respectively.

According to the fuzzy logic concepts, "x is A" is called the antecedent, while "y is B" is known as the consequent. This provides strong constructs for fuzzy inference. Fuzzy inference is the process of formulating the mapping from a given input to an output based on some fuzzy logic set of rules (Sugeno 1985; Mamdani 1975). The mapping provides a basis from which decisions can be made based on a set of linguistic control rules obtained from experienced decision makers. The process of fuzzy inference involves the following constructs: membership functions, logical operations, as well as "if-then" rules. The fuzzy inference process involves crisp (non-fuzzy) inputs, linguistic (fuzzy) rules, a defuzzifier, and the crisp output.

Fuzzy logic is built on top of the experience of experts who already understand the system under study. It is built on the structures of qualitative description used in the everyday natural language, which makes it easy to use. This is because, oftentimes, systems do not have enough precise data to allow statistical analysis, which normally demands data collection over a long time. Fuzzy logic, being tolerant of imprecise data, builds this understanding into the process rather than tacking it onto the end. Moreover, fuzzy logic can model nonlinear functions of arbitrary complexity. In general, a fuzzy logic system can be defined in three steps: fuzzification, fuzzy rules, and defuzzification (Labibi et al. 1998).

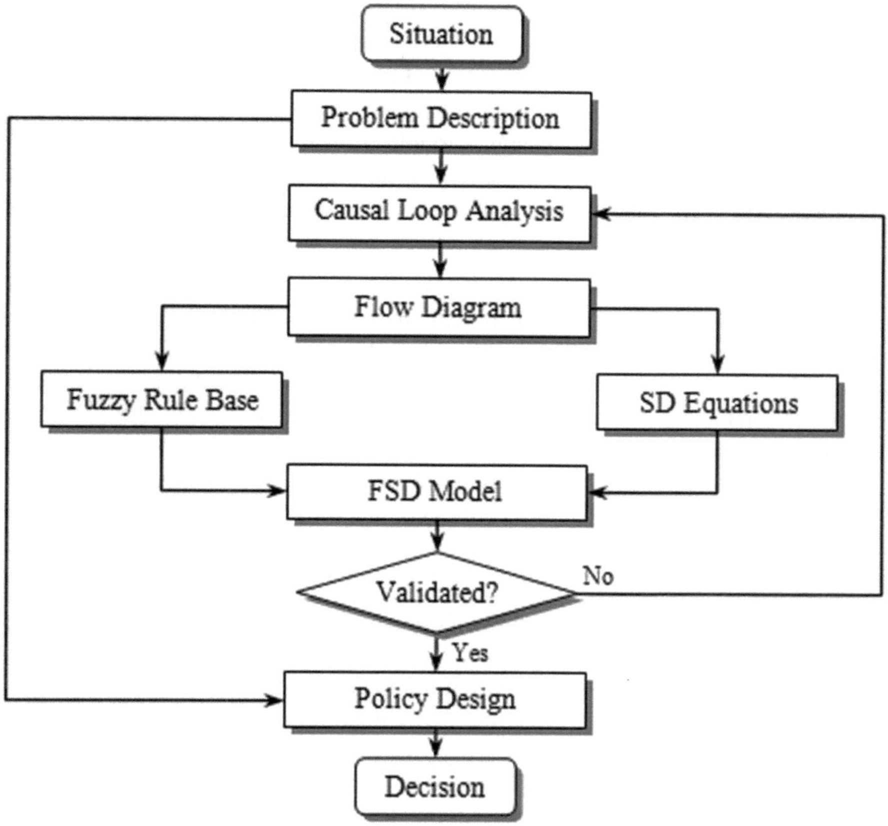

Fig. 1 Steps in an fuzzy system dynamics (FSD) study

Fuzzy System Dynamics Framework

FSD inherits its concepts from system dynamics and fuzzy set theory. Figure 1 shows a set of steps to guide a systems analyst in a thorough and sound dynamic simulation study in a fuzzy environment. The steps are categorized into six phases as follows:

Fuzzy System Dynamics Phases

The proposed FSD simulation methodology generally follows through six phases: (1) identification of problem situation, (2) causal loop analysis, (3) model formulation and development, (4) verification and validation, (5) policy analysis and improvement, and (6) implementation. Descriptions of each phase are presented, following the structure shown in Fig. 1.

Phase 1: Identification of Problem Situation This phase is concerned with the identification and understanding of the problem situation, which leads to a clear problem statement. Trends in the key variables relating to the identified problem are investigated. It is important for the modeler to make any necessary observations on the behavior of the actual system. Variables relating to the problem are filtered out while investigating their possible impact on the behavior of the actual system. This leads to system conceptualization stage in the next phase, known as causal loop analysis.

Phase 2: Causal Feedback Loop Analysis This stage involves system conceptualization, which is concerned with identification of the causal linkages and interactions between the main variables of the problem, based on the principles of cybernetics and feedback analysis. The main variables are those that are expected to have significant influences on the overall behavior of the actual system in the context of the observed problem. A causal link is indicated by an arrow that connects the causal variable at the tails of the arrow, to the effect variable at the head of the arrow. A "+" sign close to the arrowhead indicates that both the causal and the effect variables change in the same direction, while a "−" sign indicates that the tail and the head variables change in the opposite direction.

Phase 3: Model Formulation and Development The end product of model formulation and development is the FSD model. First, a stock flow diagram is developed using a suitable simulation tool such as Simulink to represent input and output flows. The block diagram should include the fuzzy logic block diagrams that model the fuzzy variables and relationships. Second, the modeling process branches into two parallel activities to produce the FSD model: (1) the development of the fuzzy rule base using suitable fuzzy logic tools, (2) the development of the SD equations.

The fuzzy rule base is constructed from expert knowledge and experience based on fuzzy logic. A set of "if-then" rules are constructed to emulate the expert in the subject area. Parallel to the construction of the fuzzy rule base, SD equations are built into the block diagram (flow diagram). The two activities yield the FSD model obtained by linking the fuzzy rule base with the SD model blocks that contain the SD equations.

Phase 4: Verification and Validation In this stage, the FSD model is verified to check for any errors in the logical flow of the model. This is followed by model validation which determines whether or not the model is an accurate representation of the real system. Validation is usually achieved through an iterative comparison of the model with the actual response of the system under study. Any discrepancies between the two are used to improve the system model. The availability of data is crucial for the success of this stage. In practice, when developing the FSD model, an appreciable set of verification and validation methods is commonly adopted with success (Forrester and Senge 1980; Sterman 2004; Barlas 1996; Saysel and Barlas 2006; Qudrat-Ullah and Seong 2010). Table 1 lists the methods that are generally accepted for validation.

Table 1 Structural validity testing methods

Number	Validation method	Brief description
1	Structural validity test	This method tests whether the model structure is consistent with relevant descriptive knowledge of the system being modeled
2	Indirect structural validity test	The indirect structure validity test method distils essential structures of the model, simplifying the model to tell the fundamental dynamics of a large-scale model
3	Extreme conditions	This method tests whether the model exhibits a logical behavior when selected parameters are assigned extreme values
4	Parameter verification of the system	This approach tests whether the parameters in the model are consistent with relevant descriptive and numerical knowledge
5	Dimensional consistency	This approach tests whether each equation in the model dimensionally corresponds to the real system
6	Boundary adequacy	This method tests whether the important concepts and structures for addressing the policy issues are endogenous to the model

Phase 5: Policy Analysis and Improvement In this phase, alternative scenarios are designed for simulation analysis in line with decisions that need to be considered. For each scenario, decisions need to be made in regards to the length of simulation runs, the run step, as well as the warm-up period. Simulation runs and their subsequent analysis are then used to estimate the performance indicators for the alternative system designs or alternative policy designs.

Phase 6: Decision Support and policy Implementation Being the last step of the simulation study, the success of the implementation phase is much dependent on how well the previous phases have been performed. The system analysis should ideally involve all the ultimate model users. The success of the implementation stage also depends on the underlying assumptions that were used in building the model.

Central to the FSD paradigm, is the development of the fuzzy logic system that can address the fuzzy variables of the system under study. The method incorporates fuzzy modeling concepts, fuzzy logic, and fuzzy logic rule base to improve realism in the modeling process. This can be implemented using system dynamics software tools such as Simulink® on a Matlab® platform and Vensim®. In this chapter, illustrations are based on Simulink applications. The next section provides an explanation of the causal loop analysis upon which the FSD model is built.

Fuzzy System Dynamics Modeling

The process of FSD modeling can be divided into two broad parts: causal feedback loop analysis and FSD model construction. A causal feedback loop analysis diagram shows the major causal linkages between the main variables of the system under

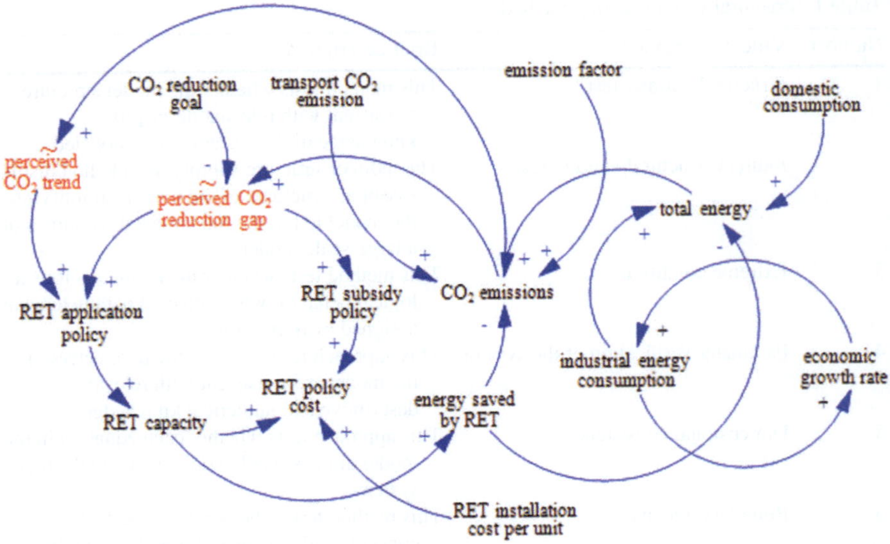

Fig. 2 The causal feedback loops for renewable energy policy with fuzzy variables

investigation. Identification of the major causal feedback loops of the system is crucial, together with the system inputs and outputs. Causal loops are used to model the causal linkages between related variables, directions of variable influences, and the system boundaries of the system. The focus of this chapter is on renewable energy policy formulation and evaluation in a fuzzy environment.

Figure 2 shows the causal feedback loops, describing the relationship between renewable energy policies and the associated carbon emissions. The inputs to the FSD system include the information on the particular RET system to be implemented, while the outputs of the system are the reduction of carbon emissions, the RET dynamic policy, and the associated cost of policy implementation. In a typical community, carbon emissions are influenced indirectly by industrial and domestic consumption of electricity generated from thermal power, and directly by transportation and domestic usage of thermal power. The main variables in the causal feedback loops are briefly described as follows:

- Perceived carbon reduction gap: the perceived difference between the carbon reduction goal and the actual emissions; the variable may take linguistic values "low", "ok", and "high"
- Perceived carbon trend: the perceived trend, i.e., increase or decrease, of the current carbon emissions; the variable may take linguistic values "decreasing", and "increasing"
- Carbon emissions: the total carbon emissions which vary in accordance with industrial, domestic, and transport energy usage
- Energy saved by RET: this variable represents the surplus energy generation saved through the application of the RET such as solar water heater systems, photovoltaic systems, and wind energy systems

- RET application policy: this variable is influenced by the perceived carbon reduction gap and the perceived carbon trend
- RET capacity: the capacity of renewable energy in use, which varies according to the RET application policy, that is, policy incentives, promotion policy, and policy control
- RET policy cost: the cost of RET policy is influenced by the subsidy policy cost, the installation costs, and the capacity of the RET
- Total energy: this is the total energy in form of electricity generated by thermal production for industry consumption and domestic consumption

Following the causal loop analysis described above, the FSD model is constructed in order to simulate and evaluate alternative RET policy scenarios. The model was developed based on a control-theoretic approach using fuzzy logic tools and Simulink in Matlab, consisting of three stocks, namely: the RET capacity, the transport, and the population. Through FSD simulation expert knowledge is built into a fuzzy rule base and simulated to see the related effects of alternative dynamic fuzzy rules on the amount of carbon emission. To capture the fuzzy variables, the perceived carbon reduction gap is converted to a fuzzy set, called perceived error. The error is defined as a function of the difference between the maximum acceptable carbon reduction gap and the perceived reduction gap. In essence, the perceived gap should be as close as possible to the maximum acceptable gap, which directly implies that the error should be as close to zero as possible. Therefore, we define the perceived error, *error*, as follows:

$$error = \frac{perceived\_gap}{perceived\_gap_m} - 1 \qquad (2)$$

Here, $perceived\_gap_m$ is the maximum acceptable perceived gap, and *perceived_gap* is the observed gap. Since *perceived_gap* and $perceived\_gap_m$ are supposed to be as close as possible, the *error* values close to zero are most preferable, and the level of preference diminishes fast as the magnitude of the error values increases. Apart from *error*, we also define perceived trend, *trend*, as a function of the observed carbon emissions, as follows:

$$trend = \frac{\mathrm{d}}{\mathrm{dt}}(CO_2\ emissions) \qquad (3)$$

The perceived *trend*, defines whether the quantity of carbon emissions is increasing or decreasing. It follows that if the *trend* is increasing, then the intensity of the corresponding energy policy initiatives should be increased. Conversely, if the *trend* is decreasing, then the desired policy efforts should be decreased. The set of these expert rules can form an effective platform for managing investment, promotional, and incentive policies that influence the adoption of renewable energy which ultimately leads to low-carbon societies. Based on the fuzzy causal loop analysis explained earlier, a fuzzy rule base is constructed to represent the fuzzy policy design for the renewable energy market place. As an illustration, let the variable *policy_change*

R1: IF (error is ok) THEN (policy change is zero);

R2: IF (error is low) THEN (policy change is reduce fast);

R3: IF (error is high) THEN (policy change is increase fast);

R4: IF (error is ok) and (trend is positive) THEN (policy change is reduce slowly);

R5: IF (error is ok) and (trend is negative) THEN (policy change is increase slowly);

Fig. 3 Fuzzy rule base for renewable energy policy

represent the desired policy adjustment. Then, a fuzzy rule base can be constructed as illustrated in Fig. 3.

According to rule R1, whenever the gap is ok, the desired policy change is "zero", meaning that the current policy remains unchanged. With reference to rule R2, whenever the error is low, it follows that the desired policy change is "reduce fast" meaning that the current policy efforts should be reduced at a faster rate since the perceived carbon reduction gap is much lower than the acceptable level. On the other hand, if the error is high, then the policy should be "increase fast". In addition, if the error is ok, that is, in the neighborhood of zero, then the actual decision depends on whether the current trend (rate) of carbon emissions is increasing or decreasing. If the trend is increasing then the policy should ideally be "reduce slowly". Conversely, if the current is decreasing, then the policy should be "increase slowly" since the trend shows that carbon emissions are somewhat on the increase. The FSD model was tested and verified using the following validation methods:

- Extreme conditions: This method tests whether the model exhibits a logical behavior when selected parameters are assigned extreme values (Qudrat-Ullah and Davidsen 2001; Qudrat-Ullah and Seong 2010)
- Indirect structure validity test: The validity test method distils essential structures of the model, via simulation, to tell the fundamental dynamics of the model (Barlas 1996; Saysel and Barlas 2006)

Following the verification of our FSD model, we present experimental simulation approaches that are essential for further evaluation and analysis of renewable energy policies in a fuzzy environment, deriving useful managerial insights. A case example is provided for further analysis and discussion in the next section.

Experimental Approaches for Simulation

Further to the formal framework suggested and outlined above for simulation and evaluation of renewable energy policies in a fuzzy environment, this section selects a case example of South Africa (SA) as a base example for analysis and discussion.

Case Example: South Africa

South Africa intends to lower its carbon emissions to 34 % below current expected levels by 2020 and to about 42 % below current trends by 2025, subject to adequate financial support from the international community (BBC 2009). Currently, the country is dependent on thermal power which accounts for 80–90 % of the total primary energy supply in the year 2010. SA's renewable sources include solar, wind, hydro, biomass, geothermal, and ocean energy. This shows that the country need to put in place an active policy to pursue RETs and set up effective policies in order to reduce carbon emissions (Winkler 2006). For instance, such policies should promote the development of solar energy industry and the utilization of solar energy products, which have an availability factor of 60 % (NER 2004). Thus, the SA government intends to promote her renewable energy policy by promoting the utilization of solar energy products, including photovoltaic systems and solar water heating systems. Several households, clinics, and schools have photovoltaic systems. There is a steady increase of solar water heater installations in households, with more than 100,000 installations every month. In addition to solar energy, wind energy is also harvested and the installations are on the increase (Winkler 2006). The government reports that at least 10,GWh per year of final energy demand should be met by renewable energy sources, including solar, wind, and small hydro (NER 2004).

The National Integrated Energy Plan for South Africa (DEES 2004) estimates that the economic growth of the country in terms of GDP is 2.8 % per annum and the population growth rate is about 1.3 % per annum. The energy demand is expected to grow by a margin of about 2–3 % per annum, as shown in Fig. 4 (Winkler 2006).

The SA energy policy has five objectives for the energy sector: (a) increased access to affordable energy services, (b) improving energy governance, (c) stimulating economic development, (d) managing energy related environmental impacts, and (e) securing diversity through diversity, which addresses the need to provide alternative sources of energy including renewable energy. It recognizes the potential of renewable energy in securing supply through diversity.

Proposed Policy Scenarios

SA endeavors to implement a renewable energy policy in form of wind and solar energy resources, with the aim of reducing carbon emissions from thermal production of electricity, industry, and domestic use (MED-SA 2003). In this connection, the policies can be matched into three possible scenarios. The first scenario, the base case, is aimed at benchmarking the carbon emissions without promoting any renewable energy policies. On the contrary, the second scenario observes the variation of carbon emissions when solar energy policies are implemented. It is important to note that policies can be deterministic, whereby the intensity of the promotion is either constant or changing periodically, or fuzzy dynamic, in which case the policies are adjusted according to the observed fuzzy trends in the system. In this framework, this

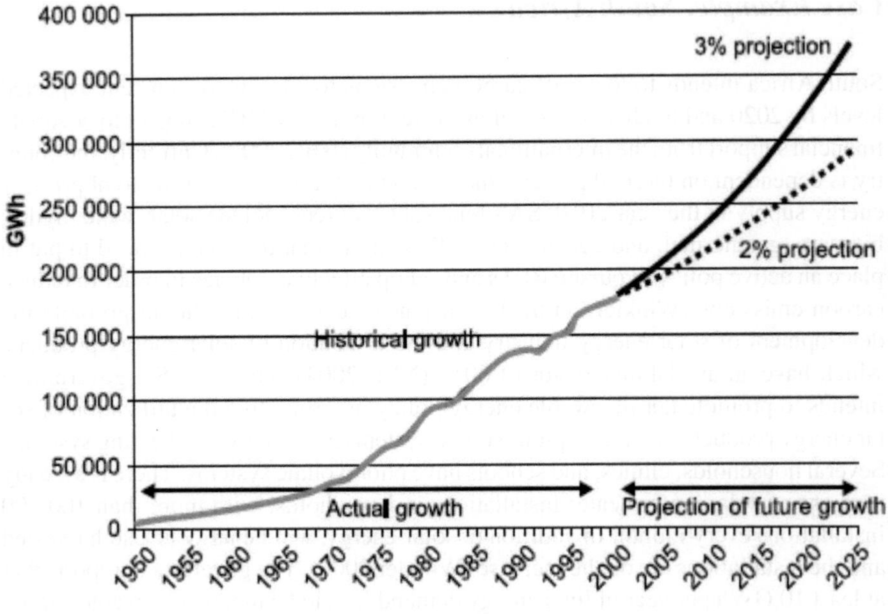

Fig. 4 Growth in electricity sales, actual and future projections. (NER 2000)

scenario is twofold: First, simulation is carried out based on the assumption that a deterministic control policy is used without incorporating the fuzzy-based dynamic policy, and second, the simulation is run assuming that a fuzzy dynamic control policy is implemented based on the two fuzzy variables: perceived carbon trend and perceived carbon reduction gap. In a similar manner, the third scenario observes the variation of carbon emissions when wind-based RET polices are implemented. The scenario is twofold; first, with deterministic promotion policies, and second, with dynamic fuzzy feedback from the market trends. Table 2 provides a summary of the policy scenarios for simulation and evaluation.

The next section presents a summary of this chapter, concluding remarks, contributions, and further research.

Concluding Remarks and Further Research

This chapter provides a formal framework for realistic formulation and evaluation of renewable energy policies. Unlike previous simulation models and frameworks, the current framework considers that real-world low-carbon energy, environment and economic systems are inundated with fuzzy variables which make the whole system complex. As such, policy makers rely on imprecise information from the renewable energy marketplace so as to formulate appropriate medium-term to long-term strategies. With this realization, the framework identifies two major fuzzy variables, namely, perceived CO_2 reduction gap and perceived CO_2 trend that are modeled as

Table 2 A summary of policy scenarios for simulation and evaluation

No.	Scenario	Description
1	Bases case—without RET policies	This scenario represents the as-is model aimed at benchmarking the carbon emissions of SA without any renewable energy promotion policies
2	Promote solar RET with and without fuzzy-based policy control	This scenario observes the variation of carbon emissions when solar RET policies are enhanced, first without fuzzy control then with fuzzy control promotion
3	Promote wind RET with and without fuzzy-based policy	This scenario observes the variation of carbon emissions when wind RET policies are promoted, first with deterministic policies, then with fuzzy-based policies

RET renewable energy technologies, *SA* South Africa

linguistic variables from a fuzzy causal loop perspective. Drawing from the fuzzy causal loop analysis, the framework provides a stepwise guide for building an FSD model based on fuzzy logic tools and control theoretic simulation on a Matlab platform. Overall, the chapter contributes to the existing body of knowledge in policy formulation and evaluation for the 3Es concept of energy, economic development, and the environment aimed at building a low-carbon society.

Contributions to Theory

The 3Es concept of energy, economy, and environment is a complex system characterized with dynamic and fuzzy variables. No doubt, the policy formulation and evaluation for such as system demands the application of system modeling tools that address both dynamic and fuzzy features of the problem. This work points to the existence of these complexities in the 3Es concept, highlighting the imperative need for developing simulation approaches that can capture the complex features of the system. Therefore, the development of an FSD model is an important contribution to the system dynamics community and to the practicing policy makers in governments and other stakeholders. In addition, this research work points out the need to build more realism into systems simulation models especially for behavioral models where essential variables involve human judgments and perceptions. Fuzzy set theory is a viable and important inclusion into system dynamics models when information is fuzzy or imprecise.

Managerial Implications

Policy formulation and evaluation for a fuzzy 3Es system of energy, environment, and economy is complex due to the presence of fuzzy and dynamic variables. As such, the policy maker needs to have in place an appropriate guide for renewable energy policy formulation. First, the policy maker needs to identify dynamic interacting variables in a causal loop form. This is followed by identification of fuzzy variables upon which the policies are anchored in order to make robust dynamic policies. The proposed approach in this chapter offers a number of advantages for the decision makers:

- FSD provides the modeler with the opportunity to model fuzzy variables upon which dynamic renewable energy policies can be anchored, that is, perceived carbon reduction gap and perceived carbon trend.
- FSD uses fuzzy logic and control-theoretic tools, that is, tools which make model building easy within a reasonable modeling time
- The FSD approach builds from the prior knowledge captured from experts in the field such that the users gain confidence and trust in the model as it is based on practical knowledge of experts, rather than theoretical assumptions
- Expert knowledge can easily be built into the fuzzy rule base and updated on time with ease

In light of the above-mentioned managerial implications, the application of FSD offers significant advantages to the policy maker concerned with renewable energy formulation and evaluation. Therefore, the FSD framework suggested in this chapter is an important contribution to the practicing policy makers concerned with low-carbon economy societies, environments, and economies.

Further Research

Further research prospects are realized in this chapter. The FSD model presented in this study can be enhanced further. For instance, when building the fuzzy rule base, the construction can be such that the rules are optimized using an optimization tool such as genetic algorithms. Given enough sample data, the rule base and the weights of the specific rules can be fine tuned and optimized using soft computing tools such as genetic algorithms in Matlab. This can further enhance policy formulation for renewable energy systems.

References

Anand, S., Vrat, P., Dahiya, R.P.: Application of a system dynamics approach for assessment and mitigation of CO_2 emissions from the cement industry. J Environ. Manage. **79**, 383–398 (2005)

Barlas, Y.: Formal aspects of model validity and validation in system dynamics. Syst. Dynam. Rev. **12**(3), 183–210 (1996)

BBC.: South Africa to cut carbon emissions by 34 %. http://news.bbc.co.uk/2/hi/africa/8398775.stm (2009). Accessed 02 June 2012

Chandrasekara, B., Tara, C.K.: An opinion survey based assessment of renewable energy technology development in India. Renew. Sustain. Energ. Rev. **11**, 688–701 (2007)

Chang, Y., Ries, R.J., Wang, Y.: The embodied energy and environmental emissions of construction projects in China: an economic input—output LCA model. J. Energ. Policy **38**(11), 6597–6603 (2010)
Chen, F., Duic, N., Alves, L.M., Carvalho, M.G.: Renewislands—renewable energy policies for islands. Renew. Sust. Energ. Rev. **11**(8), 1888–1902 (2007)
Coyle, RG.: System Dynamics Modeling—A Practical Approach. Chapman and Hall, New York (1996)
DEES.: Draft energy efficiency strategy of the Republic of South Africa. Department of Minerals and Energy (April 2004)
Demiroren, A., Yilmaz, U.: Analysis of change in electric energy cost with using renewable energy sources in Gokceada, Turkey: an island example. Ren. Sust. Energ. Rev. **14**, 323–333 (2010)
Dyner, I., Smith, R., Pena, G.: System dynamics modeling for energy efficiency analysis and management. J. Oper. Res. **46**(10), 1163–1173 (1995)
Ernest, F.B., Matthew, A.B.: Feasibility of solar technology (photovoltaic) adoption—a case study on Tennessee's poultry industry. Ren. Energ. **34**, 748–754 (2009)
Ford, J.W.: System dynamics and the electric power industry. Syst. Dynam. Rev. **13**(1), 57–85 (1997)
Forester, J.: Industrial Dynamics. Pegasus Communications, Waltham (1961)
Forrester, J. W., Senge, P. M.: Tests for building confidence in system dynamics models. Time Stud. Manage. Sci. **14**, 209–228 (1980)
GPO.: Annual energy outlook, with projects to 2030. USA Government Printing Office (2009)
Han, J., Hayashi, Y.: A system dynamics model of CO_2 mitigation in China's inter-city passenger transport. Transport. Res. D. Transport Environ. **13**(5), 298–305 (2008)
Huang, L.M.: Financing rural renewable energy: a comparison between China and India. Renew. Sust. Energ. Rev. **13**(5), 1096–1103 (2009)
IPCC—Intergovernmental Panel on Climate Change.: IPCC third assessment report, IPCC. http://www.ipcc.ch/ipccreports/tar/index.htm (2001). Accessed 22 Nov 2010
IPCC—Intergovernmental Panel on Climate Change.: IPCC fourth assessment report, IPCC. http://www.ipcc.ch/pdf/assessment-report/ar4/syr/ar4_syr.pdf (2007). Accessed 02 June 2012
Jan, T.S., Hsiao, C.T.: A four-role model of the automotive industry development in developing countries: a case in Taiwan. J. Oper. Res. Soc. **55**(11), 1145–1155 (2004)
Jin, W., Xu, L., Yang, Z.: Modeling a policy making framework for urban sustainability: incorporating system dynamics into the ecological footprint. Ecolog. Econ. **68**(12), 2938–2949 (2009)
John, B., Bo, S., William, W.: The economics of sustainable energy for rural development: a study of renewable energy in rural China. Energ. Policy **26**, 45–54 (1998)
Kosko, B.: Fuzzy systems as universal fuzzy approximation. In: Proceedings of the First IEEE International Conference on Fuzzy Systems, pp. 1153–1162 (1992a)
Kosko, B.: Neural Networks and Fuzzy Systems. Prentice-Hall, Englewood Cliffs (1992b)
Kosko, B.: Fuzzy Thinking. Harpner Collins, London (1994)
Kosko, B.: Combining fuzzy systems. IEEE FUZZ-95, Yokohama, Japan, March, 1995
Krushna, M., Leif, G.: An adopter-centric approach to analyze the diffusion patterns of innovative residential heating systems in Sweden. Energ. Policy **36**, 577–579 (2008)
Labibi, A.W., Williams, G.B., O'Conor, R.F.: An intelligent maintenance model system: an application of the analytic hierarchy process and a fuzzy logic rule-based controller. J. Oper. Res. Soc. **49**, 745–757 (1998)
Levary, R.: Systems dynamics with fuzzy-logic. Int. J. Syst. Sci. **21**(8), 1701–1707 (1990)
Li, J.F., Zhu, L., Hu, R.Q.: Policy analysis of the barriers to renewable energy development in the people's republic of China. Energ. Sustain. Dev. **6**, 11–20 (2009)
Liu, H.Y., Wu, S.D.: An assessment on the planning and construction of an island renewable energy system: a case study of Kinmen Island. Renew. Energ. **35**(12), 2723–2731 (2010)
Mamdani, E.H., Assilian, S.: An experiment in linguistic synthesis with a fuzzy logic controller. Int. J. Man Mach. Stud. **7**(1), 1–13 (1975)

MED-SA.: White paper on renewable energy. Department of minerals and energy, Republic of South Africa (2003)
Morecroft, J.D.W.: Strategic modeling and business dynamics: a feedback systems approach. Wiley, Chichester (2007)
Mutingi, M., Matope, S.: System dynamics of renewable energy technology adoption. IEEE International Conference on Industrial Technology, South Africa (2013)
Mutingi, M., Mbohwa, C.: Fuzzy system dynamics simulation for manufacturing supply chain systems with uncertain demand. In: CIE42 Proceedings of the International Conference on Computers and Industrial Engineering, South Africa, pp. 1–12 (2012)
Naill, R.: A system dynamics model for national energy policy planning. Syst. Dynam. Rev. **8**(1), 1–19 (1992)
NER.: National Electricity Regulator. Electricity Supply Statistics for South Africa.NER, Pretoria (2004)
Oikonomou, E.K., Kilias, V., Goumas, A., Rigopoulos, A., Karakatsani, E., Damasiotis, M.: Renewable energy sources projects and their barriers on a regional scale: the case of wind parks in the Dodecanese islands, Greece. Energ. Policy **37**(11), 4874–4883 (2009)
Peter, W.: Biogas production: current state and perspectives. Appl. Microbiol. Biotechnol. **85**, 849–860 (2010)
Peters, G.P.: From production-based to consumption-based national emission inventories. Ecolog. Econ. **65**(1), 13–23 (2008)
Qudrat-Ullah, H.: MDESRAP: a model for understanding the dynamics of electricity supply, resources, and pollution. Int. J. Global Energ. **23**(1), 1–4 (2005)
Qudrat-Ullah, H., Davidsen, P.I.: Understanding the dynamics of electricity supply, resources and pollution: Pakistan's case. Energy **26**, 595–606 (2001)
Qudrat-Ullah, H., Seong, B.S.: How to do structural validity of a system dynamics type simulation model: the case of an energy policy model. Energ. Policy **38**, 2216–2224 (2010)
Raja, P., Laurence, D., Vasanthi, M.P.: Adoption of photovoltaic power supply systems: a study of key determinants in India. Ren. Energ. **31**, 2272–2283 (2006)
Roger, F., Naill, R., Belanger, S., Klinger, A., Petersen, E.: An analysis of the cost effectiveness of U.S. energy policies to mitigate global warming. Sys. Dynam. Rev. **8**(2), 111–128 (1990)
Saysel, A. K., Barlas, Y.: Model simplification and validation testing. Sys. Dynam. Rev. **22**(3), 241–262 (2006)
Sterman, J. D.: Business Dynamics: Systems Thinking and Modeling for a Complex World. Irwin/McGraw-Hill, Boston (2004)
Sugeno, M.: Industrial applications of fuzzy control. Elsevier Science Pub. Co.(1985)
Tessem, B., Davidsen, P.I.: Fuzzy system dynamics: an approach to vague and qualitative variables in simulation. Sys. Dynam. Rev. **10**(1), 49–62 (1994)
Trappey, C.V., Trappey, A.J.C., Lin, G.Y.P., Chang, Y-S.: System dynamics evaluation of renewable energy policies. In: Frey, D.D., Fukuda, S., Rock, G. (eds.) Improving Complex Systems Today. Advanced Concurrent Engineering, pp. 133–140. Springer, New York (2011)
Trappey, A., Trappey, C., Hsiao, C.T., Ou, J., Chang, C.T.: System dynamics modeling of product carbon footprint life cycles for collaborative green supply chains. Int. J. Comp. Integ. M. **25**(10), 934–945 (2012a)
Trappey, A.J.C., Trappey, C.V., Lina, G.Y.P., Chang, Y.S.: The analysis of renewable energy policies for the Taiwan Penghu island administrative region. Renew. Sustain. Energ. Rev. **16**, 958–965 (2012b)
Trappey, A.J.C., Trappey, C., Hsiao, C.T., Ou, J.J.R., Li, S.J., Chen, K.W.P.: An evaluation model for low carbon island policy: the case of Taiwan's green transportation policy. Energ. Policy **45**, 510–515 (2012c)
Uemura, Y., Kai, T., Natori, R., Takahashi, T., Hatate, Y., Yoshida, M.: Potential of renewable energy sources and its applications in Yakashima Island. Renew. Energ. **29**, 581–591 (2003)
United Nations.: Kyoto protocol to the united nations framework convention on climate change. UNFCCC (1998)

Vicki, G., Tomas, M.K.: Breaking the cycle producer and consumer perspectives on the non-adoption of passive solar housing in the US. Energ. Policy **36**, 551–566 (2008)
Vizayakumar, K., Mohapatra, P.K.J.: Modeling and simulation of environmental impacts of coalfield: system dynamics approach. J. Environ. Syst. **22**(1), 59–73 (1993)
Winkler. H.: Energy policies for sustainable development in South Africa: options for the future. Energy Research Centre, University of Cape Town (2006)
Zadeh, L.A.: Fuzzy sets. Inform. Control **8**, 338–353 (1965)
Zadeh, L.A.: Fuzzy sets as a basis for a theory of possibility. Fuzzy Set. Syst. **1**, 3–28 (1978)

The Diffusion of Eco-Technologies: A Model-Based Theory

Matthias Otto Müller, Ruth Kaufmann-Hayoz, Markus Schwaninger and Silvia Ulli-Beer

Introduction

On the road toward ecological sustainability, technological change in general and innovations in particular play a crucial role. In fact, in many domains progress toward sustainability has been brought about by the replacement of a relatively inefficient, pollution-intensive technology with a more efficient, less polluting technology. In line with other contributions in the literature (Giannetti et al. 2004; Kemp and Pearson 2007; Kemp 2009; Kemp 2011), we refer to such technologies as "eco-technologies," or as "eco-innovations." Specifically, an eco-innovation may be defined as "the introduction of any new or significantly improved product (good or service), process, organizational change, or marketing solution that reduces the use of natural resources (including materials, energy, water, and land) and decreases the release of harmful substances across the whole life-cycle" (Eco-Innovation Observatory 2012, p. 8). In line with this definition, we identify a technology as an "eco-technology" if it is cleaner or more efficient compared to most commonly used

M. O. Müller (✉)
Greenwood Strategic Advisors AG, Zugerstrasse 40, 6314 Unterägeri, Switzerland
e-mail: matthiasottomueller@gmail.com, matthias.mueller@greenwood-ag.com

R. Kaufmann-Hayoz
Interdisciplinary Centre for General Ecology (IKAÖ), University of Bern,
Schanzeneckstr. 1, 3001 Bern, Switzerland
e-mail: ruth.kaufmann-hayoz@ikaoe.unibe.ch

M. Schwaninger
Institute of Management (IfB), University of St. Gallen,
Dufourstrasse 40a, 9000 St. Gallen, Switzerland
e-mail: markus.schwaninger@unisg.ch

S. Ulli-Beer
Dynamics of Innovative Systems (DIS), Paul Scherrer Institut,
5232 Villigen, Switzerland
e-mail: silvia.ulli-beer@bluewin.ch

H. Qudrat-Ullah (ed.), *Energy Policy Modeling in the 21st Century*,
Understanding Complex Systems, DOI 10.1007/978-1-4614-8606-0_4,

technology.[1] Examples of eco-technologies are highly energy-efficient buildings (compared to buildings without insulation) and solar-warmed water (compared to electric boilers powered by coal-generated electricity). In this chapter, we have in mind mainly low-emission and energy-efficient technologies. Yet, the arguments we present are likely to apply to many other types of eco-innovations, such as water-saving innovations or exhaust filters.

This leads to the questions "How do eco-technologies diffuse?" and "What can be done to accelerate their diffusion?" Starting with the seminal work by Rogers (2003) on the diffusion of innovations, several streams of research have provided insights into these issues. Rogers defines the diffusion of innovations as a process "in which an innovation is communicated through certain channels over time among the members of a social system" (Rogers 2003, p. 5). Rogers's work has inspired a large body of research and influenced other disciplines such as marketing, sociology, communication sciences, and computer sciences. However, classical innovation diffusion studies have an implicit focus on individual adopters (such as persons or organizational units). Recent contributions on sustainability transitions, such as studies based on the multilevel perspective (Geels 2004; Geels and Schot 2007), provide stronger integration of microlevel with macrolevel developments.

The recent literature on sustainability transitions is interesting and inspiring, and we acknowledge its crucial contributions to the understanding of transitions toward sustainability. There are, however, two domains in which we depart from current thinking. First, we think the role of public policy in the successful diffusion of eco-technologies is crucial, and that it has been underestimated in the literature. Second, we propose that the application of modeling and simulation methodologies enhances current theorizing based on "natural language" (Hanneman 1988). In what follows, we further elaborate on both domains.

a) The central role of public policy
Promoting eco-technologies is a key aspect of public policy,[2] both as a solution to specific environmental problems as well as an approach to increasing the sustainability of a country as a whole. Technology is a preferred policy lever, as the other two main policy levers for sustainable development (reducing the size of the population and its affluence) are of questionable value: The forced reduction of the size of the population is outside consideration from a moral and ethical point of view. Voluntary birth control policies are slow to take effect, and they also raise difficult moral and ethical questions. Promoting reductions

[1] In this conceptualization it does not matter whether the eco-technology is a completely new technology or improvement of an existing technology. What is important is that there is a "better" configuration that has to overcome a "worse" configuration (also see "Model Sectors and General Setup"; in particular Table 1).

[2] In line with Knoepfel et al. (2007, p. 24), we define public policy as "a series of intentionally coherent decisions or activities taken or carried out by different public and sometimes private actors whose resources, institutional links and interest vary, with a view to resolving in a targeted manner a problem defined politically as collective in nature." Relying on this abstract term allows us to ignore the question which specific institutions are involved.

in a population's affluence for environmental reasons would not be a winning platform for any politician. Hence, promoting sustainable development by way of fostering eco-technologies is arguably the most favored approach to decreasing the environmental impact of economic activities.

This means that while most technological innovations need to stand the test of a free market environment in order to diffuse successfully, eco-innovations are likely to get public policy support that accelerates their diffusion. In this chapter, we present an analysis of the relationships between the market, technology, policy change, and public policy interventions. The theory described in this chapter results from generalizing findings of an in-depth study of the diffusion of energy-efficient renovations (Müller 2012, 2013). A preliminary version of this framework was described in Müller (2012, p. 350–353). In this chapter, we follow up and provide a more elaborated framework. The chapter's second contribution is to reflect on the strengths and limitations of building theories with the assistance of modeling and simulation methodologies.

b) Methodology

Research on sustainability transitions typically uses descriptive theorizing and relies mostly on natural language rather than simulation methods (Ulli-Beer, in press). Therefore, such research is generally not well equipped to link structures of causalities and behaviors in the presence of interacting feedback loops that cause dynamic complexity (e.g., nonlinear behaviors). In consequence, Ulli-Beer (in press, Chap. 2) concludes that this literature does not determine "which and how causal structures influence system behavior."

An alternative conceptual-theoretical viewpoint proposes that the link between structure and behavior can be better understood by employing semimathematical languages (Hanneman 1988), such as those offered by modeling and simulation methodologies (e.g., discrete-event simulation, agent-based modeling, and System Dynamics). In several recent studies, we have applied System Dynamics modeling and simulation to the analysis of specific eco-technologies (e.g., Müller 2013; Ulli-Beer et al. 2006; and the research documented in Ulli-Beer in press). System Dynamics[3] is more appropriate to our research endeavor than alternative methodologies for the following reasons:

- System Dynamics has a special strength in the analysis and synthesis of causal relationships and in the handling of delays. In these respects, System Dynamics is superior to the other two methodologies mentioned.
- Compared to discrete-event simulation methods, System Dynamics conceives of the systems modeled as continuous processes. This is in line with our

[3] System Dynamics is an interdisciplinary, scientific methodology that is used to describe the structure of causality driving change processes and to elicit the resulting behavior produced by that structure. Specifically, change processes are represented mathematically by differential equations. In order to obtain behavior, these equations are solved, by way of computer simulation. Any kind of change process can be represented as a simulation model, regardless whether it stems from the physical, ecological, or social domain. The methodology was developed by Jay W. Forrester in the late 1950s and early 1960s by applying principles of control (from electric engineering) to the management of real-world problems (Lane and Oliva, 1998, p. 219; Sterman 2000).

project, which addresses the patterns of behavior exhibited by the system under investigation.
- Compared to agent-based modeling, System Dynamics adopts a top-down view rather than a bottom-up view. For analyzing high-level aggregates, System Dynamics is more appropriate than a methodology that focuses on the behavior of single agents.
- Finally, System Dynamics strongly encourages an "endogenous point of view" (Richardson 2011, p. 219). Based on such a systemic perspective, more effective policies can be developed.

Summarizing this discussion, the following research questions can be formulated:

- How should the diffusion of eco-technologies be described in a generic way that goes beyond the particulars of specific technologies? In particular, in what generic way do the market, technological change, and public policy trigger the diffusion of eco-technologies?
- How should System Dynamics modeling and simulation support research and public policy activities in the diffusion of a specific eco-technology?

The remainder of this chapter is structured as follows. In "A Generic Theory of the Diffusion of Eco-Technologies", we present the main elements of our generic theory, in the form of a System Dynamics simulation model. With this model, we provide the cornerstones of a *middle-range theory*[4] *of the diffusion of eco-technologies* (Merton 1957; Schwaninger and Groesser 2008). In "Discussion: How Can System Dynamics Modeling Support Research and Policy Making in Support of the Diffusion of Eco-technologies?", we discuss how our results and the System Dynamics methodology can be used to conduct research and support policy making, and we outline how such a process could actually be implemented. In "Conclusions", we summarize our findings and offer a brief reflection on the benefit of using formal models in research and policy making.

A Generic Theory of the Diffusion of Eco-Technologies

In the social sciences, the result of theory building is often stated in natural language. In contrast, we present our theory in the form of a System Dynamics simulation model. In line with Schwaninger and Groesser (2008), we see the model itself as the theory. In general, it is good practice to discuss simulation models on the level of the equations as well as on the level of the feedback loops implemented by the

[4] The concept of middle-range theory was introduced by Merton (1957). It refers to theories that are located between universal theories ("grand theories") and micro theories. They integrate theoretical and empirical research. They consolidate different hypotheses or findings. Instead of all-inclusive efforts to develop a unified theory, they are limited to specific types of contexts, which allow for the formulation and testing of specific hypotheses. A middle-range theory is generic in that it holds for a whole class of systems. In contrast, a micro theory is less abstract, deals with relatively small slices of time, and covers small numbers of objects, e.g., individuals, interactions, or families.

Table 1 Initial key characteristics of the two technologies

	Role	Market share	Technological maturity	Production costs	Environmental impact
Conventional	Incumbent	High	High	Low	High
Eco-technology	Innovation	Low	Low	High	Low

equations. However, limited space as well as the nontechnical focus of this chapter motivated us to provide a more graphically oriented, high-level description.[5]

Model Sectors and General Setup

We assume that there are two generic types of technology (see Table 1). *Conventional technology* is the incumbent technology with a large market share. Due to learning effects and economics of scale and scope, it has achieved a high degree of technological maturity and low production costs. What makes such technology problematic from a sustainability perspective is its high environmental impact. In contrast, *eco-technology* represents innovations with low environmental impact. However, due to the recent advent of such technology, it has not had any chance of moving down the learning curve or benefiting from economies of scale and scope. Therefore, it initially has a low degree of technological maturity, rather high production costs, and hence a small market share.

Should a substitution of an eco-technology for a conventional technology take place, the environmental impact of the installed base could be reduced. However, at the beginning of the substitution process, the eco-technology faces high-diffusion barriers. As it has a low market share, there is only limited potential for industrial learning, and hence it remains at a low market share. In order to promote the diffusion of the eco-technology, public policy needs to support the diffusion process until it can compete with the conventional technology based on market incentives alone.

In order to organize our generic model of the diffusion of eco-technologies, we use six distinct model sectors (see Fig. 1). Sectors 1, 2, and 3 represent the extended economic subsystem. Through the dynamics of the market (sector 1), technological quality (sector 2), and production costs (sector 3), the economic system controls the installed base of the technology, and eventually determines the environmental impact (sector 6). However, in the diffusion of eco-technologies, public policy plays a crucial role. Hence, we model how public policy changes (sector 4) and how public policy intervenes in the economic system (sector 5). Note that while major drivers of policy change may be exogenous, there is still some feedback between the economic subsystem and public policy. For example, the impact of the current technological quality of the eco-technology influences public policy interventions.

[5] The model is electronically available in the Vensim model format from Matthias Müller.

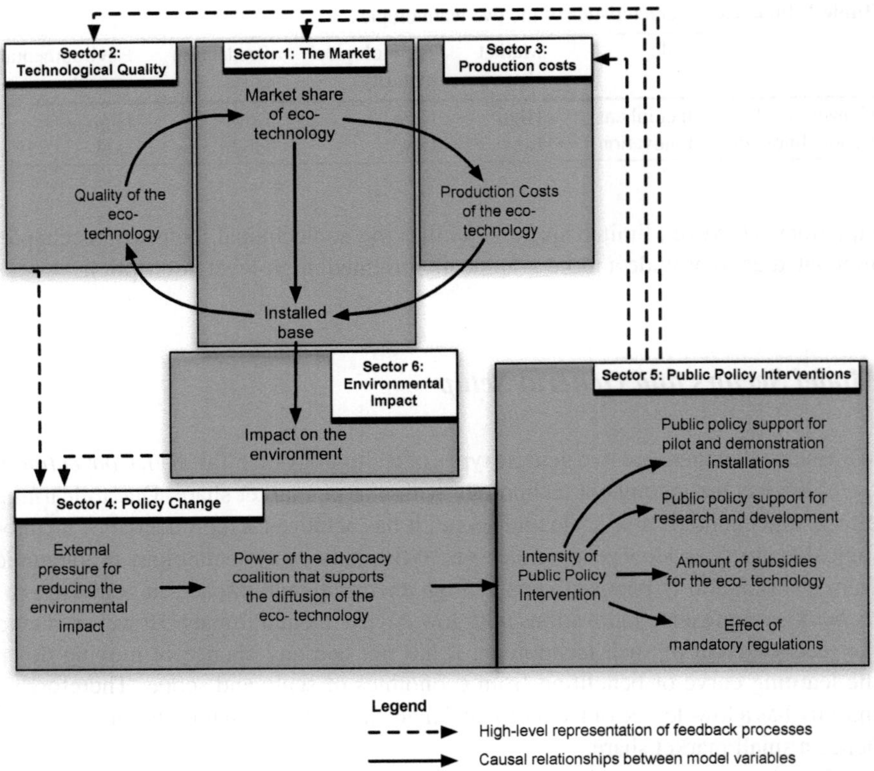

Fig. 1 Model sector diagram

In the remainder of this section, we provide a high-level description of these model sectors. Note that the actual model is subscripted. This means that in Fig. 2, for example, the equations are calculated once for the eco-technology and once for the conventional technology. In order to facilitate the communication of key model structures, the visualization contains only the most important structures. Structurally uninteresting calibration parameters and switches are omitted from this high-level description and need to be inspected in the actual model source code. We follow the convention of using <*brackets*> in the text to refer to variable names. <*Brackets*> in figures indicate that a variable was calculated in another model sector.

The Market and Its Effect on Installed Base

In our model, the market consists of two major feedback loops, demand and supply, that interact with one another. Through their interaction, these feedback loops control the installed base of the two technologies. Figure 2 shows a stock-and-flow diagram

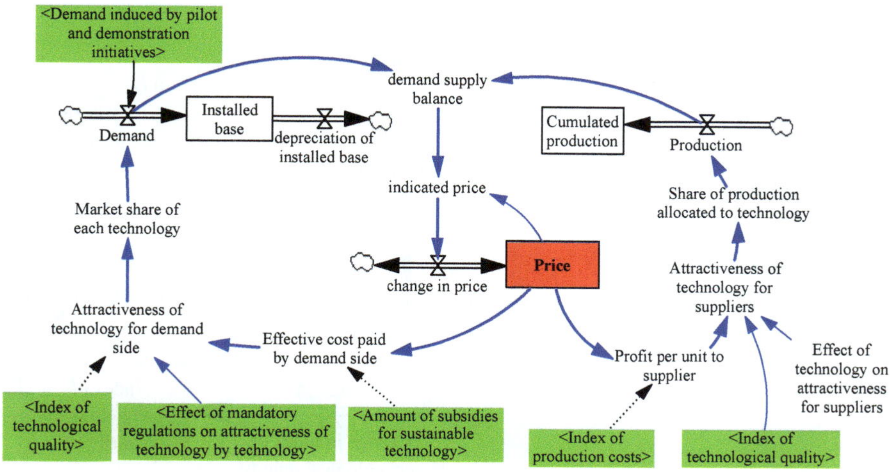

Fig. 2 Main feedback loops in the market sector. *Valves* represent flows and *boxes* represent stocks (integrations of flows). *<Variables set in single brackets>* in figures indicate that the variable is calculated in another model sector

of the two loops. In the middle, the price-setting mechanism is shown.[6] Whenever demand exceeds supply, that structure increases price. Whenever supply exceeds demand, that structure decreases price.

The basic logic of the two loops is as follows: When the *<price>* of a technology is reduced, then the *<effective cost paid by demand side>* is decreased as well, which in turn increases the *<attractiveness of technology for demand side>*. As the *<attractiveness of technology for demand side>* rises relative to the other technology, the *<market share>* of the corresponding technology rises too, eventually leading to increased *<demand>*. The*<installed base>* of each technology is modeled as a stock. This means that it is increased by the rate of *<demand>* and depleted by *<depreciation of installed base>*. As *<demand>* is increased, pressure mounts for prices to rise, thereby eventually dampening the whole demand loop.

On the supply loop, a similar basic pattern is shown. When the *<price>* of a technology is increased, the *<profit per unit to supplier>* is increased as well, thereby increasing the *<attractiveness of technology for suppliers>*. This in turn leads to an increased *<share of production . . . >* allocated to that specific technology and then to an increased *<production>* of that technology. Should*<production>* exceed *<demand>*, then the price mechanism will decrease *<price>* and thereby dampen the supply loop.

Figure 2 shows several other variables influencing the two market loops. For example, a measure of technological maturity (*<index of technological quality>*) influences both the *<attractiveness of technology for demand side>* as well as the *<attractiveness of technology for suppliers>*. Both the demand and the supply loop

[6] See Ventana Systems (2012) for a discussion of strengths and limitations of various formulations of allocation.

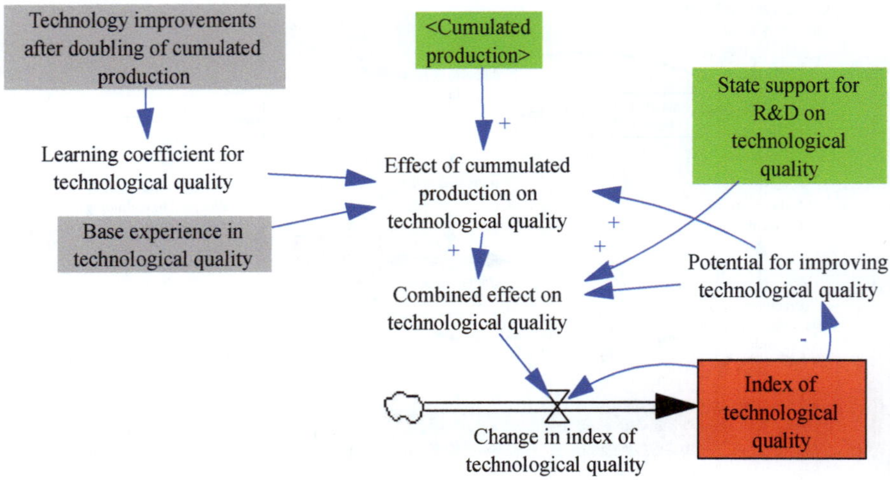

Fig. 3 Main elements of the model structure used to model improvements in technological quality

are also affected by the *<effect of mandatory regulations on attractiveness of technology>*. In order to model the effect of changes in production costs, the *<index of production costs>* is used to influence the *<profits per unit to supplier>*. Further, the *<effective cost paid by demand side>* (for the eco-technology) is influenced by the *<amount of subsidies for eco-technology>*. *<Demand>* (for the eco-technology) is increased by *<demand induced by pilot and demonstration initiatives>*. Note that these influences are not exogenous ones. Instead, they result from additional feedback loops. This means they are an integral part of the feedback structure that drives the diffusion of eco-technologies.

Changes in Technological Quality and Production Costs

Above, in "The Market and Its Effect on Installed Base", we argued that improvements in the technological quality of the eco-technology and decreases in its production costs increase the attractiveness of the eco-technology on both the demand and the supply side. In what follows, we argue that it is industrial learning and research and development (R&D) that bring about these changes. Figure 3 shows the main model structure used to model improvements in the *<index of technological quality>* as a function of the *<cumulated production>* and the *<state support for R&D on technological quality>*. As *<cumulated production>* is increased, experience with the technology is increased. This in turn increases the attractiveness of the technology on the market (see Fig. 2), thereby increasing demand and supply, which in turn contributes to technological improvements.

Initially, however, the technological quality of the eco-technology is so low that it has no chance to successfully diffuse based on market mechanisms alone. Therefore,

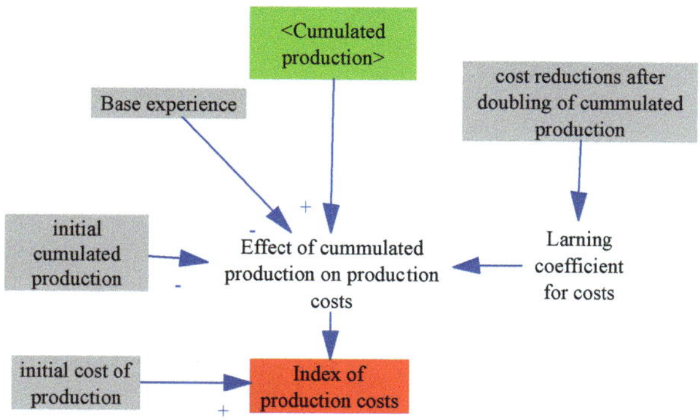

Fig. 4 Main elements of the model structure used to model reductions in production costs

we argue that a crucial initial step, *<State support for R&D on the technological quality>*, is an important contributor to technological progress. It helps to start up the industrial learning feedback loop that eventually makes eco-technology marketable (Fig. 4).

Dynamics of Policy Change

In the Introduction, we argued that the diffusion of eco-technology is frequently the outcome of public policy. In particular, we argued that public policy promotes the diffusion of eco-technology as a way of achieving environmental policy goals. What has not been discussed is why public policy goes from ignoring a particular technology domain to actively promoting the diffusion of more sustainable alternatives. This can be addressed based on theories from political science. In fact, a range of explanations of policy change have been proposed (Easton 1957; Sabatier 2007). We rely on the advocacy coalition framework (Sabatier and Jenkins-Smith 1988; Sabatier and Jenkins-Smith 1993; Sabatier 1998) in order to explain and model long-term policy change, typically lasting a decade or longer.[7] In the advocacy coalition framework, actors in a policy subsystem are aggregated into different coalitions. Members of a coalition "(a) share a set of normative and causal beliefs and (b) engage in a non-trivial degree of coordinated activity over time" (Sabatier 1998, p. 103).

Typically, there are about one to five coalitions in a given policy subsystem (Sabatier and Weible 2007, p. 196). Further, such coalitions are typically highly stable, both internally and in terms of the power they hold relative to one another. Policy change is typically brought about by external events (Weible et al. 2009, p. 124), which may cause a decline in a formerly dominant coalition's power.

Figure 5 shows how we operationalize the advocacy coalition framework into a dynamic simulation model. We start by assuming the existence of two coalitions,

[7] This description of the advocacy coalition framework substantially draws on previously published work reported in Müller (2012, Chap. 5.4.4.1; 2013).

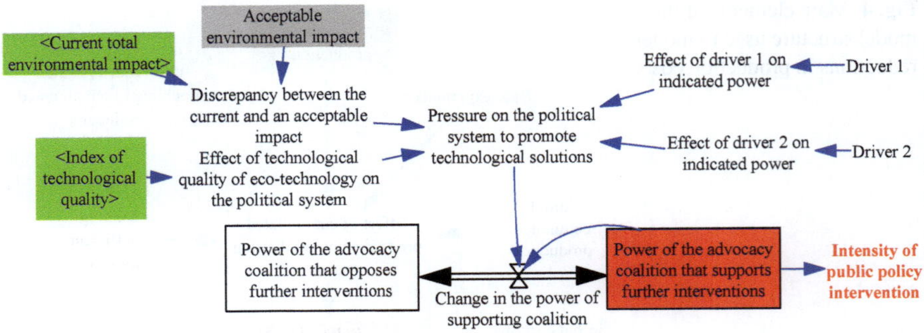

Fig. 5 Main elements of the model structure used to model policy change

one in support of and one opposed to further interventions. Empirically, we drew on studies of policy change in Switzerland's climate and energy politics (Ingold 2007, 2010; Jegen 2003; Kriesi and Jegen 2000, 2001; Lehmann and Rieder 2002). More specifically, we model the power of each of the two coalitions as a stock (e.g., *<power of the advocacy coalition that supports further interventions>*), and we model *<changes in the power of the supporting coalition>* as a flow connecting the two stocks. In line with the advocacy coalition framework, we assume that the drivers of policy change are mostly exogenous. Typical examples of exogenous drivers are the emergence of climate change concerns or energy security issues (see Müller 2012, p. 56 ff., p. 216 ff.). Such trends emerge from outside the policy subsystem, but they lead to the creation of a societal problem situation and create pressure on the political system. Technically, we implemented the effect of exogenous drivers by including two exogenous variables to operationalize such effects (*<driver 1>*, *<driver 2>*). These two variables are the main drivers of policy change, and cause the *<pressure on the political system to promote technological solutions>* to rise in the beginning.

In addition, we include two endogenous drivers of policy change. First, we assume that with a large discrepancy between the *<current total environmental impact>* and an *<acceptable environmental impact>* the *<pressure on the political system to promote technological solutions>* is large, thereby driving policy change. Second, we assume that a rising *<index of technological quality>* will reinforce the pressure on the political system. The rationale behind this is that in the context of a societal problem situation, the availability of better technology will weaken opposition to further regulations.

In the next section, we argue that in conjunction with a rising *<power of the advocacy coalition that supports further interventions>*, the *<intensity of public policy intervention >* rises.

Public Policy Interventions

Due to limitations of space, we refrain from describing the structures used to model public policy interventions in the same detail as the structures described above.

Nevertheless, the logic implemented in this model sector can be derived by examining Fig. 1 (sector 5). As the *<intensity of public policy interventions>* rises, four distinct types of public policy interventions are implemented:

- Public policy supports pilot and demonstration installations of the eco-technology, with the goal of increasing demand, eventually speeding up the industrial learning loop (as shown in Fig. 2).
- Public policy supports research and development, with the goal of directly improving the technological maturity of the technology (as shown in Fig. 3).
- Public policy subsidizes market actors that implement the eco-technology, with the goal of increasing demand, eventually speeding up the industrial learning loop (as shown in Fig. 2).
- Public policy implements mandatory regulations, with the goal of regulating environmental impacts (as shown in Fig. 2).

The effects of these interventions can be seen in Figs. 2 and 3. The underlying logic of the simulation model now becomes more evident: Technologies that are of no particular interest to public policy would be subjected to the interplay of the demand, supply, technological progress, and production cost loops. This means that what is often called "free market economics" would determine whether a technology diffuses successfully—or not. In the case of eco-technologies, however, public policy actors are interested in ensuring that the process of replacing a conventional technology (with high environmental impacts) with an eco-technology (with low environmental impacts) actually takes place, and without avoidable delays.

Behavior of the Simulation Model in a Base Calibration

Generic theories cannot be calibrated against specific data. Nevertheless, we calibrated the model such that it yields a plausible behavior and generally reproduces model behaviors documented in Müller (2012). Figure 6 shows the behavior of key variables in a base calibration. Several issues should be highlighted:

- Demand and production are closely related in this model. Further advanced models might contain model structures that allow for stock keeping by the producing company. In its current version, however, the price mechanism assures that demand and production are in balance (Figs. 6a and 6b).
- For the base calibration, we assumed that the technological quality of the eco-technology rather quickly approaches the quality of the conventional technology (Fig. 6c). The production cost of the eco-technology, however, takes much longer to catch up. This is because it takes a long time to accumulate experience and the eco-technology's low-emission characteristics cause higher production costs (Fig. 6d).
- The intensity of public policy intervention rises more slowly than technological quality (Fig. 6e). This is because political processes take a great deal of time. Eventually, however, public policy rises to levels that allow the implementation

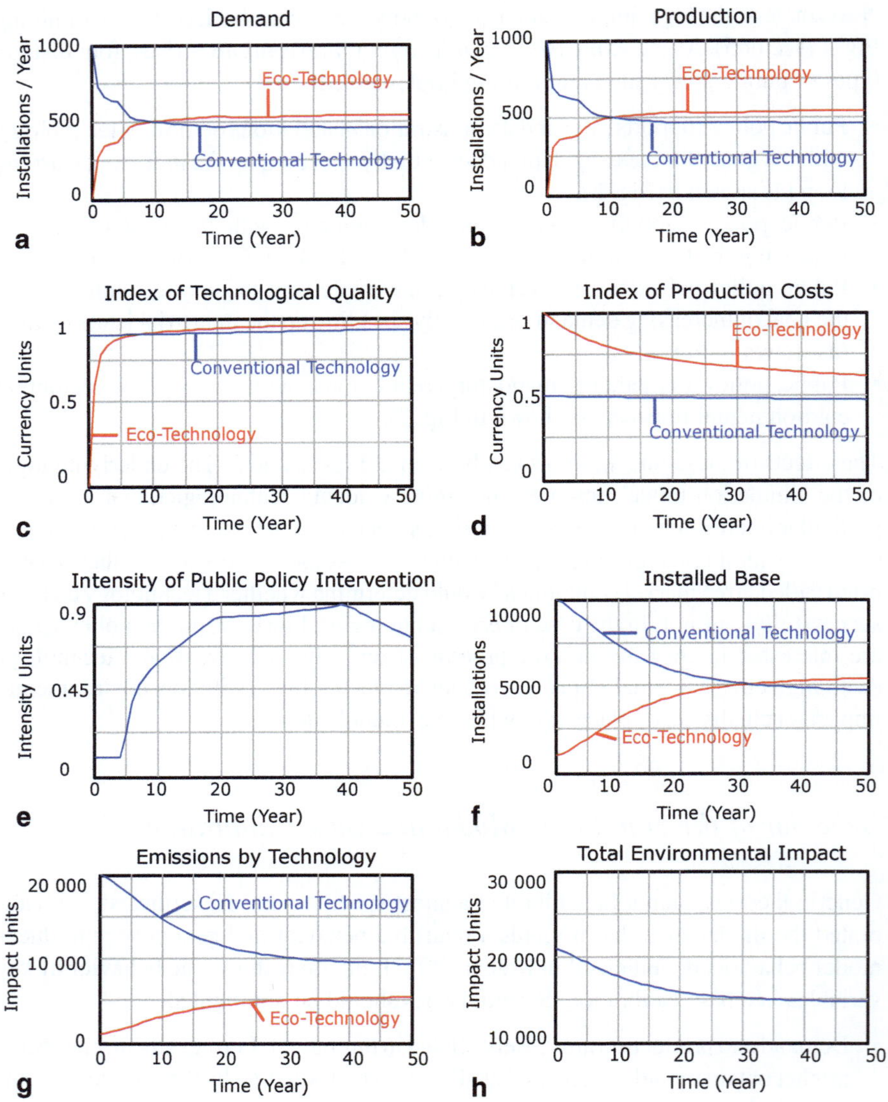

Fig. 6 Behavior of the simulation model in base calibration

of ambitious policies and instruments. Later in time, once the eco-innovation has reached a substantial market share and no longer needs public policy support to compete with the conventional technology, the intensity of public policy interventions starts to fall.

- The dynamics of the installed base (Fig. 6f) show that even when the eco-technology has the larger market share, it may take many years for the eco-technology to become the most frequently installed technology. The longer the service life of the conventional technology, the longer it will take to eventually replace it in the installed base.

- As the conventional technology gets replaced by the eco-technology in the installed base, total emissions are lowered (Fig. 6g) and the negative environmental impact is mitigated (Fig. 6h).

Model Validation

Model validation may be summarized as a systematic way of testing whether the structure and behavior of a simulation model provide a sufficiently accurate representation of the real system under investigation. Whenever a model fails a test, it needs to be changed to enhance its "fit" with the available information about the real system. A canon of tests and procedures has been proposed in the literature (Barlas 1996; Sterman 2000; Schwaninger and Groesser 2009). Hence, it is likely that the model presented here needs to be adapted and further developed when applied to the study of a specific eco-technology.

However, only a limited set of tests (e.g., unit tests, integration time step tests, etc.) could be carried out on the model in its current form. This is because we propose a generic theory rather than a substantive theory that addresses a specific eco-technology. Nevertheless, both model structure and model behavior were substantially derived from a study of the diffusion of energy-efficient renovations in Switzerland. The validation of that model is documented in Müller (2012). The tests described in detail there were applied equally to the model presented here.

Applying our generic model to a specific eco-technology will also reduce some of the challenges posed by using "qualitative" or "soft" variables (Luna-Reyes and Andersen 2003, p. 274). For example, while our generic model uses a qualitative variable called "index of quality", an applied model could rely on a more specific measure of quality and use empirical data to operationalize it. Modeling and validating more hard-to-observe variables (such as "power of the advocacy coalition that supports further interventions") still remains a challenge. Yet, "omitting structures or variables known to be important because numerical data are unavailable is actually less scientific and less accurate than using your best judgment to estimate their values" (Sterman 2010, p. 854). Also, the social sciences have a whole set of methods (e.g., expert interviews, literature analysis, discourse analysis) that can be applied to increase the empirically grounding of qualitative variables (McLucas 2003).[8]

---

[8] Also see Müller (2012, Sect. 2.2.3 and 9.5.1) for further reflections on designing System Dynamics research that includes variables that are hard to measure.

Discussion: How Can System Dynamics Modeling Support Research and Policy Making in Support of the Diffusion of Eco-Technologies?

In "Using System Dynamics Modeling to Support Research and Public Policy Initiatives Aimed at Accelerating the Diffusion of Eco-Technologies", we discuss how System Dynamics modeling in general and our model in particular could guide research and policy making in support of the diffusion of eco-technologies. We do so by elaborating on a series of propositions. In "Implementing Diffusion-Support Processes Based on the System Dynamics Methodology", we then discuss how diffusion-support processes based on the System Dynamics methodology should be implemented.

Using System Dynamics Modeling to Support Research and Public Policy Initiatives Aimed at Accelerating the Diffusion of Eco-Technologies

Proposition 1: Our Model Is Valuable As a Starting Point for Studying the Diffusion of Specific Eco-Technologies Research into the diffusion of specific eco-technologies faces somewhat of a starting problem. Without theoretical knowledge about the diffusion process under investigation, it is not clear what empirical data should be collected. Yet, without empirical grounding, it may not be clear what theories are adequate. By using our generic theory as a starting point, this starting problem can be overcome by conceptualizing the causalities embodied in the model as hypotheses that need to be empirically tested. Insights from falsification testing may then be used to enhance the model's structure as well as its calibration. Further benefits of relying on our theory in the initial phases of research derive from its focus on a system rather than on isolated elements. Hence, using our theory as a starting point helps to overcome an overly narrow focus and supports the integration of different perspectives. When sufficient insights into the system governing the diffusion of a specific eco-technology have been assembled, the model can be expanded to include further effects, such as word-of-mouth effects or network effects.

Proposition 2: System Dynamics Modeling Is Well Suited to Integrate Findings from Research Projects Using Different Methods, and It Promises to Be a Valuable Tool in Managing Entire Research Programs Research aimed at providing action knowledge for accelerating the diffusion of an eco-technology may often be organized as a research program, consisting of several dedicated research projects. Particularly in an academic context, it may be very challenging to organize the research projects such that results from different projects can be synthesized with one another. In such a situation, there is great potential in using the System Dynamics methodology to synthesize the results from different research projects into a simulation model and in using insights from the simulations to guide individual research projects.

To illustrate this point, let us imagine a research program in which individual projects provide insights into the following issues: What is the installed base of the conventional technology and the eco-technology, and how does the ratio change over time? What is the environmental impact of the conventional technology and the eco-technology? What actors influence the diffusion process, and how should they be categorized? How do producers, consumers, or investors make decisions? What preferences do they have, and what attributes do they value? How does the current institutional framework (laws, regulations, government policies, etc.) influence the diffusion process? What are the causes of policy change? In order to integrate insights from different research projects, the development of a formal simulation model like the one presented above may prove valuable. For example, developing the simulation model is likely to indicate areas in which current knowledge is insufficient and areas in which further research needs to be undertaken to enhance the understanding of the diffusion process. Close coordination and communication between a System Dynamics modeling team and the more content-related research teams could yield both meaningful research into specific aspects of the diffusion of an eco-technology as well as an empirically and theoretically well-grounded simulation model that embodies a systems perspective.

Proposition 3: A Fully Developed, Empirically Well-Grounded Simulation Model Can Be Used to Identify Policy Levers and Investigate the Dynamic Implications of Policies Directed at Such Policy Levers Once a simulation model has been developed and tested to establish its consistency with the available knowledge, it can be used to support policy makers. In a first step, policy levers—variables that have a strong effect on the diffusion process—can be identified by systematically reviewing the simulation model. In a second step, policies and instruments can be identified by which such policy levers can be influenced in the real world. Third, the simulation model can be used to analyze the effect of policies and instruments over time. What is more, the simulation model can be used to estimate the intensity of the policies and instruments to be implemented. For example, a simulation model may be used to analyze the magnitude of a tax on fossil fuels or the optimal size of a subsidy. We have not yet tested the application of our model in a concrete decision situation with policy makers. However, we have shown how that model could, in principle, be used to support decision making (Müller 2012).

Proposition 4: Simulation Models May Become Part of a Joint Learning Process and Facilitate the Emergence of Consensus Across Different Advocacy Coalitions In essence, System Dynamics simulation models are "white box" models that can be inspected and understood by anybody. The ability to visualize complex systems can help policy makers' perspectives evolve toward a shared systemic perspective and away from a narrow, nonintegrated view of the diffusion process. This can be illustrated by findings from our recent study of the diffusion of energy-efficient renovations of buildings: In our research projects, we found that actors from industry, public policy, and civil society typically know a great deal about their narrow fields of specialization. Yet even with decades of experience, they were not particularly skilled in developing a systems perspective that could take in the whole diffusion process.

Finally, simulation models promise to facilitate joint learning processes among actors from different backgrounds. To avoid debating fragmented perspectives and vague fundamental values, simulation models could be used to debate policies and instruments based on a systemic understanding of the diffusion process.[9] This holds the potential for identifying win-win solutions and moving toward consensus on policies that perform well in the simulation model.

Implementing Diffusion-Support Processes Based on the System Dynamics Methodology

How could a diffusion-support process based on the System Dynamics methodology be implemented? In our experience, it is good practice to start by modeling the installed base of the conventional technology and the eco-technology and by capturing their dynamics over time. This may entail modeling car fleets or building stocks and tracking various characteristics such as fuel consumption, heating systems, and CO_2 emissions. Based on a rather small yet empirically well-grounded model of the installed base, preliminary policy recommendations can become evident.

In a second step, researchers should focus on the feedback loops that control the installed base and drive the technological substitution process. We expect that all the feedback loops included in our generic model are present in most technological diffusion processes. Yet, including additional feedback loops in a simulation model of the diffusion process may prove insightful and rewarding. Identifying additional feedback loops may entail both empirical research (e.g., face-to-face interviews, desktop research, analysis of quantitative data) and a review of theories that shed light on particular aspects of the system under investigation.

Along with the analysis of feedback loops driving the diffusion process, actors should be analyzed. Which actors are relevant? What part of the feedback structure do they control? What are the interests of the various actors, and how can they be influenced to contribute to the diffusion of the eco-technology rather than block it?[10]

As the understanding of the system under investigation deepens, a quantitative simulation model should be built that adequately represents the diffusion process of a specific eco-technology. Through iterations of model testing and subsequent model improvements, the quality of the model will be improved (see Barlas 1996; Schwaninger and Groesser 2009). When there is a lack of knowledge, further empirical research should be conducted.

In a next step, the simulation model should be used to identify policy levers that have a substantial impact on the diffusion rate of the eco-technology. We expect that such sensitivity analysis will rule out many potential policy levers. However, a set

[9] A pertinent example for the case of municipal waste management is documented in Ulli-Beer (2006).

[10] See Müller et al. (2011) and Müller (2012, Chap. 5) for further insights on identifying and representing relevant actors.

of potentially powerful policy levers will remain. Together with representatives of relevant actors, a dialogue on policies and instruments should be started. The goal of such a dialogue should be to identify pragmatic policies and instruments that make use of highly sensitive policy levers. As a result of collaboration with representatives of relevant actors, the quality of public policies would be improved and policy resistance would be reduced, thereby enhancing the effectiveness of diffusion-support policies.

Conclusions

In the Introduction to this chapter, we presented two research questions. First, we asked how eco-technologies diffuse. In particular, we were interested in what generic causalities associated with the market, technological change, and public policy drive the diffusion of eco-technologies. In the main part of this chapter ("A Generic Theory of the Diffusion of Eco-Technologies"), we argued that changes in technological quality and production cost influence supply and demand on the market. The diffusion process of technologies that are of no special interest for public policy is controlled by the technology and market loops that we described (see "The Market and Its Effect on Installed Base" and "Changes in Technological Quality and Production Costs"). However, technologies that are promoted as a means of achieving public policy goals are not subjected to the interplay of technology and market forces alone. In addition, public policy provides diffusion support by way of interventions like pilot and demonstration installations, support for research and development, subsidies, and mandatory regulations. We described this logic of intervention and argued that our model could be used as a starting point for research and policy initiatives aimed at supporting the diffusion of specific technologies. We claimed that our model is, in principle, applicable not only to one kind of technology but to a whole class of eco-technologies. If that assertion is justified, our model embodies a theory of the middle range.

This leads to our second research question, which asks how modeling and simulation could support the diffusion of a specific eco-technology. We showed that the System Dynamics methodology is well suited to integrate a broad range of insights and perspectives into a more complete, systemic perspective. Furthermore, we argued that a fully developed, empirically well-grounded simulation model can be used to derive robust policy recommendations.

Finally, we indicated how System Dynamics modeling might guide policy making in settings characterized by multiple actors as well as value and interest conflicts. Future research might take our generic model as a starting point and attempt its application to different eco-technologies. Furthermore, it would be promising to explore the potential of using System Dynamics modeling to guide and integrate larger research networks.

Acknowledgments The authors thank Susanne Bruppacher, Franz Schultheis, Heidi Hoffmann, and Stephan Walther for contributing valuable discussions and comments toward improving this article. The research underlying this chapter was supported by the Swiss National Science Foundation within the context of the National Research Program 54 (Sustainable Development of the

Built Environment), Novatlantis—Sustainability at the ETH Domain, the Swiss Federal Office of Energy, the City of Zürich, the Interdisciplinary Centre for General Ecology at the University of Bern, the University of St. Gallen, and Paul Scherer Institut.

References

Barlas, Y.: Formal aspects of model validity and validation in system dynamics. Syst. Dynam. Rev. **12**(3), 183–210 (1996)

Easton, D.: An approach to the analysis of political systems. World Polit. **9**(3), 383–400 (1957)

Eco-Innovation Observatory: Methodological report. Website of the Eco-Innovation Observatory. http://www.eco-innovation.eu/images/stories/Reports/eio_methodological_report_2012.pdf (2012). Accessed 14 April 2013

Geels, F.W.: From sectoral systems of innovation to socio-technical systems Insights about dynamics and change from sociology and institutional theory. Res. Policy **33**, 897–920 (2004)

Geels, F.W., Schot, J.: Typology of sociotechnical transition pathways. Res. Policy **36**, 399–417 (2007)

Giannetti, B.F., Bonilla, H., Almeida, C.M.V.B.: Developing eco-technologies: a possibility to minimize environmental impact in Southern Brazil. J. Clean. Prod. **12**, 361–368 (2004)

Hanneman, R.A.: Computer-Assisted Theory Building. Modeling Dynamic Social Systems. Sage, Newbury Park (1988)

Ingold, K.: The influence of actors' coalition on policy choice: the case of the swiss climate policy. In: Friemel, T. (ed.) Applications of social network analysis, Proceedings of the 3rd Conference on Applications of Social Network Analysis 2006. UVK Medien, Konstanz (2007)

Ingold, K.: Apprendre pour le futur: une analyse de la politique climatique suisse. Swiss Polit. Sci. Rev. **16**(1), 43–76 (2010)

Jegen, M.: Energiepolitische Vernetzung in der Schweiz. Analyse der Kooperationsnetzwerke und Ideensysteme der Energiepolitischen Entscheidungsträger. Helbing & Lichtenhahn, Basel (2003)

Kemp, R.: From end-of-pipe to system innovation. Paper for DRUID Summer Conference. Copenhagen, 17–19 June 2009 http://www2.druid.dk/conferences/viewpaper.php?id=500475&cf=32 (2009). Accessed 14 April 2013

Kemp, R.: Ten themes for eco-innovation policies in Europe. Sapiens **4**(2) http://sapiens.revues.org/1169 (2011). Accessed 14 April 2013

Kemp, R., Pearson, P.: Final Report MEI Project About Measuring Eco-Innovation. UM MERIT, Maastricht (2007)

Knoepfel, P., Larrue, C., Varone, F., Hill, M.: Public Policy Analysis. Policy, Bristol (2007)

Kriesi, H., Jegen, M.: Decision-making in the swiss energy policy elite. J. Public Pol. **20**(1), 21–53 (2000)

Kriesi, H., Jegen, M.: The Swiss energy policy elite: the actor constellation of a policy domain in transition. Eur. J. Polit. Res. **39**, 251–287 (2001)

Lane, D.C., Oliva, R.: The greater whole: towards a synthesis of system dynamics and soft systems methodology. Eur. J. Oper. Res. **107**(1), 214–235 (1998)

Lehmann, L., Rieder, S.: Wissenschaftliches Wissen in der politischen Auseinandersetzung. Fallstudie zur Genese des CO_2-Gesetzes im Auftrag der Arbeitsgruppe Transdisziplinarität der Energiekommission der Schweizerischen Akademie der Technischen Wissenschaften. SATW Bericht Nr 4, Schweizerische Akademie der Technischen Wissenschaften, Zürich (2002)

Luna-Reyes, L.F., Andersen, D.L.: Collecting and analyzing qualitative data for system dynamics: methods and models. Syst. Dynam. Rev. **19**(4), 271–296 (2003)

McLucas, A.C.: Incorporating soft variables into system dynamics models: a suggested method and basis for ongoing research. In: Proceedings of the 21st international conference of the System Dynamics Society, pp. 20–24 (2003)

Merton, R.K.: Social Theory and Social Structure. Free Press, New York (1957)
Müller, M.O.: How can the diffusion of energy-efficient renovations of buildings be accelerated? Dissertation, University of St. Gallen, School of Management, Economics, Law, Social Sciences and International Affairs to obtain the title of Doctor Oeconomiae. http://verdi.unisg.ch/www/edis.nsf/SysLkpByIdentifier/4030?OpenDocument&,lang=en(2012). Accessed 7 Apr 2013
Müller, M.O.: Diffusion Dynamics of Energy-Efficient Renovations. Causalities and Policy Recommendations. Lecture Notes in Energy, 14. Springer, Heidelberg (2013)
Müller, M.O., Grösser, S.N., Ulli-Beer, S.: How do we know who to include in collaborative research? Toward a method for the identification of experts. Eur. J. Oper. Res. **216**(2), 495–502 (2011)
Richardson, G.P.: Reflections on the foundations of system dynamics. Syst. Dynam. Rev. **27**(4), 219–243 (2011)
Rogers, E.M.: Diffusion of Innovations. Free Press, New York (2003)
Sabatier, P.A.: The advocacy coalition framework: revisions and relevance for Europe. J. Eur. Public Policy **5**(1), 98–130 (1998)
Sabatier, P.A. (ed.): Theories of the Policy Process. Westview, Cambridge (2007)
Sabatier, P.A., Jenkins-Smith, H.: An advocacy coalition model of policy change and the role of policy oriented learning therein. Policy Sci. **21**, 129–168 (1988)
Sabatier, P.A., Jenkins-Smith, H.: Policy Change and Learning: An Advocacy Coalition Approach. Westview, Boulder (1993)
Sabatier, P.A., Weible, C.M.: The advocacy coalition framework: innovations and clarifications. In: Sabatier, P.A. (ed.) Theories of the Policy Process, pp. 189–220. Westview, Boulder (2007)
Schwaninger, M., Groesser, S.: System dynamics as model–based theory building. Syst. Res. Behav. Sci. **25**(4), 447–465 (2008)
Schwaninger, M., Groesser, S.: System dynamics modeling: validation for quality assurance. In Meyers, R.A. (ed.) Encyclopedia of Complexity and Systems Science, vol. 9, pp 9000–9014. Springer, New York (2009)
Sterman, J.D.: Business Dynamics. Systems Thinking and Modeling for a Complex World. McGraw-Hill, Boston (2000)
Sterman, J.D.: Business Dynamics. Systems Thinking and Modeling for a Complex World. Tata McGraw Hill Education, New Delhi (2010)
Ulli-Beer, S.: Citizens' Choice and Public Policy. A System Dynamics Model for Recycling Management at the Local Level. Shaker, Aachen (2006)
Ulli-Beer, S.: Conceptual grounds of socio-technical transitions and governance. In Ulli-Beer, S. (ed.) Governance Dynamics of Energy Technology Change Towards Sustainability. Analyzing and Substantiating Socio-Technical Transitions (chapter 2). Sustainability and Innovation series. Springer, Heidelberg (in press)
Ulli-Beer, S., Bruppacher, S., Groesser, S.N., Geisshüsler, S., Müller, M.O., Mojtahedzadeh, M., Schwaninger, M., Ackermann, F., Andersen, D., Richardson, G.P., Stulz, R., Kaufmann-Hayoz, R.: Understanding and accelerating the diffusion process of energy-efficient buildings: Introducing an action science venture. Proceedings of the 24th international conference of the system dynamics society. Nijmegen, The Netherlands http://www.systemdynamics.org/conferences/2006/proceed (2006). Accessed 21 Nov 2010
Ventana Systems: Allocation by priority. Website of Ventana Systems Inc. http://vensim.com/allocation-by-priority-alloc-p (2012). Accessed 7 April 2013
Weible, C.M., Sabatier, P.A., McQueen, K.: Themes and variations: taking stock of the advocacy coalition framework. Policy Stud. J. **37**(1), 121–140 (2009)

Managing the Energy Basket in the Face of Limits

A Search for Operational Means to Sustain Energy Supply and Contain Its Environmental Impact

Khalid Saeed

Introduction

The term sustainability is best defined by a homeostasis in which stocks of resources are maintained by counterbalancing flows. Mill (1848) called this a stationary state, while Forrester equates it to an equilibrium in a dynamical system (Hopkins 2009). Sustaining energy supply, however, has often meant meeting the rising demand to the policy makers while its environmental impact still remains a subject of debate, which is fueled by a plethora of models that are often tied to specific viewpoints. Such models have led to recommendations which are largely normative statements rather than operational policy instruments. The implementation of such recommendations beyond a moral appeal often calls for powerful command and control infrastructure at all, local, national, and global, levels for which institutional mechanisms often do not exist. System dynamics models are no exceptions to this pattern. The most influential system dynamics model addressing environmental agenda is World3 that was created for the famous Limits to Growth Study (Meadows et al. 1972, 1974). The main recommendations of the Limits study were to drastically limit resources use, control population, and reduce pollution rate. How to accomplish those ends cannot be inferred from experimenting with the model since it is tied to the environmentalist viewpoint and does not have the policy space in it to explore interventions for changing the behavior of the human actors in the system to realize the manifestations of the alternative viewpoints.

This chapter revisits the important agenda of sustaining energy supply and containing environmental impact of energy use with a simple model adapted from Saeed (1985), which attempted to resolve the debate on the supply potential of earth's resource system and explored policy options to sustain its yield. It furthermore addresses the problem of containing environmental impact of energy consumption using the concepts developed in Saeed (2004), which investigated the possibility of

K. Saeed (✉)
Social Science and Policy Studies Department, Worcester Polytechnic Institute,
Worcester, MA 01605, USA
e-mail: saeed@wpi.edu

H. Qudrat-Ullah (ed.), *Energy Policy Modeling in the 21st Century*,
Understanding Complex Systems, DOI 10.1007/978-1-4614-8606-0_5,

internalizing environmental restoration effort into a market economy. It is important to note that the models of both above papers attempt the design of operational means to attain sustainability using the experimental system dynamics protocol without forecasting the future demand for energy resources or estimating the carbon volume that energy consumption would add to the environment. Those objectives, although widely pursued, neither represent a valid use of system dynamics modeling nor add much value to policy formulation that might avoid a problematic future. The logic of this chapter is, instead, built on the following premises:

a. A system dynamics model is a stylized representation of a slice of a complex system whose boundary is determined by the problem pattern being analyzed. Behavioral patterns generated by such a model can qualitatively represent both a stylized historical pattern and its future extrapolations. However, numerical values in the simulation of a stylized model are only manifestations of the numerical integration process used and may neither replicate an actual time history nor give a valid point forecast of future.
b. Stylized models highlighting a problematic pattern of behavior or a specific viewpoint may not have adequate policy space in them to design operational means for intervention to alleviate that problem.
c. The incorporation of an adequate policy space in a model requires that we look at multiple manifestations of or opposing viewpoints about the future before building the model of the pattern being investigated, so there is a potential structure for making a transition from one pattern to the other. It also requires including decision rules related to the possible interventions by an identifiable agent or institution so plausible mechanisms of change can be experimented with.
d. While a policy design process might also use stylized models, they can be separated from the models that highlight the problem. Thus, the growth of demand for resources projected by the Limits to Growth study can be represented as an exogenous input in a model focused on policy design pertaining to supply.
e. Different aspects of policy design can be addressed in different models.

Premises of an Operational Policy

The approaches to policy design can be placed in two broad categories: normative and descriptive. The normative decision theory is concerned with how people should act in order to achieve better results. It provides rules that will improve the consequences of actions. The policies formulated with an orientation of normative decision theory involve imposition of prescriptions about social behavior decided exogenously and often without taking into account the compatibility of such prescriptions with the existing circumstances. Due to the very nature of the premises behind the policies of this class, intervening through command and control or moral appeals is the most common strategy posited for the implementation of normative policies. The descriptive decision theory, on the other hand, is concerned with how people actually go about handling a problem irrespective of whether or not the outcomes are admirable.

This theory describes the patterns of behavior that characterize action, so it provides a simple picture of how organizations work, which is the basis for improving organizational performance (Bauer 1968; Bower 1968; Rapoport 1989).

In either approach, the process of policy formulation involves several distinct steps, such as setting goals, formulating general policy directives and guidelines, identifying appropriate policy leverages, and, finally, selecting policy instruments. Although the nature of the formulated policy might depend on its underlying decision theory orientations, if it fails to define operational instruments for affecting the day-to-day decisions of the pertinent actors in the system, the implementation of the policy would necessarily require powerful intervention through command and control. Unfortunately, interventionist designs are prone to failure. Firstly, it is not an easy task to achieve the needed level of centralization to implement most command and control regimes. Secondly, even when decision making can be centralized, the actors entrusted with making the decisions may not empathize with the objectives of the design. Finally, centralization may conflict with a prevalent management ideology, may be unacceptable to the members of organization in which the design is to be implemented, and may invoke much conflict that is destructive (Acharya and Saeed 1996; Saeed 1996).

I have pointed out in Saeed (1994) that while it is possible to design operational policies by employing the heuristic protocol of system dynamics, this is not attempted in a large number of cases. An operational policy design should aim at mobilizing the internal forces of the system into creating functional patterns and avoiding dysfunction by influencing motivations of the actors that guide their decisions. However, if this design is conceived in terms of changing a few sensitive parameters of a system dynamics model representing social rather than individual behavioral characteristics, its implementation may still require a powerful intervention by the leadership who may neither have the motivation nor the means to commit to such an intervention, especially when the context is public interest rather than personal gain.

Policy design for public agendas must, therefore, be conceived in terms of either new feedback loops that are created to modify the anatomy of critical decisions of the concerned actors or the way the influence structure of the existing feedback loops is changed so that the dominance of insidious mechanisms is minimized and the role of benign mechanisms enhanced. Since action in a feedback creates an iterative process that is driven by a discrepancy, the magnitude of an intervention need not be precisely specified, as it would be dynamically regulated by the discrepancy. I have also suggested in Saeed (1992) and Saeed (2003) that a model intended for exploring policy options for system change must subsume multiple manifestations of problem behavior and viewpoints that are separated by time and geography since only then its underlying structure would contain the mechanisms of change from one manifestation to the other. This means differing theoretical perspectives that are often based on selected empirical evidence should be considered a part of the behavioral variety subsumed in a model addressing controversial issues.

Contribution of the Limits Models to Sustainability Agendas

The energy-environment issues cut across natural resources, society, economy, and technology domains creating some of the most complex systems of the present day world whose management is a challenge. However, while the intertwined nature of energy and environment is widely recognized today, policy actions pertaining to the two domains remain detached. In the energy domain, the emphasis is on increasing efficiency and managing demand while in the environmental domain a search for an abundant and economical yet nonpolluting energy source and the denial of the environmental impact of energy use continue to be pitted against each other.

Although system dynamics modeling is often expected to resolve debate by creating a shared understanding of issues and identify policies that might change system behavior by influencing the day-to-day decisions of the actors, in many cases this may not be realized. Especially, when a stylized modeling exercise aims mainly at raising issues rather than designing an operational means for intervention, it may in fact fuel debate. A case in point is the Limits to Growth study that was sponsored by the Club of Rome to extrapolate the future consequences of the current economic growth policies (Meadows et al. 1974) and which was widely criticized by mainstream economists after its publication (Nordhaus 1973). This study developed a detailed system dynamics model based on Forrester's World Dynamics (Forrester 1971) that created insightful future scenarios to articulate the environmentalist viewpoint but was not designed to resolve the environmentalist–technologist controversy (Boyd 1972) and create an operational policy framework to change the future it predicted. When the interventions it suggested are literally translated into policy, they appear to call either for a powerful exogenous intervention or a miraculous value change to limit population, abate pollution, and drastically reduce resource use, for which neither an appropriate institutional structure is currently in place nor can it be created without gravely contradicting the existing systems of commerce and governance. Yet, these interventions never really alleviate the limits the underlying viewpoint of the study propounds.

The Limits models, however, made the important connections between the size of the consumption base and the environmental capacity that resides in the intertemporal domain, which cannot be addressed by the intratemporal process of price adjustment and its subsequent impacts on technology, supply, and demand, included in the contemporary models of resource economics (Nordhaus 1979). The reference mode of the World3 model commissioned by the Limits study is however based largely on an environmental perspective rather than subsuming the environmentalist–technologist controversy prevalent at the time it was developed (Tietenberg 2003). Hence, its characteristic behavior is hard to change at the outset. The policy prescriptions of the World3 model are also based on sensitive parameters representing social rather than individual characteristics; hence, their implementation appears to require powerful exogenous intervention. The revised model of its sequel, Beyond the Limits (Meadows et al. 1992), indeed replaces some of the sensitive parameters with self-regulating feedback structure, but is still unable to deliver adequately operational policy guidelines since, like the original model, it does not subsume multiple manifestations implicit in the contradicting theoretical perspectives on resource policy, which are critical to creating policy space for a productive line of policy experimentation. Finally, "Limits to Growth—The 30 year update" reviewed various proposals on sustainability to understand their efficacy and to suggest guidelines for avoiding an impending catastrophe (Meadows et al. 2004). These guidelines, though insightful, continue to reside in a normative rather than an operational domain.

When run for an extended period of time, the Limits models only defer the impending catastrophe, even when their policy recommendations are fully implemented. Hayes (1993) simulated the Beyond the Limits model (with all prescribed policies) from 1900 through 2400. The policies, which appeared to be effective in ensuring a sustainable world over the time frame of the original study, could only postpone the collapse until the middle of the twenty-second century. The nature of the policy prescription of the Limits models arises from the way the resource sectors (i.e., natural resources and arable land) have been modeled. The homogenous stocks of these resources have only outflows, which make the ultimate collapse inevitable since these outflows continue as long as there is any production supported by the remaining resources. These models exclude any energic inputs into the global resource system that may create regeneration of resources or land and do not consider any possibility of long-term sustenance through regeneration, which is a widely recognized natural process that is fueled by the energic inputs from the sun (Miller 1982). Through this process, earth's ecosystem can regenerate all spent resources, even though the regeneration time is very long in some cases (Cook 1976; Ourisson 1984). The Limits models also do not allow for changing the composition of the resources in use for matching consumption and regeneration, which rules out the consideration of policy options that might avoid intergenerational transfers by adopting a flexible *resource basket* proposed in Saeed (1985) that I will discuss in the next section.

A Simple Model of the Resource Ecosystem Subsuming Multiple Viewpoints

Georgescu-Roegen (1971) pioneered the famous hourglass model of the earth's ecosystem that draws its energic inputs from the sun, which I have tried to interpret in Saeed (1985) as shown in Fig. 1 showing how the solar energic inputs drive the renewal process in earth's ecosystem. The energy resources of the earth can be placed in four aggregate categories: (1) Usable Resources, which can be expended using currently available technologies; (2) Exploitable Resources, which become usable after they have been exploited; (3) Potentially Usable Undiscovered Resources, which would later become exploitable; and (4) Spent Resources, which must be regenerated by the ecosystem to become potentially usable. The total mass of energy resources in the system represented by the large rectangle might remain constant, but the proportion of usable energy resources within this stock will depend on the

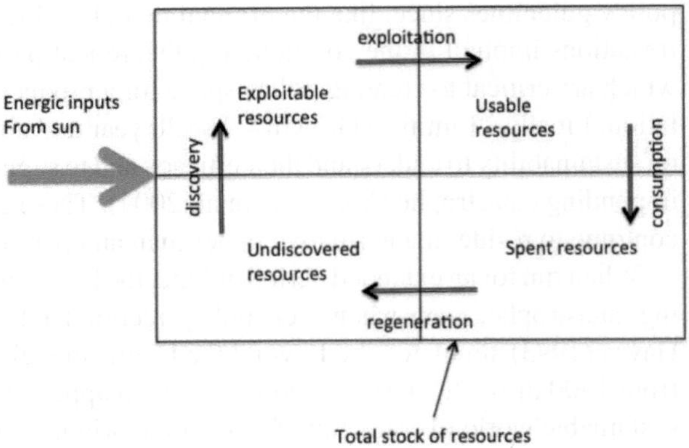

Fig. 1 An interpretation of Georgescu-Reogen's entropy hourglass

speed of circulation within the resource system, which is determined by the technological and management practices rather than being given for all times (Abelson and Hammond 1974; Brooks and Andrews 1974).

Figure 2 shows how energy resources move between the four categories contained in the large rectangle of Fig. 1. The expenditure rate converts usable resources into spent form and is primarily determined by the demand made on the resource ecosystem but is limited by the available inventory of the usable resources. The regeneration rate converts spent resources into the potentially usable form. Regeneration is made possible because of the energic inputs continuously received by the resource ecosystem from the sun, but regeneration time depends on which materials are included in the *resource basket* in use. For energy, the aggregate regeneration time that should be applied to this model may range from a few instants for direct use of solar energy to millennia for fossil fuels and radioactive materials depending on the composition of the *energy resource basket* in use.

The discovery rate allows transfer of potentially usable resources to the exploitable category. Both discovery and exploitation rates are speeded up if the inventory of usable resources declines below a desirable level, as a condition of resource scarcity would raise prices, which would draw investment into research and development for new resources. A persisting condition of scarcity would also provide motivation for developing technologies for reclassifying spent resources into exploitable ones. Model equations are given in the Appendix.

Implicit Assumptions and Energy Supply Scenarios

When the demand profile is based on criteria exogenous to this model (such as a simple trend), the resource expenditure patterns produced by it will depend on the implicit assumptions made about technologies that determine the regeneration time of the resource basket in use and the rates of regeneration and reclassification it yields. Figure 3 compares the expenditure patterns generated with the pessimist and optimist viewpoints when the demand profile is a simple trend.

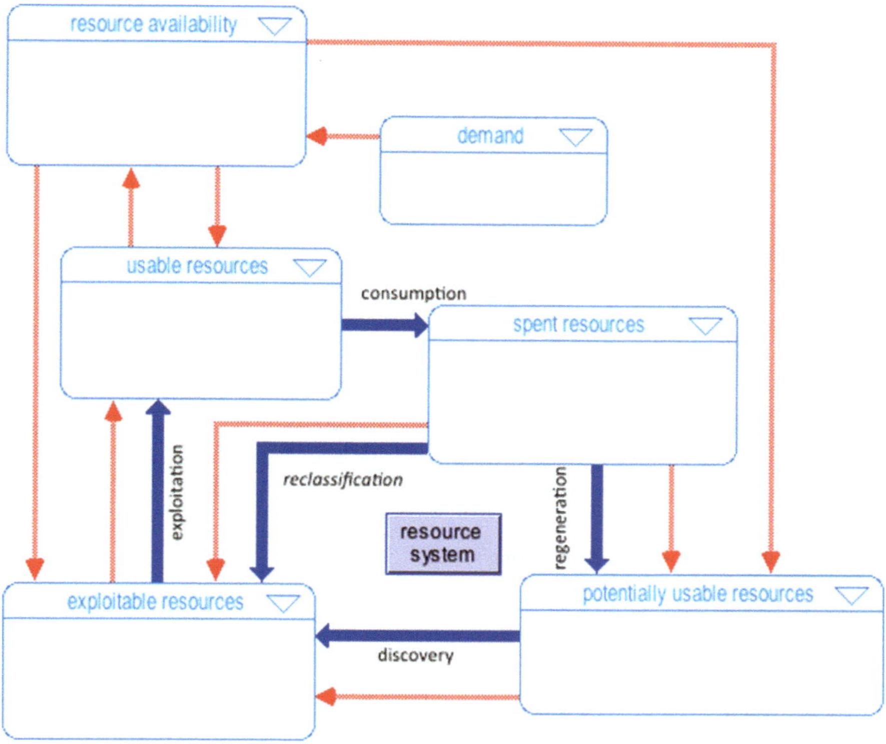

Fig. 2 A system dynamics model of the resource system of Fig. 1

The pattern associated with the pessimist view results from the assumption that the regeneration time of the resource basket in use is infinitely long and there is no possibility of reclassifying or regenerating spent resources into potentially usable ones. These assumptions allow a temporary increase in expenditure when demand rises, but this is followed by a catastrophic decline when usable, exploitable, and potentially usable resource inventories decline. At the other extreme is the pattern representing the optimist view positing an unlimited supply of backstop or exploitable resources (Nordhaus 1979), which results from the heroic assumption that even the spent resources may always be reclassified as exploitable through technological advances when demand rises, while the composition of the resource basket is of no consequence. These two patterns incorporate implicit assumptions of the technological progress made by the environmentalist and the technologist models of resource use, respectively. In between these views lie the patterns corresponding to the revisionist perspectives calling for use of fast renewable resources like wind, waves, and sun. As shown in Fig. 4, these strategies result in some increases in the inventory of usable resources and thus help to alleviate a catastrophic decline in their expenditure rate, although they are unable to match an ever-increasing exogenous demand.

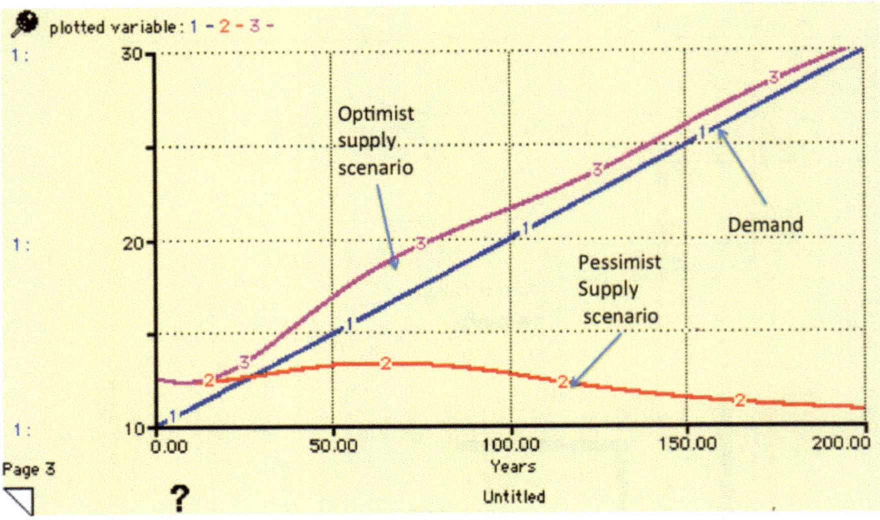

Fig. 3 Supply scenarios based on opposing views

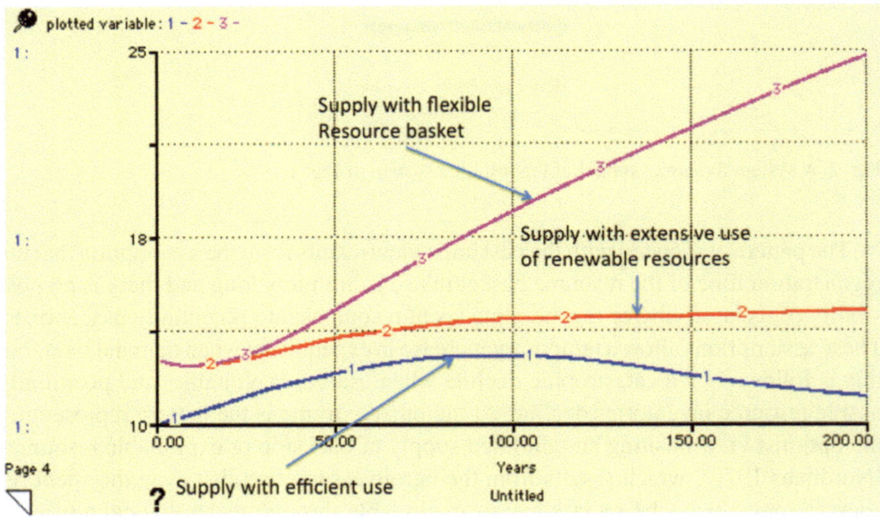

Fig. 4 Revisionist and flexible basket supply scenarios

When usage is confined to mostly fast renewable resources, the expenditure rate is limited by the quantity and the frequency of resources in circulation. Thus, limiting energy sources to a narrow group of fast renewable resources may not necessarily allow society to take full advantage of the potential of the resource environment. Another viewpoint calls for increasing efficiency of use of energy that should limit its rate of expenditure. A scenario implementing this policy is also shown in Fig. 4.

The expenditure rate is indeed lower at first, but the ensuing rise in availability allows meeting more of the rising exogenous demand, which soon depletes any initial advantage. The initial advantage is also defeated when an increased availability arising from efficient usage suppresses exploration and discovery rates that replenish depleting stocks.

Sustaining Energy Supply with a Flexible Resource Basket

None of the resource use scenarios discussed above appear to be satisfactory. If we expand consumption in the hope that future technologies would always make it possible to reclassify some of the spent resources into usable ones, we would be making heroic assumptions about technology and, possibly, penalizing future generations. If we make conservative assumptions about technology and show an overwhelming concern about the maintenance of the resource ecosystem, we may not only limit the benefits to the human society but also generate much conflict while implementing conservationist policies. Fortunately, the resource ecosystem of the earth contains a very large variety of substances from which we can obtain materials for our consumption. Several sources for a single raw material can often be identified, although, not all of these can be exploited simultaneously since the prevalent economic criteria call for consuming the cheapest source first. The cheapest source to exploit is often the one that is richest in the materials we need for our consumption, which is especially true for energy resources. Such resources have usually undergone the longest regeneration processing in the resource ecosystem.

It should also be noted that the distinction between renewable and nonrenewable resources is a superficial one. Given enough time, all resources in the ecosystem could be renewed. Articles made from clay break and change back into clay. Metals can either directly be recycled or re-extracted from the oxides, which are formed when metals deteriorate. Metal ores are also continuously created and enriched through long-term geological processes (Cook 1976). Plastics and man-made fibers may not be easily biodegradable, but they do not remain stable indefinitely. Eventually, they deteriorate into their simpler components, which can be assimilated by nature. There might remain an unconverted residue in a single regeneration cycle, but in each subsequent cycle, a fraction of the residue remaining from the last cycle would again be regenerated together with a fraction of a more recent batch of spent materials. Thus, most of the spent materials from a given period may ultimately be regenerated while many vintages of them may be undergoing the process of regeneration at a given time.

Similar regeneration processes also exist for energy sources other than the sun wind, and waves. Felled trees clear space for growing more trees. Residues from burning wood and coal fertilize land. Carbon dioxide and moisture generated from burning are used by growing plants and contribute towards the development of their cellular structure. Coal and oil are formed by nature by the destructive distillation of plant and animal cellulose. Burning of oil also deposits carbon dioxide, moisture, and waste heat in the air that help to nourish plants which, in turn, nourish animals and microbial organisms that provide cellulose for making oil (Ourissonet al. 1984). Radioactive metals can also be regenerated by the tremendous heat and pressure of the earth's inner core. In some of these cases, however, the regeneration process may take an incredibly long time.

The survival of human society depends not on the life of the universe but on the balancing of the consumption and regeneration of resources. If all resources we use are converted into the spent form and their regeneration takes a few million years, human society may not live to see the regenerated resources, while the ecosystem of the earth that would eventually regenerate all spent resources lives on. If we could wait for nature to complete its regenerative process on materials, it would perhaps make sense to use only the richest sources. However, such consumption could be sustained only as long as expenditure does not exceed regeneration rate. Otherwise, expenditure and regeneration will be separated by delays which human society may not survive. Thus, ideally, we ought to select a *flexible resource basket* from our environment whose aggregate regeneration rate matches our consumption. When consumption rises, resources with a shorter renewal time should be added to the basket in use and those with a longer renewal time dropped from it. The remaining plot in Fig. 4 illustrates implications of such a policy. As the stock of usable resources is depleted, more and more materials with a shorter regeneration time are introduced, which increases the aggregate rate of circulation of materials through the regeneration cycle of the resource ecosystem. Consequently, the stock of spent resources is more rapidly converted into the stock of usable resources. Thus, it becomes possible to sustain a higher expenditure rate. Periods of minor shortages may still be experienced, but these shortages also provide the driving force for changing the composition of the resource basket. Since the substitution process is iterative, a high degree of precision is not needed to drive this process.

While some older studies have suggested that considerable slack exists between this ultimate limit and the current levels of consumption, provided we are able to take advantage of the variety in the resource base (Brooks and Andrews 1973; Ravelle 1973), an ultimate limit dictated by the absolute amount of resources in the ecosystem and the maximum speed at which these can be circulated would still exist and perhaps some measure for moving towards a *steady state economy* would be in order when this limit is approached (Daly 1974). The immediate need, however, is to facilitate technological developments which may allow to substitute the energy resources that have a long regeneration time and that are being currently rapidly exhausted, with those that are in abundant supply and that also have a shorter regeneration time. A *flexible energy basket* is indeed the key to realizing this substitution process.

The neoclassical economic theory advocates using natural resources to maximize the present utility determined by market situation, discount rates, and technology in use, which are subsumed in the price responses. The price mechanisms are, however, good only for assuring intratemporal efficiency of resource use, and they cannot address the issue of intertemporal equity (Pearce et al. 1989; Page 1977). Because, according to the theory of market economy, reserving resources for future use makes sense only when the expected future price of the resources is increasing, at least, at a rate equal to the market rate of interest, which generally exceeds the rate at which the society wishes to discount future. Hence, the market mechanisms always favor the present use of resources over the future use, which does not serve the societal interest in terms of intertemporal equity (Solow 1974). They may achieve intratemporal efficiency, but not intertemporal equity.

To address the problem of intertemporal (or intergeneration) distribution of natural resources, variable severance taxation also based on current resource availability must be used as a proactive policy lever that Solow (1974) and Page (1977) have favored. This taxation may be driven by geological information rather than prices. The policy levers affecting the resource basket must, therefore, be based both on the principles of neoclassical economic theory that should yield intratemporal efficiency and the physics of the resource ecosystem that should yield intertemporal equity.

I have proposed in Saeed (1985) the formation of a national resource board that constantly monitors the consumption and regeneration rates of known energy resources to adjust the severance tax rates to ensure intergenerational equity. The taxation structure modifies prices that lead to selection of resources for use on the basis of matching their regeneration rate with their consumption rate rather than their economic prices. It dispenses with an antagonistic comparison of the present with the future and assures intergenerational equity, as each generation may make the best possible use of the resources available to it without shifting the burden to future generations.

Containing the Environmental Impact of Energy Use

As long as the scale of human settlements was small, and mostly locally found renewable resources constituted the resource basket used, the resource limits remained easily recognizable. It is not surprising that indigenous knowledge enabled traditional societies to live in a way that maintained a balance between development and environment. For example, ancient agricultural methods such as slash-and-burn farming were restricted to small ranges, desert cultures adopted nomadic ways to assure regeneration of the oases that sustained them, planting trees was believed to earn spiritual merit, and fallow practice and diversity of crops were widely used as standard farming practices that sustained land fertility.

Small size also allowed the wastes to remain within the absorption capacity of the ecosystem we live in. The abundant forests and plant resources could easily regenerate emissions created by human use of energy. The growth of human society and its consumption has, however, increased emissions concomitantly with a drastic reduction in the regeneration capacity as more and more of the forests are cleared for accommodating and fueling human activity. Wealth, as we define it, manifests in accumulation of capital whose creation, maintenance, and gainful employment must consume energy and create emissions that nature is no more able to absorb. Thus, economic activity, as we define it, creates goods and services through gainful employment of man-made capital, while the regeneration of the natural capital we destroy in the process is left to nature and helping it has no commercial value.

Extending Market to Subsume Environmental Restoration Activity

Many institutional concepts have been proposed to restore environmental responsibility in society once its need was recognized. Examples of these include the creation of private national trusts that would purchase and maintain historical heritages and reserves; the imposition of environmental taxation on the production of commodities so their price is modified in accordance with the environmental burdens they create; the trading of emission rights so the cost of environmental degradation can be borne by the responsible parties with the help of the market; and Mitigation banking so environmental degradation is off-set by a compensatory restoration effort while the cost of mitigation is borne by the parties who consume environmental resources (Lindell et al. 1996).

The compensatory mitigation concept supports the notion that the net loss of natural capital resulting from production of goods and services is maintained at zero. Private mitigation banks can be formed to carry out the environmental restoration work and sell the credits so earned to companies engaged in production of goods and services for the regular economy that consume natural capital. Mitigation banking creates a trading system whereby deposits can be credited in advance of natural capital consumption by means of ecosystem creation or restoration. Also, since the regulation accompanying mitigation banking creates a cost for production that degrades the environment, it would lead to minimizing the degradation either by limiting production or by reducing its environmental impact. Allowing the market to determine the price of credits earned by restoration of natural capital and mitigation of environmental damage creates a balance between the production and restoration sectors of the economy.

Modeling the Environmental Mitigation Banking System

Whether or not the institutional designs for internalizing environmental costs into goods production can reinstate the environmental responsibility function in society cannot be ascertained, since those designs have not been adequately tested to allow us to guarantee their success. Mitigation banking however appears to be a promising way to internalize environmental restoration activity into the economy as I have investigated in Saeed (2004) on basis of experimentation with a model of a market subsuming environmental mitigation and banking processes into an extended market. An abstract view of this model is shown in Fig. 5.

While a normal market is focused on goods production that consumes natural capital, an extended market also includes mitigation production for restoring natural capital. Restoration production yields mitigation credits that are banked and must be purchased by the goods production sector before the natural capital-consuming activity is undertaken.

A mitigation banking system may function under a variety of organizational and regulatory arrangements. It can be established in the public or private sector. The

Fig. 5 An abstract view of the extended market subsuming restoration activity. (Adapted from Saeed 2004)

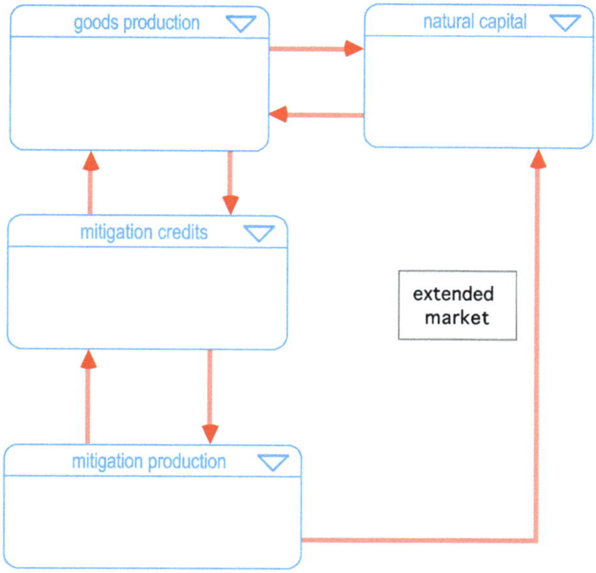

price of the mitigation credits it creates can be fixed, tied to costs using engineering methods, supported by subsidies, determined by the market, or influenced by combinations of all of these factors. Furthermore, the regulations governing the requirement of mitigation credits for the formation and operation of the built environment may be fixed or tied to the condition of the environment. Many views exist on what might be an appropriate way for a mitigation bank and mitigation regulation to function. Currently, the establishment and use of mitigation banks are being promoted in many countries. In the United States, active mitigation banking systems are in place in Minnesota, Florida, and California for preserving forests and wetlands. In all cases, the implementation of the concept is in a nascent stage and its efficacy under a variety of arrangements needs to be carefully evaluated (Wildlandsinc.com) before the scope of its use can be expanded to cover a variety of environmental contexts.

Mitigation Banking for the Energy Market

The mitigation banking concept is ideally suited to aligning the size of the energy basket in use to the restoration activity that aims at mitigating its carbon impact, since both the carbon impact of energy use and its mitigation are not location specific, which has been posited as a limitation of a mitigation banking system applied to maintenance of forests and wetlands. Mitigation credits must be required for all energy use creating carbon emissions in proportion to the volume of emissions. They can be earned and banked by enterprises creating infrastructure that fixes or absorbs carbon. This infrastructure can subsume a wide range of activities including the plantation of forests and wetlands to technologies that might directly sequester carbon from atmosphere and water (Herzog 2011).

I have shown in Saeed (2004) that a mitigation banking system operating in a market can function well without any need to regulate credit prices or provide subsidies when the market is able to recognize the mitigation of environment as a profitable activity. A global regulation requiring mitigation credits for all types of energy use creating carbon emissions can help to promote carbon sequestering industry that is an integral part of the extended economy and that assures a healthy balance between energy use and its impact. Such a regulation can, however, be easily implemented at local levels.

Conclusion

Meeting the rising demand for energy and containing the environmental impact of energy use are intertwined issues. They arise from market-based policies that have implicitly assumed infinite supplies and unlimited environmental capacity for absorbing emissions. The Limits models made a valuable contribution to the recognition of the finite nature of our resource system both in terms of sustaining supply of resources and absorbing the impact of their use. They should also be credited with bringing to fore those issues almost a quarter of a century ago, when little awareness existed about them. The policy agenda they raised, however, could be considered only in the context of a global command and control order or a radical value change, both of which are difficult to realize, which calls for further exploration of interventions in the operational domain.

This chapter has explored ways to deal with the intergenerational transfers related to the depletion of so called nonrenewable energy resources together with containing the impact of energy use by internalizing the management of the resource system into the working of a market. This is done through using different models that take the growth of demand shown by the Limits models as given and focus on operational means for intervention. The creation of a severance tax structure that should minimize intergenerational transfers together with the creation of a system to commercialize mitigation of the damage to the environment are posited as key interventions for sustaining concomitantly energy supply and environment. Both interventions require creation of new institutions. The first of these should help to maintain a flexible resource basket for energy supply whose regeneration balances the rate of use. Since any resource basket in use will lead to some emissions, the second institution aims to internalize the mitigation of the environmental damage into the market by creating a reward system that would make it feasible for private enterprise to pursue environmental restoration for profit. Stylized system dynamics models are used to develop and test designs for both institutions.

An important moral of the story is that models addressing policy design should take into consideration pertinent structure that should allow manipulating parameters relating to individual behavior rather than to social characteristics in the management of environment. Furthermore, multiple modes of behavior subsuming opposing views should be considered while constructing models for policy intervention so mechanisms for changing from one mode to the other could be explored. Thus, both technologist and environmental views of future should be considered as multiple modes constituting the reference mode. Only then a model will allow exploration of operational policies for change. Last, the design of operational policy can be attempted using models that need not replicate the problem pattern, which can be taken as given. Further research is needed on exploring the design and testing of interventions, especially on public policy.

Appendix: Model Equations

Demand

exog_demand_f = (initial_expenditure)*(1 + RAMP(slope, T))
DOCUMENT: RESOURCE DEMAND SCHEDULE (EXOGENEOUS)
initial_expenditure = 10
DOCUMENT: INITIAL EXPENDITURE RATE
slope = 0.01
DOCUMENT: SLOPE OF EXOGENOUS DEMAND SCHEDULE
$T = 0$
DOCUMENT: TIME PARAMETER IN EXOGENOUS DEMAND SCHEDULE
Exploitable Resources
exploitable_res(t) = exploitable_res(t – dt) + (reclassification + discovery – exploitation) * dt
INIT exploitable_res = initial_expenditure*normal_exploit_delay
DOCUMENT: EXPLOITABLE RESOURCES
INFLOWS:
reclassification = spent_res*fr_reclassified*reclass_sw
DOCUMENT: RECYCLING RATE
discovery = potentially_usable_res/discovery_delay
DOCUMENT: DISCOVERY RATE
OUTFLOWS:
exploitation (IN SECTOR: usable resources)reclass_sw = 1
Potentially Usable Resources
potentially_usable_res(t) = potentially_usable_res(t – dt) + (regeneration – discovery) * dt
INIT potentially_usable_res = initial_expenditure*normal_discovery_delay
DOCUMENT: POTENTIALLY USABLE RESOURCES
INFLOWS:
regeneration = spent_res/regen_time
DOCUMENT: REGENERATION RATE
OUTFLOWS:
discovery (IN SECTOR: exploitable resources)
resource availability
availability = usable_res/desired_usable_res
DOCUMENT: RESOURCE AVAILABILITY

av_availability = SMTH1(availability, time_to_smooth_av)
DOCUMENT: AVERAGE RESOURCE AVAILABILITY
desired_usable_res = res_demand*res_coverage_time
DOCUMENT: DESIRED USABLE RESOURCES
discovery_delay = normal_discovery_delay*effect_of_av_on_discovery_relay
DOCUMENT: DISCOVERY DELAY
exploitation_delay = normal_exploit_delay*effect_of_res_av_on_expl_delay
normal_discovery_delay = 50
DOCUMENT: NORMAL DISCOVERY DELAY
normal_exploit_delay = 20
DOCUMENT: NORMAL EXPLOITATION DELAY
normal_regen_time = 10000
DOCUMENT: NORMAL REGENERATION TIME 10000
regen_time = normal_regen_time*(1-res_basket_sw) + normal_regen_time*effect_of_shortage_on_reg_time*res_basket_sw
DOCUMENT: REGENERATION TIME
res_basket_sw = 1
res_coverage_time = 20
DOCUMENT: RESOURCE COVERAGE TIME
res_demand = exog_demand_f
DOCUMENT: RESOURCE DEMAND
time_to_smooth_av = 50
DOCUMENT: TIME TO SMOOTH RESOURCE AVAILABILITY
effect_of_av_on_discovery_relay = GRAPH(availability)
(0.00, 0.4), (0.5, 0.6), (1.00, 1.00), (1.50, 1.60), (2.00, 2.00)
DOCUMENT: EFFECT OF RESOURCE SHORTAGE ON DISCOVERY DELAY
effect_of_av_on_exp = GRAPH(availability)
(0.00, 0.00), (0.2, 0.29), (0.4, 0.51), (0.6, 0.71), (0.8, 0.87), (1.00, 1.00), (1.20, 1.10), (1.40, 1.18), (1.60, 1.22), (1.80, 1.24), (2.00, 1.25)
DOCUMENT: EFFECT OF RESOURCE AVAILABILITY ON EXPENDITURE
effect_of_res_av_on_expl_delay = GRAPH(availability)
(0.00, 0.4), (0.5, 0.6), (1.00, 1.00), (1.50, 1.60), (2.00, 2.00)
DOCUMENT: EFFECT OF RESOURCE AVAILABILITY ON EXPLOITATION DELAY
effect_of_shortage_on_reg_time = GRAPH(availability)
(0.00, 0.01), (0.1, 0.05), (0.2, 0.095), (0.3, 0.16), (0.4, 0.23), (0.5, 0.315), (0.6, 0.42), (0.7, 0.54), (0.8, 0.665), (0.9, 0.815), (1, 1.00)
DOCUMENT: EFFECT OF RESOURCE SHORTAGE ON REGENERATION TIME
fr_reclassified = GRAPH(av_availability)
(0.00, 0.005), (0.1, 0.0033), (0.2, 0.0022), (0.3, 0.00143), (0.4, 0.000975), (0.5, 0.00065), (0.6, 0.000425), (0.7, 0.000275), (0.8, 0.00015), (0.9, 7.5e-05), (1, 0.00)
DOCUMENT: FRACTION SPENT RESOURCES RECLASSIFIED.0005 OR 0

fr_recycled = GRAPH(av_availability)
(0.00, 0.4), (0.1, 0.255), (0.2, 0.165), (0.3, 0.105), (0.4, 0.07), (0.5, 0.045), (0.6, 0.03), (0.7, 0.02), (0.8, 0.01), (0.9, 0.005), (1, 0.00)
DOCUMENT: FRACTION EXPENDED RESOURCES RECYCLED
Spent Resources
spent_res(t) = spent_res(t – dt) + (expenditure – reclassification – regeneration) * dt
INIT spent_res = initial_expenditure*normal_regen_time
DOCUMENT: SPENT RESOURCES
INFLOWS:
expenditure (IN SECTOR: usable resources)OUTFLOWS:
reclassification (IN SECTOR: exploitable resources)regeneration (IN SECTOR: potentially usable resources)
usable resources
usable_res(t) = usable_res(t – dt) + (exploitation – expenditure) * dt
INIT usable_res = initial_expenditure*res_coverage_time
DOCUMENT: USABLE RESOURCES
INFLOWS:
exploitation = exploitable_res/exploitation_delay
DOCUMENT: EXPLOITATION RATE
OUTFLOWS:
expenditure = res_demand*effect_of_av_on_exp/efficiency_of_use
DOCUMENT: EXPENDITURE RATE
av_expenditure = SMTH1(expenditure, time_to_smooth_av_exp)
DOCUMENT: AVERAGE EXPENDITURE RATE
time_to_smooth_av_exp = 5
DOCUMENT: TIME TO SMOOTH EXPENDITURE RATE
Not in a Sector
efficiency_of_use = 0.8
DOCUMENT: EFFICIENCY OF USE
plotted_variable = expenditure*plot_SW + (1-plot_SW)*res_demand
plot_SW = 0

References

Abelson, P.H., Hammond, A.L.: The new world of materials. Science **191**(4228), 683–689 (1974)

Acharya, S.R., Saeed, K.: An attempt to operationalize the recommendations of the "Limits to Growth" study to sustain future of mankind. Syst. Dynam. Rev. **12**(4), 281–304 (1996)

Bauer, R.A.: The study of policy formation: an introduction. In: Bauer, R.A., Gergen, K.J. (eds.) The Study of Policy Formation, The Free Press, London (1968)

Bower, J.L.: Description decision theory from the administrative view point. In: Bauer, R.A., Gergen, K.J. (eds.) The Study of Policy Formation, The Free Press, London (1968)

Boyd, R.: World dynamics, a note. Science **177**(4048): 516–519 (1972)

Brooks, D.B., Andrews, P.W.: World Population and Mineral Resources: Counterintuitive or Not? UN Symposium on Population, Resources and Environment. United Nations, Stockholm (1973)

Brooks, D.B., Andrews, P.W.: Mineral resources, economic growth, and world population. Science **185**(4145), 13–19 (1974)
Cook, E.: Limits to exploitation of non-renewable resources. Science **191**(4228), 677–682 (1976)
Daly, H.E.: The economics of steady state. Am. Econ. Rev. **64**(2), 15–21 (1974)
Forrester, J.W.: World Dynamics. Wright-Allen, Cambridge (1971)
Georgescu-Roegen, N.: The Entropy Law and the Economic Process. Harvard University Press, Cambridge (1971).
Hayes, B.: Balancing on a pencil point. Am. Sci. **81**(6), 510–516 (1993)
Herzog, H.: Scaling up carbon dioxide capture and storage: from megatons to gigatons. Energ. Econ. **33**(4), 597–604 (2011)
Hopkins, M.S.: The loop you can't get out of, an interview with Jay Forrester. Sloan Manage. Rev. **50**(2): 9–12 (2009)
Lindell, L., Porter, D.R, Salveen, D.A. (eds.): Mitigation Banking, Theory and Practice. Island, Washington (1996)
Meadows, D.H., Meadows, D.L., Randers, J., Berhens III, W.W.: The Limits to Growth. Universe Books, New York (1972)
Meadows, D.L., Meadows, D.H., et al.: Dynamics of Growth in a Finite World. Wright-Allen, Cambridge (1974)
Meadows, D.H., Meadows, D.L., Randers, J.: Beyond the Limits. Chelsea Green, Post Mills (1992)
Meadows, D.H., Meadows, D.L., Randers, J.: Limits to Growth: The 30-Year Update. Chelsea Green, Post Mills (2004)
Mill, J.S.: Of the stationary state. In: Mill, J.S.: Principles of Political Economy with Some of Their Applications to Social Philosophy, book 4, chap. 6. Parker, London (1848)
Miller, J.G.: The Earth as a system. Behav. Sci. **27**(4), 303–322 (1982)
Nordhaus, W.D.: World dynamics: measurement without data. Econ. J. **83**(332), 1156–1183 (1973)
Nordhaus, W.D.: The Efficient Use of Energy Resources. Cowels Foundation, Yale University Press, New Haven (1979)
Ourisson, G.: The microbial origin of fossil fuels. Sci. Am. **251**(2), 34–41 (1984)
Page, T.: Conservation and Economic Efficiency: An Approach To Materials Policy. Johns Hopkins University Press, Baltimore (1977)
Pearce, D.W., Turner, R.K.: Economics of Natural Resources and Environment. Johns Hopkins University Press, Baltimore (1989)
Rapoport, A.: Decision Theory and Decision Behavior: Normative and Descriptive Approaches. Kluwer, Norwell (1989)
Ravelle, R.: Will Earth's Land and Water Resources Be Sufficient for Future Populations? UN Symposium on Population, Natural Resources and Environment. United Nations, Stockholm (1973)
Saeed, K.: An attempt to determine criteria for sensible rates of use of material resources. Technol. Forecast Soc. **28**, 311–323 (1985)
Saeed, K.: Slicing a complex problem for system dynamics modeling. Syst. Dynam. Rev. **8**(3), 251–261 (1992)
Saeed, K.: Development Planning and Policy Design: A System Dynamics Approach. Avebury, Aldershot (1994)
Saeed, K.: Sustainable development, old conundrums, new discords, Jay Wright Forrester award lecture. Syst. Dynam. Rev. **12**(1), 59–80 (1996)
Saeed, K.: Articulating developmental problems for policy intervention: a system dynamics modeling approach. Simulat. Gaming **34**(3), 409–436 (2003)
Saeed, K.: Designing an environmental mitigation banking institution for linking the size of economic activity to environmental capacity. J. Econ. Issues **38**(4), 909–937 (2004)
Solow, R.: Economics of resources or resources of economics. Am. Econ. Rev. **64**(2), 1–14 (1974)
Tietenberg, T.H.: Environmental and Natural Resource Economics, 6th edn. Addison Wesley, Boston (2003)
Wildlandsinc.com.: Mitigation Banks. http://www.wildlandsinc.com. Accessed: 20 Aug 2013

Power Plant Relocation Policy versus Investments in Transmission Network Infrastructure: A Study on the Italian Energy Market

Silvano Cincotti and Giulia Gallo

Introduction

Transmission investment opportunities and methodologies are among the most debated topics in the electricity market sector. Before the transition towards a liberalized electricity sector, Hogan pointed out that locational price differences defined the opportunity cost of transmission and that the potential to arbitrage these same price differences were able to provide a market incentive for transmission investments (Hogan 1999). Moreover, Joskow argued that economic and reliability-based criteria for transmission investment were fundamentally interdependent and ignoring these interdependencies might have adverse effects on the efficiency of investment in transmission infrastructure and would have undermined the success of electricity market liberalization (Joskow 2005).

While deregulated electricity markets were initializing their operations, many practitioners started studying approaches to increment transmission investments in order to reduce the exertion of market power and to give other economic signals to the different actors. However, in 2006 Stoft suggested that it was too early to begin policy initiatives, inasmuch as there was no deep knowledge of the new market structure as yet and, more important, the failure experienced in California was not entirely forgotten (Stoft 2007). During the following years, as deregulated markets were established, empirical market data were collected and some positive outcomes were accepted by the scientific community, the topic of transmission investments witnessed a new increase in importance, and several approaches have been proposed. Game theory, computational economics, and artificial intelligence as well as electrical engineering methodologies were used to design innovative solutions. In (Siddiqui and Gupta 2007) Siddiqui used a real options approach to determine both optimal investment timing and line capacity under uncertain congestion rents. Leou et al.

S. Cincotti (✉) · G. Gallo
DIME-CINEF, University of Genoa, via Opera Pia, 15-16145 Genova, Italy
e-mail: silvano.cincotti@unige.it

G. Gallo
e-mail: giulia.gallo@unige.it

H. Qudrat-Ullah (ed.), *Energy Policy Modeling in the 21st Century*,
Understanding Complex Systems, DOI 10.1007/978-1-4614-8606-0_6,

proposed a method combining Monte Carlo simulations and greedy algorithms to find optimal transmission expansion plans (Leou and Teng 2011). Skoteinos et al. proposed a methodology to assess the economic evaluation of alternative transmission expansion plans, based on measures of market performance using IEEE 24-bus reliability test systems to determine cost–benefit scenarios and to test their performance (Skoteinos et al. 2011).

In the framework of zonal pricing, the market convergence rate (i.e., a measure indicating the hours during which the market actually operates as uniform) has been suggested as a proxy for grid investment opportunities by Makkonen et al. (Makkonen and Viljainen 2012). Again, investments in transmission capacity were generally considered a viable way to market efficiency and thus to increase consumers' social welfare.

During the last two decades, generation capacity investments have been discussed in parallel with transmission network investments. As described by Ventosa et al., the most commonly used modeling techniques follow three main trends: optimization models, dynamic simulation models, and equilibrium models (Ventosa et al. 2005). Within the first context, Schroeder et al. presented equilibrium models that incorporated long-term uncertainty and multistage decision making, accounting for the real option character of investments, in order to quantify how fuel and carbon price risk affect investment incentives of thermal power plants (Skoteinos et al. 2011). Moreover, Burger et al. studied game-theoretic models for generation capacity investment decisions in deregulated electricity markets by means of S-adapted Cournot equilibrium in the German electricity market (Burger and Ferstl 2008). Within the context of dynamic simulation models, Joskow argued that market imperfections and institutional constraints might have the effect of keeping wholesale prices for energy and operating reserves below their efficient levels during hours when prices should be very high and possibly lead to underinvestment in generating capacity (Joskow 2006). At the junction between dynamic and equilibrium models, Botterud et al. studied how uncertainty influences the optimal timing of investments in new power generation capacity, implementing a stochastic dynamic optimization model to solve the problem for a decentralized and profit-maximizing investor in the electricity market (Botterud and Korpas 2004).

More recently, a huge number of renewable power plants have been installed worldwide and understanding their impact on the level of prices has become a crucial problem. In (Smith et al. 2010), Smith et al. examined the design and operation of a cross-section of electricity markets in the United States giving insights into the needs of markets necessary to accommodate significantly higher levels of variable renewable energy in the future. Boerema et al. studied how key characteristics of the underlying wind and solar resources may affect their energy value within the Australian National Electricity Market (Boerema et al. 2010). Their analysis showed that these energy resources have key characteristics that could have a marked impact on their energy value within the wholesale electricity market. A summary of policy best practices that energy ministers and other stakeholders can pursue to ensure that electricity markets and power systems can effectively coevolve with increasing penetration of variable renewable energy has been compiled by Cochran et al. (Cochran et al. 2012).

Irrespective of the concerted attention and efforts of the scientific community, the debate on the localization of renewable energy sources is still open, thus offering opportunities for unconventional approaches.

In these respects, this chapter presents a comparison between investments in transmission networks and generation capacity relocation policy under a zonal pricing mechanism. The aim is to understand if the existing Italian power mix allows possible solutions that increase consumers' social welfare, by taking advantage of the existing constraints in the transmission network and the zonal pricing mechanism. It is worth remarking that comparing relocation policy with investment in transmission capacity is strongly counterintuitive as classical literature considers investments in transmission infrastructure the only viable path to an efficient electricity market. Thus, the purpose of this chapter is to provide an innovative framework for discussing and proposing policy design mechanisms that are able to alleviate high zonal prices and that can better integrate renewable generation.

First, an empirical analysis on day-ahead market prices is performed in order to evaluate such opportunities in the case of the Italian electricity market. In this respect, a computational framework that solves the market and replicates the Italian day-ahead market has been used. In this context, investments in transmission capacity have been evaluated as well as the presence of over- and undergeneration capacities in the different areas. This led to the opportunity to evaluate possible scenarios of generation relocation, in particular for renewable power plants. The different solutions arising from investments in transmission capacity and from power plant relocation have been compared by means of the daily average PUN (i.e., the consumers' social welfare proxy defined as the average unit cost paid daily by consumers). Results have shown that with a proper localization of power plants it is possible to increase consumers' social welfare within a zonal splitting mechanism with transmission grid limits.

The chapter is organized as follows: a description of the Italian electricity sector, with attention to the day-ahead market, is provided together with an empirical analysis on market outcomes. The analyses on investments in transmission network and power plant relocations are presented separately. The comparison between the policies and the conclusions are reported at the end of the chapter.

The Italian Electricity Sector

The Italian electricity market, called the Italian Power Exchange (hereafter IPEX) is the fundamental instrument for creating a competitive electricity market in Italy. The electricity market arose in Italy from Legislative Decree no. 79/99 of March 16th, 1999 as part of the EU Directive 96/92/EC concerning common rules for the domestic energy market in electricity. The Italian wholesale market started to operate as a pool in April 2004 and became an exchange in 2005 with the liberalization of demand-side bidding. In 2011 there were 195 operators and volumes traded on the exchange were equal to180.4 TWh against 67.3 TWh in 2004 (AEEG 2012). The presence of new independent power producers alongside the old (ex) monopolist, introduced

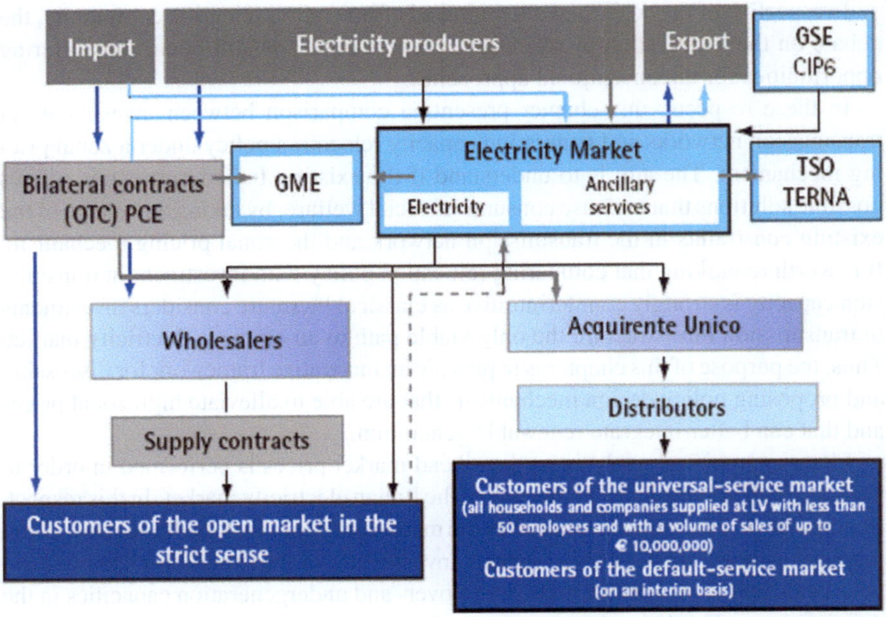

Fig. 1 Organization of the electricity market in Italy. (For further details, see GME 2009)

a problem of coordination between time-varying demand and supply of electricity. Coordination was not an issue in the old integrated industry where the sole producer was also responsible for transmission and distribution and was endowed with all the relevant information about demand and supply. Therefore the new liberalized market structure requires a central mechanism in order to match demand and supply continuously.

The Regulatory Authority for Electricity and Gas is the independent body that regulates, controls, and monitors the electricity and gas sectors and markets in Italy. The Authority's role and purpose is to protect the interests of users and consumers, promote competition, and ensure efficient, cost-effective, and profitable nationwide services with satisfactory quality levels. Its mission includes defining and maintaining a reliable and transparent tariff system, reconciling the economic goals of operators with general social objectives, and promoting environmental protection and the efficient use of resources.

Since January 1st, 2005, the market has been opened to full demand-side participation: all interested operators may trade the electricity that they need directly on the power exchange, under the obligation of hourly scheduling their withdrawal and injection profiles. The organizational structure and the actors involved in the functioning of the electricity market are shown in Fig. 1. In particular, the main entities contributing to the operation of the power system are:

- "Gestore dei Servizi Energetici" (GSE), which buys the electricity generated by CIP-6 power plants and sells it in the market
- "Gestore dei Mercati Energetici" (GME), which organizes and manages the electricity market under principles of neutrality, transparency, objectivity, and competition among producers
- Terna S.p.A., which manages the national transmission grid under security conditions, as well as the power flows thereon through the dispatching activity, that is, by balancing supply and demand of electricity 365 days a year and 24 h a day
- "Autorità per l'Elettricità e l'Energia" (AEEG), which guarantees the promotion of competition and efficiency in the sector and has regulation and monitoring tasks

GME organizes and manages the energy markets, which consist of the day-ahead market (called "Mercato del Giorno Prima"), the intra-day market (called "Mercato di Aggiustamento"), and the forward electricity market (called "Mercato a Termine dell'Energia"), as well as the platform for physical delivery of financial contracts stipulated on the Italian Derivatives Power Exchange (IDEX). Therefore, GME does not organize merely financial markets but real physical markets, where physical injection and withdrawal commitments are scheduled.

One of the peculiar aspects of the Italian electricity market—and especially of the Italian power system—is the presence of market zones. The zones play a crucial role in the splitting of the market in the case of congestion and lead to a zonal pricing algorithm for clearing the market.

Almost all European countries have adopted zonal pricing as market model. In the Nordic countries (i.e. Finland, Sweden, Norway, Denmark, and Estonia) the wholesale markets were combined into a single market and the choice of market model to be applied was zonal pricing with 10 price areas with one of these being the entire Estonian price area. The zonal pricing model was a natural choice as the transmission capacity both between the countries and within them was sufficient for the formation of price areas. Nowadays the number of price areas has changed with the creation of a new larger one which includes Sweden, inasmuch as the congestion inside the country had been causing problems in the entire market. Moreover, in the Central West European area, France, Belgium, the Netherlands, and Germany constituted a unified market, which adopted zonal pricing with a market coupling mechanism; that is, an area price is calculated for each country, but when the transmission capacity is sufficient, each area has the same price.

It is worth remarking that the main difference between the Italian electricity market and the other European markets consists of the economic mechanism for allocating transmission capacity. Indeed, instead of settling the transport capacity for each participant to the market before starting the day-ahead market session with an explicit auction, the Italian mechanism adopts an implicit transmission capacity auction: the hourly transport capacity and the related fees are implicitly calculated by the market resolution algorithm.

For the sake of power system security, the Italian power system consists of portions of transmission grids linked by connections characterized by the physical limits of electricity transmission. The identification of the zones of the critical grid (so-called

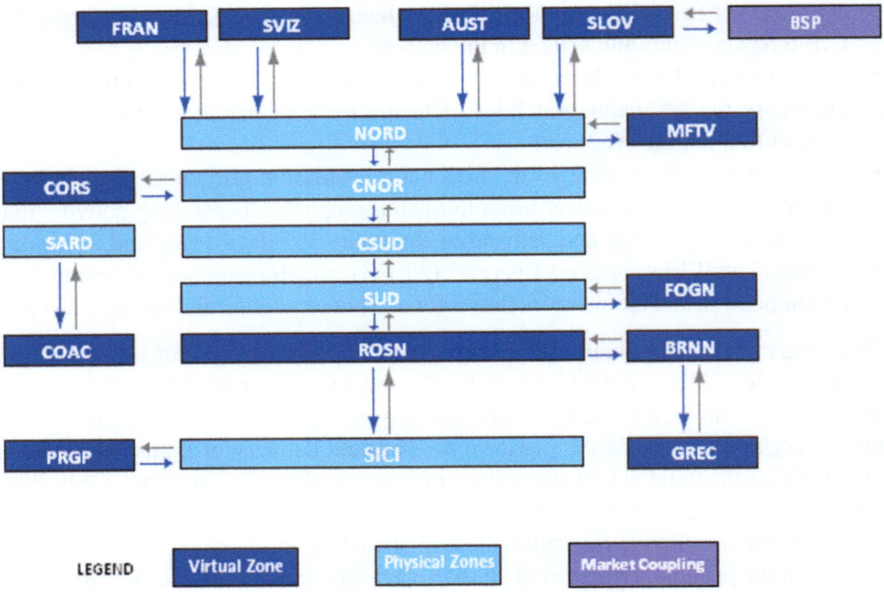

Fig. 2 Zonal representation of the Italian power system network. The market coupling with the Slovenian market is represented with the BSP zone

"rete rilevante") is reviewed every two or three years by Terna to take into account the three-year National Transmission Grid Development Plan (TERNA 2009). The zones of the critical grid may correspond to physical geographical areas, virtual areas (i.e., without a direct physical correspondence), or to constrained zones (i.e., virtual zones whose generation is subject to constraints in terms of management of the power system due to security conditions).

The transmission grid is directly considered by the Italian day-ahead market mechanism, and GME uses the simplified map of the grid shown in Fig. 2 with a DC optimal power-flow optimizer (i.e., a representation comprising the most significant transmission limits in the transmission grid linking the zones), in order to determine the locational marginal price.

The Italian Day-Ahead Market and the Italian Power Mix

The day-ahead market (hereafter DAM) in Italy is a market organized under the implicit capacity auction model. It hosts most of the transactions of purchase and sale of electricity. Indeed, the DAM is a wholesale electricity market, where hourly blocks of injection and withdrawal commitments are negotiated for the next day.

Participants submit price/quantity offers for each hour separately to the DAM which are aggregated by the market operator in order to determine the hourly supply and demand curves. Producers and consumers are allowed to engage in bilateral contracts for the short- and long-term exchange of electricity. The quantity traded bilaterally is mandatorily included in the total demand and supply recorded in the exchange as price–taker offers, as they contribute to the implicit transmission capacity auction.

In a DAM session, the economic merit order criterion and the transmission capacity limits between zones are considered in order to accept offers and bids in a uniform nondiscriminatory auction with zonal splitting. The zones in the Italian market correspond to virtual zones representative of foreign neighboring markets (Austria, Corsica, France, Greece, Slovenia, and Switzerland), limited production poles (Brindisi, Foggia, Monfalcone, Priolo, Gargano, and Rossano) and also physical national zones (Northern Italy, Central Northern Italy, Central-Southern Italy, Southern Italy, Sardinia, and Sicily).

Hourly marginal prices can differ across zones due to transmission limits and the presence of different locational marginal prices denotes congestion. It is worth remarking that the accepted supply offers are evaluated at the clearing price of the zone to which they belong, whereas the accepted demand bids are evaluated at the unique national price (hereafter PUN also called "Prezzo Unico Nazionale") which is the average of the zonal prices weighted by zonal consumption. Such price for hour h is given by

$$PUN_h = \frac{\sum_{z=1}^{N} LMP_{z,h} \cdot D_{z,h}}{\sum_{z=1}^{N} D_{z,h}} \tag{1}$$

where $z = 1, \ldots, N$ denotes zone z in the Italian market, $LMP_{z,h}$ is the locational marginal price of the zone z at hour h, and $D_{z,h}$ is the total demand accepted in zone z at hour h. Moreover, in addition to PUN_h it is also useful to define the aggregate accepted demand Q_h at hour h which is given by

$$Q_h = \sum_{z=1}^{N} D_{z,h} \tag{2}$$

It is worth noting that the quantities in Equations (1) and (2) also include the energy delivered through bilateral contracts and the power imports from foreign countries.

The difference between the zonal prices paid to producers and the PUN paid by consumers results in a complex economic system. Indeed, the presence of a differentiated zonal price should, in principle, provide a correct localization of power plants. Producers would have an incentive to build production facilities in areas with less efficient generation and a limited ability to interconnect with the national transportation network. The purchase price, on the contrary, is unique in the whole national territory even in the presence of congestion. Therefore, the presence of a single national price (PUN) should not penalize areas of the country characterized by a less-efficient generation set.

Energy efficiency policies imposed by governmental agencies are appropriate means of capturing efficiencies that the market alone cannot assure (Gillingham et al. 2009). For example, demand response and energy efficiency can help improve electric-system operations by reducing the demand peak and driving peak prices to a lower level (Sahraei-Ardakani et al. 2012).

In 2007, the European Council adopted ambitious energy and climate change objectives for 2020, consequently adopted in Italy and in most European countries, both to reduce the general level of prices especially during peak hours and to mitigate the rise of congestion. The aim of these policies was to integrate and promote large-scale investments in renewable generation and to decommit old and high-cost power plants.

A major drawback of these pan-European actions is that there is yet no evidence whether the new power plants have been strategically placed. Moreover, investments in generation should be carried out with a careful and close look at the transmission network infrastructure, given the scarcity of the transmission capacity shown thus far.

Indeed, having a coal-based energy mix has many effects, both on electricity prices and, in general, on the social environment. Conversely, the more renewable sources are participating in the market (typically offering their energy with a quasi-null price) the lower are energy prices, assuming a constant and quasi-inelastic demand curve. Moreover, these social and economical outcomes can be obtained by prompting consumers to adapt their behavior to dynamic prices and to shift their peak usage of electricity when it is more convenient, that is, late-day hours and night. Many countries have adopted schemes and guidelines to stimulate a more efficient use of electricity, especially with the growing penetration level of renewable energy in the market.

It is worth noting that the stochastic nature of this type of resource requires increased deployment of operating reserves to balance the system and demand response can play an important role in reducing overall system costs, especially if a price-responsive demand mechanism is set up and customers can shift their behavior. In this context, Italy's energy mix has been characterized by a strong dependence on fossil fuels, which has always satisfied the demand for electricity for more than its 70 % (TERNA 2012).

To adhere to the EC policy target, Italy had to move towards a less coal-based electricity market and to stimulate a shift towards a more "green" energy mix. During the last four years, several steps have been taken in this direction, especially to boost new photovoltaic plants, as stated by (European Commission 2009a, b). "Conto Energia" subsidy regulations had a very ambitious aim, and had both success stories and some drawbacks, which we investigate further on. Indeed, "Conto Energia" was introduced in Italy with the EC Directive for renewable sources (European Commission 2001). This subsidy mechanism, which gives premiums for all the energy produced with photovoltaic plants, had several modifications introduced to boost investments and the production of energy derived from solar plants. The subsidy schemes started in 2005 and were well received by generation companies, small firms, and also private entities, and led to an enormous development of photovoltaic (hereafter PV) installations.

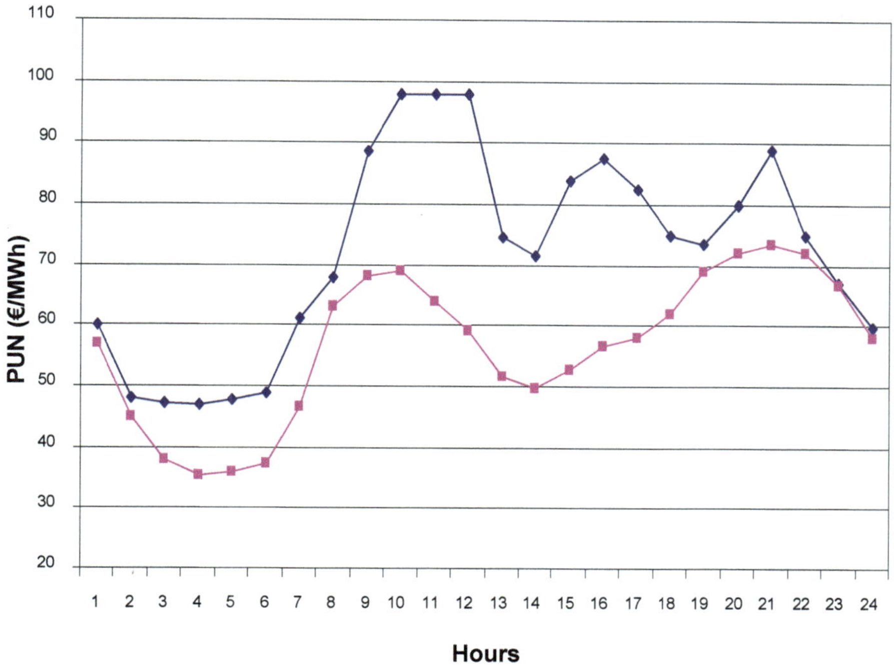

Fig. 3 PUN time series during May 3rd 2011 (*in blue*) and May 3rd 2012 (*in purple*)

It is a matter of record that this boost had a huge economic impact on the actors of the electricity market, which had to review some instances of the market in order to exploit this new blast of green electricity. To this aim, the so-called "Ritiro Dedicato" policy has been adopted by GSE, and had the objective of transferring the rights of operating PV plants to itself and selling their electricity directly on the day-ahead market. This technological policy also had economic drawbacks that can be observed either by looking at the changes in the market architectures and at price levels in some market zones.

In order to exploit the growing penetration level of renewable sources in the Italian electricity market, some infrastructure changes appear to be necessary either to the transmission network or to the market mechanisms also to provide dispatching security and network aggregate security.

Therefore, with an increased share of energy in the market bid at a null price, owners of high-cost power plants gradually decreased their intervention in the market because their bids were rejected. Moreover, they had to start reprogramming their production in subsequent markets in order to be able to sell their electricity. Furthermore, the new share of nondispatchable renewable sources in the day-ahead market increased the need for more ancillary services to guarantee security of the transmission system when renewable power plants were not producing electricity.

It is worth noting that day-ahead market prices in IPEX have always shown a peculiar shape due to the lack of efficient power plants able to lower prices during peak hours. Examples of the PUN time series are shown in Fig. 3, where PUNs during May 3rd, 2011 and 2012 are reported. It can be observed that the average

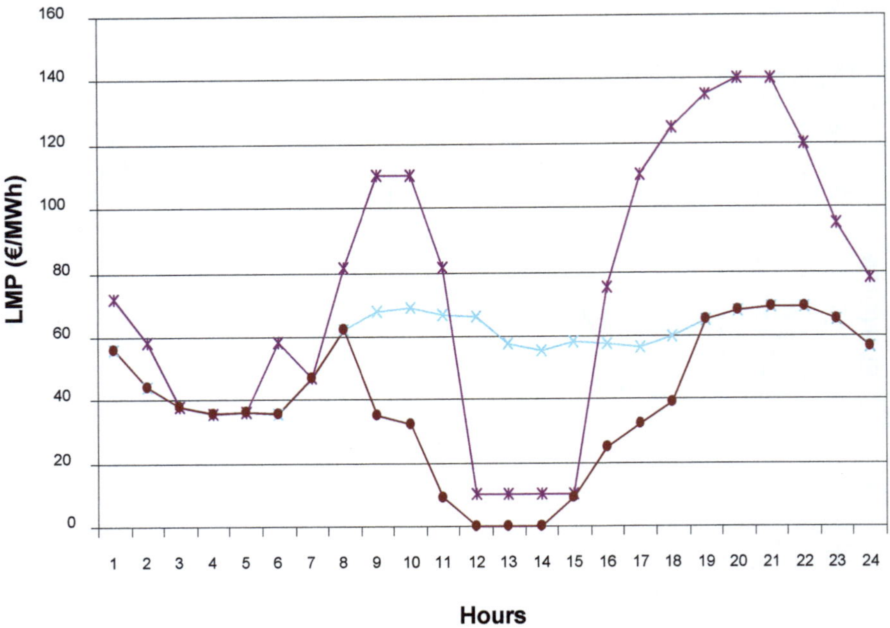

Fig. 4 Zonal prices time series during May 3rd 2012, Sicily (*in cyan*), Southern Italy (*in brown*) and Northern Italy (*in dark purple*)

PUN price is lower in 2012, especially during peak hours where there is a large valley instead of a peak. This is a direct effect of the increased share of renewable power plants and it is confirmed also by the broader shape of the remaining peaks.

As a result of the increasing share of PV plants connected to the grid and operating in the market, the peaks of PUN prices gradually decrease. Moreover, due to good weather conditions during May 2nd and 3rd, 2012, some LMPs had a value of 0 € / MWh, as shown in Fig. 4.

This analysis clearly points out the crucial contribution given by the renewable power plants to the peak-shaving action and which led to a smoother profile during peak and off-peak hours.

Nowadays, the event of zonal prices equal to zero was isolated to few days in May 2012 but new questions arise. When thinking about the incentive schemes for boosting renewable energy sources, did the regulator and the authorities expect that the policy could also give an economic outcome? Second, is the presence of zonal prices equal to zero opening an opportunity for investments in transmission capacity and power plant relocation policy? Third, which solution should we adopt?

One of the aims of having a high share of RES in the energy mix is to decrease the average prices on the wholesale market and to reduce peak prices, which in the case of Italy are among the highest in Europe. This might also lead to a decrease in the burden paid by consumers, if the effect on consumer prices were a direct reflex of this mechanism. However, due to the incentive schemes and some of the decisions taken by the Italian population after the referendums,[1] the electricity bill is affected by the high cost of fossil fuels and is composed of some tariffs that are not going to decrease, even with a greener energy mix.

Answering the questions is rather complex and requires one to look at many aspects: the transmission network infrastructure and the scarcity of transmission capacity, the presence of under- and overcapacity, and the combination of all these elements in the Italian DAM mechanism.

The contribution given to the generation side by the renewable sources was significant regarding consequent network congestion. Moreover, the combination of network congestion and zonal prices equal to zero demonstrates the need for policies that improve the use and especially the flow of energy produced by renewable sources in the zones with higher prices. To solve these issues only two approaches are possible:

- Incrementing network transmission capacity investing in those lines more subject to congestion
- Performing a zonal reconfiguration of a subset of renewable power plants based on the position of power plants, zonal prices, and congestion

Generally speaking, classical economic theory suggests that investments in infrastructure represent a viable path towards an efficient market. Indeed, eliminating any friction to the free flowing of goods (i.e., transmission network constraint in the electricity market) is the most efficient way to reduce prices and thus to increase consumers' social welfare. This approach has strongly influenced our society and economy and is driving the idea of a pan-European electricity market comprised of a single area with a uniform price from Norway to Greece without transmission constraints. However, this approach does not consider the complexity of the system under investigation together with the opportunity offered by the power mix and its possible relocation within a specific zonal splitting market clearing mechanism. Indeed, such frictions might result in either unexpected weakness or strength of the possible scenarios that might lead to unconventional and counterintuitive solutions, as discussed in the following sections.

Investments in Transmission Infrastructures

Starting from the public data published by GME and from our computational framework that replicates exactly the Italian MGP mechanism, two different market scenarios have been employed to calculate LMPs and PUN prices.

In the first scenario, by using the transmission limits adopted by GME and the real bids and offers, the locational marginal price LMPs and PUN prices (i.e., PUN_{con}) have been calculated under constrained interzonal capacity. On the other hand, in the second scenario the same real bids and offers have been considered by suppressing the transmission limits among zones, thus leading to hourly unique market

---

[1] http://en.wikipedia.org/wiki/Italian_referendums,_2011.

Table 1 LMPs (in € /MWh) in the Italian geographical zones on May 3rd, 2012 and PUN

Hour	PUN_{con}	CNORD	CSUD	NORD	SARD	SICI	SUD	BRNN	FOGN	PRGP.	ROSN
1	57.06	56.01	56.01	56.01	56.01	72	56.01	56.01	56.01	72	56.01
2	44.9	44	44	44	44	58	44	44	44	58	44
3	38.02	38.02	38.02	38.02	38.02	38.02	38.02	38.02	38.02	38.02	38.02
4	35.34	35.34	35.34	35.34	35.34	35.34	35.34	35.34	35.34	35.34	35.34
5	36.02	36.02	36.02	36.02	36.02	36.02	36.02	36.02	36.02	36.02	36.02
6	37.34	35.87	35.87	35.87	35.87	58	35.87	35.87	35.87	58	35.87
7	46.65	46.65	46.65	46.65	46.65	46.65	46.65	46.65	46.65	46.65	46.65
8	63.11	62	62	62	62	81.39	62	62	62	81.39	62
9	68.3	68.01	68.01	68.01	68.01	110	35	35	35	110	35
10	69.01	69	69	69	69	110	32	32	32	110	32
11	63.9	66.49	66.49	66.49	66.49	81.26	9	9	9	81.26	9
12	59.21	66	66	66	66	10	0	0	0	10	0
13	51.71	57.9	57.9	57.9	57.9	10	0	0	0	10	0
14	49.67	55.35	55.35	55.35	55.35	10	0	0	0	10	0
15	52.78	58	58	58	58	10	9	9	9	10	9
16	56.82	57.75	57.75	57.75	57.75	75.29	25	25	25	75.29	25
17	58.14	56.75	56.75	56.75	56.75	110	32	32	32	110	32
18	62	59.69	59.69	59.69	59.69	125.01	38.8	38.8	38.8	125.01	38.8
19	69.02	65	65	65	65	135.01	65	65	65	135.01	65
20	72.27	67.9	67.9	67.9	67.9	140.01	67.9	65	67.9	140.01	67.9
21	73.65	69	69	69	69	140.01	69	69	69	140.01	69
22	72.18	69	69	69	69	120.01	69	69	69	120.01	69
23	66.82	64.9	64.9	64.9	64.9	95.01	64.9	64.9	64.9	95.01	64.9
24	58.01	56.65	56.65	56.65	56.65	78.01	56.65	56.65	56.65	78.01	56.65

clearing prices (i.e., PUN_{uncon}). It is worth remarking that these two scenarios differ only in interzonal transmission capacity whereas all other elements have been kept unchanged. This condition is important as it allows determining both required investment in transmission infrastructure and effects of such investments on market results.

Table 1 summarizes the results of the first scenario showing LMPs and PUN_{cons} prices during the 24 h. As clearly shown, three macro zones can be identified (i.e., aggregation of zone characterized by the absence of a market splitting due to energy flows in transmission interconnections within capacity limits):

1. SICI and PRGP (Macro Zone 1, hereafter MZ1, highlighted by orange color in Table 1)
2. SUD, ROSN, FOGN, and BRNN (Macro Zone N°2, hereafter MZ2, highlighted by light blue color in Table 1)
3. NORD, CNOR, SARD, and CSUD (Macro Zone N°3, hereafter MZ3, highlighted by light yellow color in Table 1)

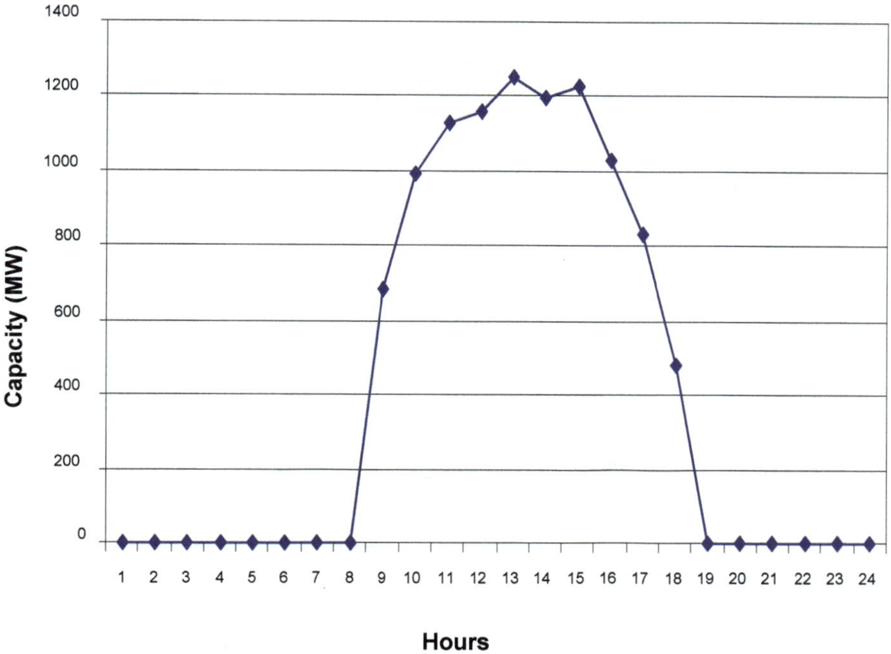

Fig. 5 Extra-transmission capacity needed in the connection between SUD and CSUD

Table 1 points out the presence of network congestion during certain hours of the day. In particular, the first congestion arises in the transmission connection between MZ3 and MZ1, whereas the second network congestion has been observed in the transmission connection between MZ1 and MZ2.

The congestion between MZ1 and MZ2 is always present, apart from hours 2, 3, and 4, in which the market is characterized by the absence of network congestion (i.e., all LMPs are identical and equal to the PUN_{con}).

During peak hours congestion is also present between MZ3 and MZ1 thus leading to an Italian power market divided into three price areas.

Furthermore, MZ1 is characterized by the highest locational zonal prices during the 24 h, whereas during the central hours of the day MZ2 has witnessed zero prices. This allows us to conclude that MZ1 and MZ2 are surely characterized by under- and overgeneration capacity, respectively. Furthermore, the presence of market zonal splitting among MZ2 and both MZ1 and MZ3 suggests the connections between MZ2 and MZ1 and between MZ2 and MZ3 as candidates for possible investments in the transmission infrastructure.

In order to understand the necessary amount of investment in the transmission network, hourly power flows on the transmission connections under unconstrained limit conditions have been computed and compared to the nominal capacity of transmission connections. This analysis points out that only the connection between SUD and CSUD is characterized by lack of capacity. In particular, Fig. 5 shows the required

Table 2 Hourly unconstrained and constrained PUN (in € /MWh) on May 3rd, 2012

Hour	PUN_{uncon} (€/MWh)	Q_{uncon} (MWh)	PUN_{con} (€/MWh)	Q_{con} (MWh)
1	58.15	27089 .50	59.07	27257 .51
2	44.01	26113 .73	44.89	26113 .73
3	38.02	25311 .31	38.02	25311 .31
4	35.34	25058 .83	35.34	25058 .83
5	36.02	25251 .54	36.02	25251 .54
6	36.65	25213 .41	37.30	25219 .63
7	46.65	27399 .80	46.65	27399 .80
8	62.00	31174 .79	63.10	31174 .79
9	66.24	36318 .67	68.17	36293 .20
10	66.00	38073 .54	68.62	38233 .20
11	64.00	37908 .50	62.88	38109 .50
12	63.00	37536 .86	58.15	37789 .19
13	50.00	36254 .65	50.62	36093 .67
14	49.50	35958 .15	49.41	35961 .31
15	50.65	36138 .39	51.74	35993 .98
16	51.65	35739 .48	56.20	35824 .54
17	53.00	35026 .83	57.87	35058 .87
18	59.25	34156 .02	61.91	34214 .99
19	65.01	33365 .20	69.00	33376 .89
20	66.88	33398 .32	71.30	33398 .32
21	71.00	33822 .00	73.64	33822 .00
22	71.00	33236 .53	72.17	33236 .53
23	66.07	30096 .49	66.81	30096 .49
24	65.00	27039 .31	58.00	27124 .40

extra transmission capacity with respect to the current real transmission limits during the 24 h. As a consequence, one can conclude that with an investment of 1,251 MW in the transmission capacity between SUD and CSUD zones, all friction to energy flow is eliminated. It is worth remarking that the network capacity of the line is 2,000 MW. Thus, such an investment requires a 60 % increment of the nominal network capacity.

In order to evaluate the effects of such investments it is useful to compare the aggregate market results in the constrained and unconstrained scenarios. Table 2 summarizes the PUN and the total aggregate accepted demand during the 24 h.

Investing in infrastructure results in a reduction of zonal prices and of the PUN, corroborating the idea that eliminating friction in the flow of energy is a positive solution towards an efficient market. However, Table 2 points out that there are some hours (i.e., 11, 12, 14, and 24) where the system is not behaving as expected. This evidence suggests that transmission limits can result in a factor that will improve consumers' social welfare. In particular, the presence of theItalian zonal pricing clearing mechanism together with constraints in the transmission grid led to a complex ecosystem that reduced the costs paid by consumers more than a free flow of energy. Therefore, scenarios where power plants are relocated are worth being investigated as alternatives to investments in transmission capacity infrastructure and this topic is addressed in the next section.

Power Plant Relocation Policy

In this section, possible reconfigurations of a subset of the existing power plants in Italy are presented and discussed. The aim is to evaluate if and how a relocation policy can result in reducing zonal prices and increasing consumers' social welfare. In order to address such analyses we need:

- To determine which macro zones are characterized by overgeneration capacity and to estimate the corresponding amounts
- To determine which macro zones are characterized by undergeneration capacity and to estimate the corresponding amounts
- To identify the potential power plants that can be relocated
- To calculate the effect of the relocation policy

In order to address the different under- and overgeneration capacity estimates, we started with the constrained and unconstrained scenarios discussed in the previous section. In particular, in each macro zone z the overgeneration capacity for each hour h is determined by considering the supply offers that have been discharged (i.e., not accepted by the market) but that were characterized by a submitted limit price less than or equal to the unconstrained PUN price at the same hour (shown in Table 2). The sum of such a quantity corresponds to the zonal overgeneration capacity, that is, energy that could not be transferred from one zone with lower supply offer prices to one with higher supply offer prices due to the limits in transmission capacity. This analysis pointed out that only macro zone MZ2 is characterized by overgeneration capacity and the aggregated hourly profile of the overgeneration capacity is shown in Fig. 5.16. Dually, in each macro zone z the undergeneration capacity for each hour h is determined by considering the demand bids that have been discharged (i.e., not accepted by the market) but that were characterized by a submitted limit price higher than or equal to the unconstrained PUN price at the same hour (shown in Table 2). In this case, the sum of such a quantity corresponds to the zonal undergeneration capacity, that is, energy demand that could not be satisfied by a zone with lower supply offer prices due to the limits of transmission capacity. This analysis pointed out that, as expected, macro zones MZ3 and MZ1 are characterized by undergeneration capacity and in particular within macro zone MZ3 it showed that zone NORD is characterized by undergeneration capacity, whereas the other zones have limited requests of additional generation capacity.

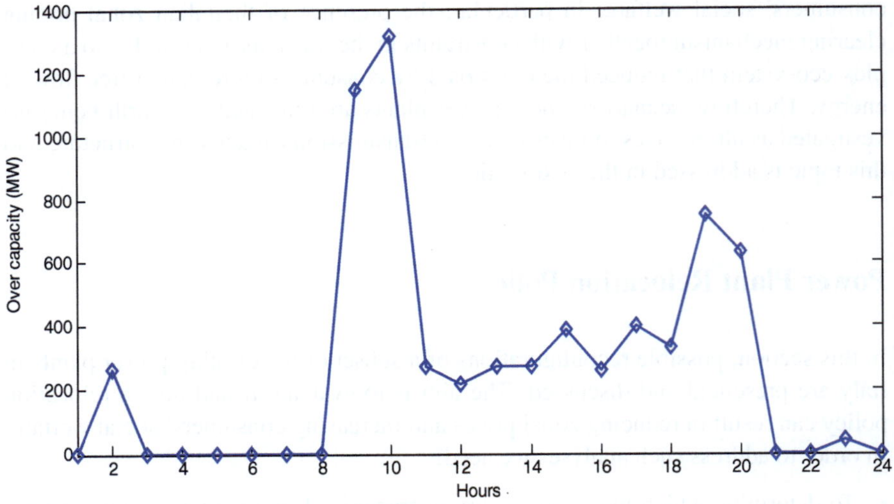

Fig. 6 Over-generation capacity profile in MZ2 during May 3rd

Based on these results, we came to the conclusion that the subset of power plants that are candidates for relocation was located in macro zone MZ2. In particular, we selected 36 power plants from the share of renewable power plants in MZ2 characterized by an offer price equal to zero. The renewable power plants were considered both because they were responsible for the presence of zero LMPs in MZ2 (see Table 2) in the constrained scenario discussed in the previous section and because they are the newer power plants installed in Italy. Thus, the proposed relocation policy can in fact also be considered a design policy for the installation of generation capacity that should have been considered by the AEEG and Italian government. The aggregated quantity offer profile of the selected 36 power plants is shown in Fig. 7 and a comparison with the overgeneration capacity shown in Fig. 6, points out quite a different profile but almost identical maxima.

Once having selected the 36 power plants, different market scenarios have been determined in order to evaluate the policy effects. In particular, if we consider that each power plant could in principle be relocated from macro zone MZ2 to either MZ1 or MZ3, the total number of possible scenarios is $2^{36} \cong 10^{11}$. A complete evaluation of all existing scenarios is not possible, thus we resorted to performing a random sampling of selected 10^5 cases. For each scenario, the Italian market clearing algorithm has been used to evaluate the corresponding LMPs and PUN for the 24 h under constrained conditions. It is worth remarking that each of these scenarios differs for the real GME solution just for the location of the selected 36 power plants whereas all other elements (i.e., interzonal transmission capacity and all the other offers) have been kept unchanged.

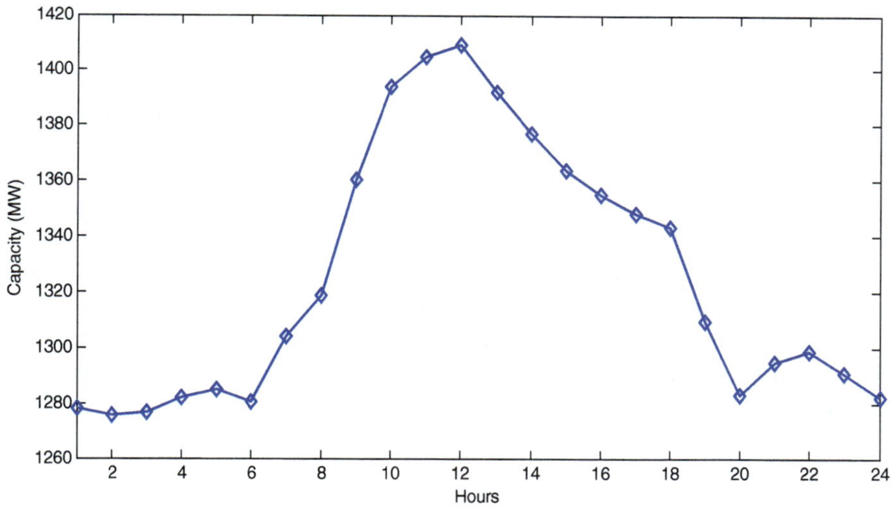

Fig. 7 Aggregated offer quantity bids of the selected set of 36 power plants (see text)

In order to compare the different solutions, we need to define an economic indicator. To this aim, we have considered the daily average unique national price $\overline{PUN}$defined as

$$\overline{PUN} = \frac{\sum_{h=1}^{24} PUN_h \cdot Q_h}{\sum_{h=1}^{24} Q_h} \quad (3)$$

where PUN_h and Q_h are given by Eqs. 1 and 2, respectively.

It is worth remarking that $\overline{PUN}$ is a valid proxy of the consumers' social welfare as it describes the average price paid by consumers during the 24 h of a day. Thus, it allows a direct and concise comparison of the different scenarios. In order to proceed with the comparison, we first calculated the daily average unique national price for the constrained (i.e., $\overline{PUN_{con}}$) and unconstrained (i.e., $\overline{PUN_{uncon}}$) cases discussed in the previous section. The results point out that $\overline{PUN_{con}}$ and $\overline{PUN_{uncon}}$ are equal to 57.599 € /MWh and to 56.700 € /MWh, respectively.

Based on classical economics, these two values represent the possible range for an effective relocation policy as one should expect. Moreover, such a policy improves the consumers' social welfare with respect to the current situation with transmission constraints (i.e., $\overline{PUN_{con}}$ is the upper boundary) and does not overperform the free-flow solution determined by the unconstrained scenario (i.e., $\overline{PUN_{uncon}}$ is the lower boundary).

In fact, the Italian energy market regulation points out a rather complex context as over the 10^5 scenarios considered, 100 and 57 % overperformed the constrained (i.e., $\overline{PUN_{con}}$) and unconstrained (i.e., $\overline{PUN_{uncon}}$) cases, respectively.

It is worth remarking that classical economic theory does not encompass the complexity of the Italian electricity system, nor does it account for the zonal splitting market clearing mechanism that might result in unexpected strength of the transmission-constrained scenarios which eventually lead to unconventional and counterintuitive solutions. In particular, the best result among the tested 10^5 scenarios leads to a daily average unique national price $\overline{PUN}$ equal to 56.001 € /MWh. Even considering the limited opportunity offered by the system, it overperforms by 47 % the free-flow solution determined by the unconstrained scenario.

Thus, these results corroborate the hypothesis that relocating a reduced subset of power plants from an overgeneration capacity macro zone to an undergeneration capacity macro zone in a limited transmission capacity context can lead to better performance for the consumers' social welfare than myopically investing in transmission network infrastructure.

Conclusions

Zonal prices and network congestion arose in Italy since the establishment of the electricity market. The existing power mix and the network infrastructure policies adopted in order to decrease congestion and reduce electricity prices did not succeed in solving the problem. This chapter presented a comparison between investments in transmission networks and generation capacity relocation policy under a zonal pricing mechanism. The aim is to understand if a relocation of the existing Italian power mix is able to maximize consumers' social welfare by taking advantage of the existing constraints in the transmission network and of the zonal pricing mechanism. The proposed policy is strongly counterintuitive as classical literature considers investments in transmission infrastructure the only viable path to an efficient electricity market. Thus, the purpose of this chapter is to provide an innovative framework for discussing and proposing policy design mechanisms that are able to alleviate high zonal prices and integrate renewable generation more efficiently. An empirical analysis on the results of the Italian day-ahead market prices together with computer determination of the solution of the scenarios in the Italian day-ahead market has been employed to address the study. In this context, investments in transmission capacity have been evaluated as well as the presence of over- and undergeneration capacity in the Italian market zones. These analyses allow us to determine the possible scenarios of generation relocation, in particular for renewable power plants. The different solutions arising from investments in transmission capacity and from power plant relocation have been compared by means of the daily average unique national price which is a valid proxy of consumers' social welfare as it describes the average price paid by consumers during the 24 h of a day. Results have shown that a proper localization of a subset of power plants allows increasing consumers' social welfare within a zonal splitting mechanism, even with respect to a classical efficient solution based on the elimination of all possible friction in the flow of energy.

This innovative and counterintuitive result suggests that (i) investing in transmission networks is first useful in order to eliminate congestion; (ii) eliminating congestion does not necessarily lead to increased consumer social welfare; and (iii) relocating existing power plants and taking advantage of the Italian market mechanism and its transmission capacity limits is an advisable policy in order to increase consumers' social welfare.

Acknowledgments This work has been partially supported by the University of Genoa, by the Italian Ministry of Education, University and Research (MUR), by the European Social Fund (ESF), and by Regione Liguria, Italy.

References

AEEG: Annual report to the European commission on regulatory activities and the state of services in the electricity and gas sectors, July (2012)

Boerema, N., Merlinde, K., MacGill, I.: Renewable energy integration into the Australian national electricity market: characterising the energy value of wind and solar generation. Environmental Economics Research Hub Research Reports EERH 88, Australian National University, http://purl.umn.edu/107740 (2010). Accessed 25 Aug 2013

Botterud, A., Korpas, M.: Modelling of power generation investment incentives under uncertainty in liberalised electricity markets. Proceedings of the Sixth IAEE European Conference, pp. 1–3 (2004)

Burger, A., Ferstl, R.: Generation capacity investment in oligopolistic electricity markets under uncertainty. EEM 2008, 5th international conference on the European electricity market, vol., no. pp. 1–10. doi: 10.1109/EEM.2008.4579118 (2008)

Cochran, J., Bird, L., Heeter, J., Arent, D.A.: Integrating variable renewable energy in electric power markets: best practices from International Experience, Summary for Policymakers. Technical Report, National Renewable Energy Laboratory (NREL) (2012). doi: 10.2172/1041368

European Commission. Directive 2001/77/EC. http://eur-lex.europa.eu/LexUriServ/LexUriServ.do?uri=OJ:L:2001:283:0033:0033:EN:PDF (2001). Accessed 25 Aug 2013

European Commission. Directive 2009/72/EC. http://eur-lex.europa.eu/LexUriServ/LexUriServ.do?uri=OJ:L:2009:211:0055:0093:EN:PDF (2009a). Accessed 25 Aug 2013

European Commission. Directive 2009/73/EC. http://eur-lex.europa.eu/LexUriServ/LexUriServ.do?uri=OJ:L:2009:211:0094:0136:en:PDF (2009b). Accessed 25 Aug 2013

Gillingham, K., et al.: Energy efficiency economics and policy. Annu. Rev. Resour. Econ., Annu. Rev. **1**(1), 597–620, 09 (2009)

GME: Vademecum of the Italian power exchange. http://www.mercatoelettrico.org/En/MenuBiblioteca/documenti/20091112Vademecumoflpex.pdf (2009). Accessed 25 Aug 2013

Hogan, W.W.: Market-based transmission investments and competitive electricity markets, mimeo. Center for Business and Government, JFK School of Government, Harvard University. http://www.whogan.com (1999). Accessed 25 Aug 2013

Joskow, P.J.: Patterns of transmission investment. Working Papers 0504, Massachusetts Institute of Technology, Center for Energy and Environmental Policy Research (2005)

Joskow, P.L.: Competitive electricity markets and investment in new generating capacity. Working Papers 0609, Massachusetts Institute of Technology, Center for Energy and Environmental Policy Research (2006)

Leou, R.-C., Teng, J.C.: A transmission plan considering uncertainties under a deregulated market. International conference on information and network technology, IPCSIT vol. 6. IACSIT Press, Singapore (2011)

Makkonen, M., Viljainen, S.: Costs of insufficient transmission network investments. The Nordic electricity market case. EEM 2012. 5th International conference on the European electricity market, vol., no., pp. 1–8, 10–12. doi:10.1109/EEM.2012.6254742. (2012)

Sahraei-Ardakani, M., Blumsack, S., Kleit, A.: Distributional impacts of state-level energy efficiency policies in regional electricity markets. Energ. Policy Elsevier **49**(C), 365–372 (2012)

Siddiqui, A., Gupta, H.: Transmission investment timing and sizing under uncertainty. Research Report No. 285, Department of Statistical Science, University College London (2007)

Skoteinos, I.I., Orfanos, G.A., Georgilakis, P.S., Hatziargyriou, N.D.: Methodology for assessing transmission investments in deregulated electricity markets. IEEE Trondheim PowerTech Conference (2011)

Smith, J.C., Beuning, S., Durrwachter, H., Ela, E., Hawkins, D., Kirby, B., Lasher, W., Lowell, J., Porter, K., Schuyler, K., Sotkiewicz, P.: Impact of variable renewable energy on US electricity markets. IEEE Power and Energy Society General Meeting, vol., no., pp. 1–12, 25–29 (2010). doi:10.1109/PES.2010.5589715.

Stoft, S.: Transmission investment in a deregulated power market, chapter 4 in F. Lévêque, Competitive Electricity Markets and Sustainability. Edward Elgar Publishing Limited, Cheltenham (2007)

TERNA: Identification of zones in the significant grid (In Italian). http://www.terna.it/LinkClick.aspx?fileticket=77563 (2009). Accessed 25 Aug 2013

TERNA: Statistical data on electricity in Italy. http://www.terna.it/default/Home/SISTEMA_ELETTRICO/statistiche/dati_statistici.aspx (2012). Accessed 25 Aug 2013

Ventosa, M., Alvaro Barıllo, A.R., Rivier, M.: Electricity market modeling trends. Energ. Policy **33**(7), 897–913. http://dx.doi.org/10.1016/j.enpol.2003.10.013 (2005)

Simulation of Greenhouse Gas Cap-and-Trade Systems with ENERGY 2020

Jeffrey S. Amlin

Introduction

ENERGY 2020 is an integrated, multi-region, energy model that has been actively used by state, provincial, and national governments and private energy companies since the early 1980s to conduct energy and emission related policy analysis and forecasting. Beginning in 1998, it has also been used to analyze several different cap-and-trade systems in the US and Canada. This chapter provides an overview of the ENERGY 2020 simulation model, describes using ENERGY 2020 to simulate various greenhouse gas (GHG) cap-and-trade systems, and reflects on the lessons learned in the modeling process.

Overview of ENERGY 2020

ENERGY 2020 is an integrated, multi-region, energy model that provides complete and detailed, all-fuel demand and supply sector simulations (SSI 1996). These simulations can additionally include macroeconomic interactions to determine the benefits or costs to the local economy in response to new facilities or changing energy prices. Greenhouse gas and criteria air contaminant (CAC) pollution emissions and costs, including allowances and trading, are endogenously determined, thereby allowing assessment of environmental risk and co-benefit impacts.

The basic implementation of ENERGY 2020 for North America now contains a user-defined level of aggregation down to the 12 provincial and 50 state (and sub-state) levels. ENERGY 2020 contains historical information on all generating units in the US and Canada. Mexico data can also be incorporated as needed. ENERGY 2020 is parameterized with local data for each region, as well as all the associated energy suppliers it simulates. Thus, it captures the unique characteristics (physical,

---

J. S. Amlin (✉)
1519 Heatherwood Trail, Xenia, OH 45385, USA
e-mail: JeffAmlin@ENERGY2020.com

H. Qudrat-Ullah (ed.), *Energy Policy Modeling in the 21st Century*,
Understanding Complex Systems, DOI 10.1007/978-1-4614-8606-0_7,

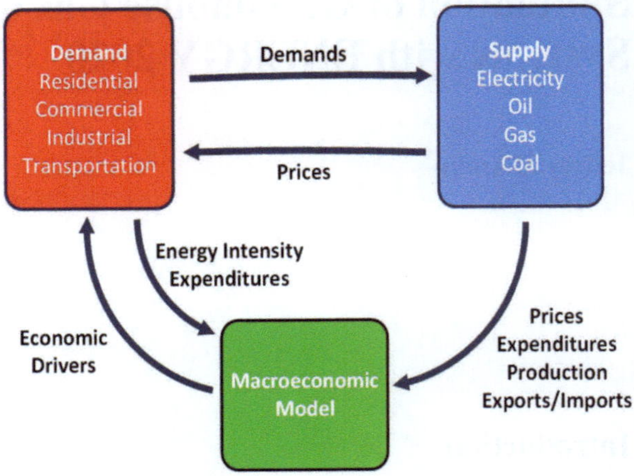

Fig. 1 Overview of ENERGY 2020. (@copyright Systematic Solutions, Inc. 2013, reprinted by permission)

institutional, and cultural) that affect how people make choices and use energy. Collections of state and provincial models are currently validated from 1986 to the latest annual numbers.[1]

ENERGY 2020 can be linked to a detailed macroeconomic model to determine the economic impacts of energy or environmental policy and the energy and environmental impacts of national economic policy. For US regional and state level analyses, the Regional Economic Models, Inc. (REMI) macroeconomic model has often been linked to ENERGY 2020.[2] The Informetrica macroeconomic model is routinely linked to ENERGY 2020 for Canadian national and provincial efforts.[3] The macroeconomic models include inter-state/provincial impacts, US trade flows, world trade flows, prices, and investments and simulate the real-time impact of energy and environmental concerns on the economy and on the energy system. Figure 1 illustrates the interactions of a macroeconomic model with the demand and supply sectors within ENERGY 2020. A macroeconomic model provides initial economic drivers for the demand sector and then receives inputs from both the demand and supply sectors to produce economic impacts caused by changes to the energy system.

The structure of the model is well tested and has been used to simulate not only US and Canadian energy and environmental dynamics but also those of several countries in South America, Western, Central, and Eastern Europe. These efforts include strategic and tactical analyses for both planning and energy industry restructuring/deregulation. The US EPA used ENERGY 2020 to perform the regional energy, environmental, and macroeconomic impacts of proposed Kyoto initiatives at the 50-state level.

[1] Energy supplier data come from the Federal Energy Regulatory Commission (FERC) and the US Department of Energy (DOE) for the US and from Statistics Canada for Canada. US and Canadian fuel and demand data come from the US DOE and Natural Resources Canada, respectively. US and Canadian pollution data come from the US Environmental Protection Agency (EPA) and Environment Canada, respectively.

[2] www.remi.com.

[3] Informetrica Limited, www.informetrica.ca.

Demand Sector

The default model simulates demand by three residential categories (single family, multi-family, and agriculture/rural), over 60 North American Industry Classification System (NAICS) commercial and industrial categories, and three transportation services (passenger, freight, and off-road). There are approximately six end-uses per category and six technology/mode families per end-use.[4] Currently the technology families correspond to six fuel groups (oil, gas, coal, electric, solar, and biomass) and 30 detailed fuel products. The transportation sector contains 45 modes, including various types of automobile, truck, off-road, bus, train, plane, marine, and alternative-fuel vehicles. More end-uses, technologies, and modes can be added as data allow. For all end-uses and fuels the model is parameterized based on historical, location-specific data. The load duration curves are dynamically built up from the individual end-uses to capture changing conditions under consumer choice and combined gas/electric programs.

Each energy demand sector includes cogeneration, self-generation, and distributed generation simulation, including mobile-generation, micro-turbines, and fuel-cells. Fuel-switching responses are rigorously determined. The technology families (which can be split, as an option, to portray specific technology dynamics) are aggregates that, within the model, change building shell, economic-process and device efficiency, and capital costs as prices, or other information that the decision makers see. ENERGY 2020 utilizes the historical and forecast data developed for each technology family to parameterize and disaggregate the model (Fig. 2).

Figure 2 shows the complex relationship between energy prices and energy demands in ENERGY 2020. The energy prices (including emission taxes) impact the economic growth, the fuel choice, the marginal process efficiency, the marginal device efficiency, the utilization of the capital stock, and the decision to invest in cogeneration. These marginal decisions are accumulated and aged through a stock and flow structure (production capacity, process energy requirements, device energy requirements, and cogeneration capacity) to determine the forecast for energy demands.

Supply Sectors

ENERGY 2020 supply sectors include electricity, oil, natural gas, refined petroleum products, ethanol, land-fill gas, and coal supply. Energy used in primary production and emissions associated with primary production and its distribution are included in

[4] End-uses include process heat, space heating, water heating, other substitutables, refrigeration, lighting, air conditioning, motors, and other non-substitutables (miscellaneous). Detailed modes include: small auto, large auto, light truck, medium-weight truck, heavy-weight truck, bus, freight train, commuter train, airplane, and marine. Each mode type can be characterized by gasoline, diesel, electric, ethanol, natural gas, propane, fuel-cell, or hybrid vehicles.

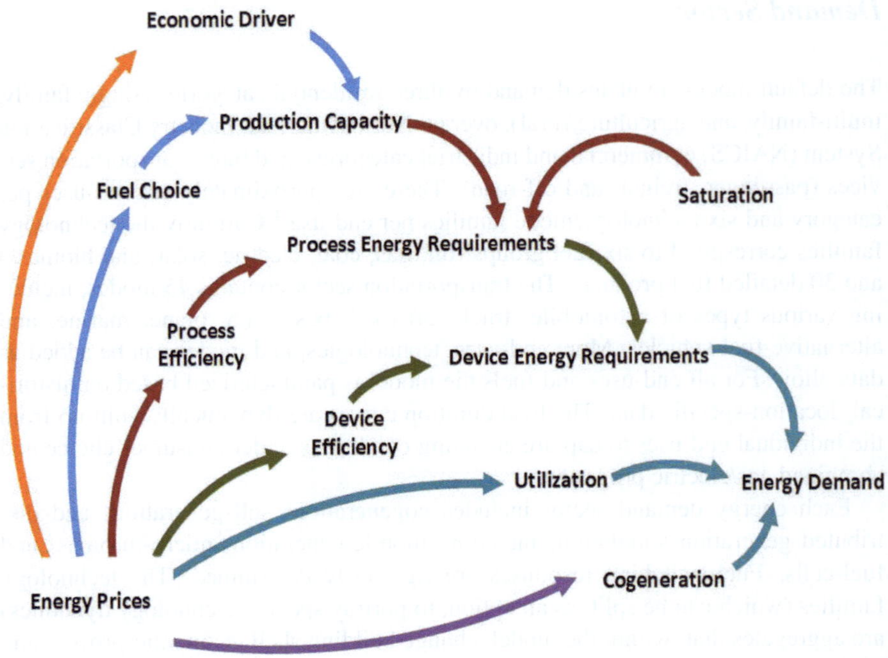

Fig. 2 Energy price and demand relationships in ENERGY 2020. (@copyright Systematic Solutions, Inc. 2013, reprinted by permission)

the model. The supply sectors included in a particular implementation of ENERGY 2020 will depend on the characteristics of the area being simulated and the problem being addressed. If the full supply sector is not needed, then a simplified simulation determines delivered-product prices. Figure 3 is an overview diagram of a generic energy supply sector in ENERGY 2020.

The electric supply sector is an endogenous detailed simulation of capacity expansion, generation, fuel usage, emissions, and electricity rates.[5] The model dispatches plants according to the specified rules, whether they are optimal or heuristic, and simulates transmission constraints when determining dispatch.[6] A sophisticated dispatch routine selects critical hours along seasonal load duration curves as a way to provide a quick but accurate determination of system generation. Peak and base hydro usage is explicitly modeled to capture hydro-plant impacts on the electric system.

[5] ENERGY 2020 does include a complete, but aggregate representation of the electric transmission system. Electric transmission data are provided by FERC, the Department of Energy, and the National Electric Reliability Council. The dispatch technologies in the basic model include: oil/gas combustion turbine, oil/gas combined cycle, oil/gas combined cycle with CCS, oil/gas steam turbine, coal steam turbine, advanced coal, coal with CCS, nuclear, base load hydro, peaking hydro, small hydro, wind, solar, wave, geothermal, fuel-cells, flow-battery storage, pumped hydro, biomass, landfill gas, trash, and biogas, but other technologies can be added.

[6] A 110 node transmission system is used in the default model, but a full AC load-flow bus representation model has also been interfaced with ENERGY 2020.

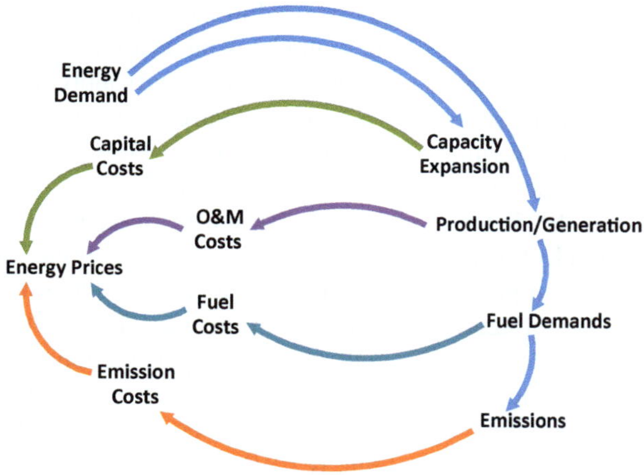

Fig. 3 Generic energy supply sector in ENERGY 2020. (@copyright Systematic Solutions, Inc. 2013, reprinted by permission)

Emissions

ENERGY 2020 includes pollution accounting for combustion (by fuel, end-use, and sector), non-combustion, and non-energy (by economic activity) for SO_2, NO_2, N_2O, CO, CO_2, CH_4, PM_T, $PM_{2.5}$, PM_5, PM_{10}, VOC, CF_4, C_2F_6, SF_6, and HFC at the state and provincial level by economic sector. Other (gaseous, liquid, and solid) pollutants can be added as desired. Pollution does not need to be determined directly by coefficients but can be recognized by the accumulation of capital investments that result in pollution emission with usage. National and international allowance trading is also included. Plant dispatch can consider emission restrictions.

The model captures the feedback among energy consumers, energy suppliers, and the economy using qualitative choice theory (McFadden and Domencich 1975) and co-integration (Granger and Newbold 1977; Granger and Engle 1987).[7] For example, a change in price affects demand that then affects future supply and price. Increased economic activity increases demand; increased demand increases the investment in new supplies. The new investment affects the economy and energy prices. The energy prices also affect the economy.

A Brief History of ENERGY 2020

ENERGY 2020 traces its roots back to the FOSSIL model developed at Dartmouth College in the mid-1970s. The FOSSIL model later became FOSSIL2 and still later the IDEAS model and was the official DOE energy analysis model until 1995.

[7] The model has used the work of Daniel McFadden and Clive Granger since its inception in the late 1970s.

ENERGY 2020 was first used in 1986 for the Kansas Gas & Electric Company (KG&E) analysis of the Wolf Creek nuclear unit. In the 1980s and early 1990s the model simulated a single energy demand and supply area and was used to generate energy plans for government agencies and electric forecasts for electric utilities. Clients included Vermont DPS, Massachusetts DOER, Canada OERD, Hawaii DBEDT, Saskatchewan Energy and Mines, Poland, Bulgaria, Estonia, Latvia, and Lithuania.

In 1995, the model was expanded to simulate multiple service areas and multiple electric utility companies. Data were compiled for all 50 US states and 14 Canadian provinces and territories and created a system to generate a model with any set of aggregated states and provinces. This enabled the simulation of electric system deregulations and other multi-jurisdictional analysis and forecast including electric transmission and regional electric capacity expansion. Clients included British Columbia Hydro, Bonneville Power Administration, PacifiCorp, Canadian Energy Research Institute (CERI), Southern California Edison, New Century Energies, Westar Energy, Houston Lighting & Power Company, Cinergy, and Ohio Edison.

In 1998, the emission sectors of the model were enhanced to enable the model to simulate all GHG emissions plus GHG sinks and offsets which was needed to support the Kyoto Protocol analysis for Canada. Since the initial GHG work, the ENERGY 2020 emissions sectors have been enhanced to support both CAC (air pollution) and GHG forecasting and analysis. The broad and comprehensive scope of the model enables it to simulate cap-and-trade systems involving both the US and Canada. The model has been used for cap-and-trade analysis by Environment Canada, the Western Climate Initiative (WCI 2009), the California ARB (ARB 2008), the Illinois Climate Change Advisory Group (Illinois 2008), BPA, and Wisconsin's Governor's Task Force on Global Warming (Wisconsin 2008).

In 2011, the oil and gas supply sectors were enhanced to better simulate the production of oil and gas (including oil sands and shale gas) in Canada and the US. These enhancements are ongoing with additional enhancements to natural gas transmission and oil refining.

A more detailed history is available in "A History of Making Energy Policy" by George Backus, Jeff Amlin, and Ottie Nabors (Amlin et al. 2009).

Simulation of GHG Cap-and-Trade Systems

ENERGY 2020 provides an excellent framework for simulating GHG cap-and-trade systems. GHG emissions are unique in that their impact is global, not local. So, reducing the GHG emissions in one part of the system does not reduce emissions overall if there is an increase in another portion of the system. Similarly, reducing GHG emissions in one sector is equivalent to reducing them in any other sector. This relationship leads to the importance of having the entire system simulated including GHG offsets. GHG offsets are mechanisms which reduce GHG emissions in sectors not included in the cap-and-trade system. These sectors often include GHG reductions from the agriculture, forestry, or landfill gas sectors.

ENERGY 2020 simulates all GHG emissions, including fugitive, process, and non-combustion, as well as, combustion related emissions, sources, and sinks for GHG emissions. The simulation of all energy demands and supplies enables ENERGY 2020 to reflect the impact of emissions switching sectors. If an industrial sector switches from fossil fuels to electricity for their energy needs, then the industry will reduce their GHG emissions. However, the electric supply sector will see an increased demand for electricity and will need to generate more electricity, often increasing electric sector GHG emissions. The net change in GHG will depend on the fuels that are displaced in the industrial sector and the source of the generation in the electric supply sector. If the industrial sector is reducing its use of coal while the electric supply sector is increasing its wind generation, then there will be an overall reduction in GHG. However, if the industrial sector is reducing its use of natural gas while the electric supply sector is increasing coal generation, then net GHG will increase. For GHG policy analysis it is important to simulate all sources of GHG emissions to ensure that the net impact can be determined.

ENERGY 2020 simulates a cap-and-trade system by establishing the GHG allowance price, allowing each sector to respond to the price. If the GHG target is not met, then ENERGY 2020 increases the price and allows each sector to respond a second time. The model continues to iterate until a solution is found. This iterative process enables a solution to be found, regardless of the methodology of each sector. The algorithms in each sector can take any reasonable, functional form, including non-linear algorithms.

GHG Cap-and-Trade Systems

GHG cap-and-trade systems are a form of regulation in which each entity that produces a unit of GHG must provide a GHG allowance (or permit) to the regulatory authority. These allowances are obtained through purchases from a GHG market or may be allocated freely to participants. The total number of allowances (either allocated freely or auctioned to the market) is controlled by the regulating authority. The regulatory authority will set the number of allowances equal to the desired GHG emission goal. A market for GHG allowances is created from which any GHG entity can buy or sell allowances as needed. This market will establish a price for allowances that will clear the market and thus meet the GHG goal.

Figure 4 is a diagram of how allowance prices are set in ENERGY 2020 and how those allowance prices feedback through energy prices to impact the energy demand and economic activity.

GHG Cap-and-Trade Concepts

A cap-and-trade system design has multiple structural concepts which must be taken into account when attempting to develop a comprehensive simulation. This section

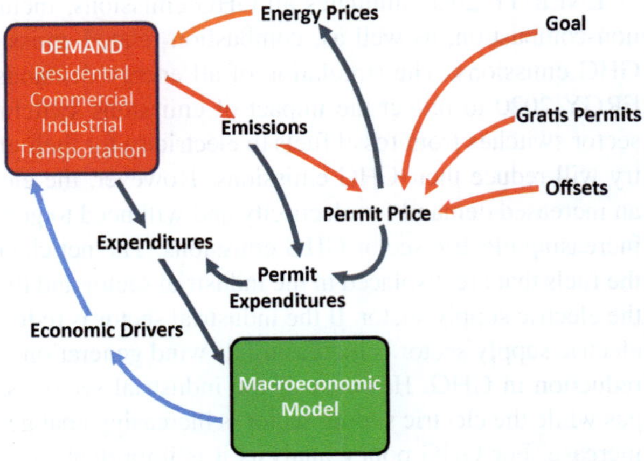

Fig. 4 Cap-and-trade system structure in ENERGY 2020. (@copyright Systematic Solutions, Inc. 2013, reprinted by permission)

discusses ENERGY 2020's approach to modeling the following cap-and-trade concepts:

- Emissions coverage—geographic, economic sectors and fuels, pollutants
- Emissions goal—historical, forecast, intensity; goals by sector
- Allocated allowances (gratis permits)—historical, forecast, intensity
- Offsets—local, domestic, international, government, offset limits, offset prices
- Allowance price limits (minimum, maximum)
- Allowance reserves
- Banking and borrowing allowances
- Allowance revenues
- Macroeconomic feedback

Emissions Coverage

Emissions coverage simply means all the geographic areas, economic sectors, and emissions are included in the cap-and trade system. The coverage is specified in the cap-and-trade system design.

Geographic Areas Through the use of model switches, ENERGY 2020 is designed to assign any area (state, province, or territory) to be included or excluded as part of a cap-and-trade system. The switches have proven to be important in order to respond to potential changes in cap-and-trade design in the middle of a study. For example, in the original Western Climate Initiative cap-and-trade analysis, Arizona, California, Montana, New Mexico, Oregon, Utah, Washington, and British Columbia were included in the cap-and-trade system. Later Manitoba, Ontario, and Québec were added to the analysis, and finally only California and Quebec were included. As the design changed, the model switches were revised to include only those areas included in the cap-and-trade design.

Economic Sectors and Fuels ENERGY 2020 assigns model switches to indicate which sectors and fuels in the model are covered under a specified cap-and-trade system. Typically, the electric power sector and the large industrial customers are the first sectors included. The residential, commercial, and transportation sectors may be included later, although their inclusion may be indirectly simulated through the entities that sell fuel to these sectors. For example, gasoline suppliers may be required to purchase an allowance for each gallon of gasoline sold within the cap-and-trade area. The model can easily simulate coverage of any or all sectors with the selection of the proper switches. Even within a sector, certain uses of fuel, such as emissions from industrial generation of electricity, may be treated differently than the emissions from other uses. In some instances in which there exists a significant amount of imported electricity, the electric generators outside the cap-and-trade area may be considered covered under the cap-and-trade. Designing the model switches to allow for a wide range of combinations is important given the changing nature of a typical design.

Pollutants There are several types of GHG with the primary ones being CO_2, N_2O, CH_4, SF_6, HFC, and PFC; however, not all these emissions may be covered for all sectors. The fluorocarbons (SF_6, HFC, and PFC) are often only covered in a limited number of sectors while even N_2O and CH_4 may not be covered for the residential or transportation sectors. Emissions are generated from combustion and non-combustion of fuels, process and fugitive sources. Again, some of these sources may not be covered for all sectors or pollutants.

In ENERGY 2020 the covered area, covered sectors, and covered emissions are specified with a single variable which ranges between 0 (not covered) and 1.0 (100 % covered). Functional values are often used to simulate systems that cover only facilities exceeded a certain level of emissions (for example facilities which emit more than 25,000 t). These values can change over time as more sectors, areas, or pollutants are incorporated into the cap-and-trade system.

Emission Goal

The emission goal may be specified in different ways, such as a reduction from a historical year emissions or a reduction from a 'business as usual' forecast year projection. An emission intensity can also be used as the goal instead of a quantitative inventory. For example:

1. The emission goal in 2025 is set to 6 % below the 2000 emissions.
2. The emission goal in 2025 is set to 30 % below the BAU forecast of 2020 emissions.
3. The emission goal in 2025 is set as a 50 % reduction in the emission intensity (tonne/US$GRP) below the emission intensity of 2010.

Variations on these methods are nearly endless including different treatment for each area, sector, and pollutant and different treatment for new versus existing facilities.

In ENERGY 2020 equations simulating the selected method for the calculation of the emission goal are incorporated into the model. In some cases these are simple equations while in other cases the equations may be quite complex.

Allocated Allowances

A simulation of a cap-and-trade system must include a method of calculating allocated allowances. The number of emission allowances is equal to the emission goal. These allowances are either allocated to participants or sold and traded in the market. Generally, some of the allowances are allocated freely to participants (gratis allowances) to reduce the economic impact of the program on the participants. For example, a paper mill may have emissions of 100 kt and be freely allocated 60 kt of allowances. The paper mill must purchase 40 kt of allowances from the market. If the paper mill could reduce its emissions to 50 kt, then it could sell 10 kt of allowances into the market. The goal of the program is to provide an incentive to reduce emissions while minimizing the financial burden on each participant.

Allowances can be allocated in many different ways including historical, forecast, and intensity based. With historical allocation, the participant is allocated a certain percentage of its historical emissions which, for example, could be 80 % of its average emissions between 2010 and 2012. With the forecasted allocations, a business-as-usual forecast of emissions will be generated by the regulatory authority. For example, a cement industry is forecasted to have emissions of 50 kt in 2020, but will only be allocated 25 kt of allowances. It will therefore need to either reduce emissions below the baseline or purchase allowances to make up the other 25 kt of emissions. With emission intensity allocations, the allocated allowances depend on the economic production and a regulated emission intensity. For example, the iron and steel industry may be allocated 75 % of their average emission intensity between 2008 and 2012. So the number of allocated allowances for each year would be the iron and steel production times 75 % of the historical emission intensity.

The allocated allowance formulas may contain any number of factors including the age of the participants (new or old facility), the type of fuel being burned (special allowance for renewable fuels or waste fuels), or the type of operations (industrial generation of electricity). The allocated allowances are often reduced over time, so initially 80 % of allowances may be allocated, but by 2025 only 15 % may be allocated freely, with the remainder being purchased at auction in the market.

In ENERGY 2020 the calculation of the allocated allowances is based on the allocation methodology of the cap-and-trade system design.

Offsets

Most cap-and-trade systems do not include all the sources and sinks for GHG. Certain sectors such as agriculture and forestry are often left out of the market even though these sectors are capable of reducing the net amount of GHG. These sectors are available for offsets. "A carbon offset is a reduction in emissions of carbon dioxide or greenhouse gases made in order to compensate for or to offset an emission made elsewhere" (Wikipedia 2013). When building a cap-and-trade system (or when simulating one), it is important to identify the level of offsets available and the expected price of those offsets. Offsets are intended to provide flexibility (and thus lower costs) in meeting the GHG goals. For example, a certain level of agricultural, forestry, and landfill gas offsets are available at a relatively modest (less than USD $20/tonne) price. The cap-and-trade system is designed to promote the development of these offsets and thus reduce GHG at a lower cost.

The offsets are often limited by type, geographic area, and size in order to encourage internal GHG reductions in as many sectors as possible. The cap-and-trade system will delineate the types of offsets allowed and set up a protocol for determining if an offset is valid. The level of offsets from outside the regulated area (but still inside the country) may be limited. In practice, offsets available from international purchases are almost always restricted since purchasing international allowances is seen as a way for rich countries to shift the burden of GHG reduction.

In ENERGY 2020 offsets are generally simulated with an offset curve. This curve has the GHG allowance price (USD $/tonne) as an input, while the output is the level of GHG reductions (tonne/year).

Offsets, however, can have a more complicated simulation. The landfill gas offset results in the construction of electric generating capacity which burns methane from landfill gas to produce electricity. Any excess methane, not used in electric generation, is flared. In both cases, the methane portion of landfill gas is burned to reduce methane emissions but increase CO_2 emissions.

Allowance Price Limits

Allowance prices are generally set through an allowance auction, but the price is constrained within a range of a minimum and maximum. The minimum price ensures that at least a modest price signal is sent to all sectors for all years, while the maximum price attempts to limit the financial burden of purchasing allowances.

The simulation of the minimum and maximum prices is straightforward; however, the model does need to know to stop trying to increase the GHG allowance price if it has hit the maximum. If the model does hit the maximum price, the system will not reach its emissions goal.

Allowance Reserves

Allowance reserves are a pool of allowances controlled by the regulatory authority that are released into the market to attempt to moderate prices. Typically, when the allowance price reaches a certain threshold, a set number of allowances will be added to the market, reducing growth in allowance prices. The regulatory authority can "fill" the allowance reserve by removing allowances from the market, creating new allowances, or purchasing allowances when the allowance prices are lower than desired.

ENERGY 2020 adds allowances to the market when the price thresholds are reached. These extra allowances will mitigate the upward pressure on prices and result in a lower price to meet goals.

Banking and Borrowing Allowances

In order to provide flexibility (and thus reduce the financial burden), participants may be allowed to bank and borrow allowances. Banking consists of storing allocated or purchased allowances. Participants may bank allowances when prices are low. They also bank allowances during periods when they are easily able to reduce emissions. Then they use the allowances later when emission reductions have become more difficult and prices have risen. Borrowing consists of using allowances before they have been allocated or purchased. The participant reduces the emission allowances submitted in the early years, while agreeing to increase the emission allowances submitted in later years. Borrowing would imply that in later years emission reductions are assumed to be easier to obtain or allowance prices are lower. A cap-and-trade design will establish rules for the levels of restriction and/or prohibition of banking and borrowing.

ENERGY 2020 uses banking and borrowing when the GHG allowance price iteration involves an entire price series (a price for every year of the analysis period). When the model is run with a single price series, some years meet the goal, some years exceed the goal, and some years fall short of the goal. The model assigns banking and borrowing to carry excess or shortfalls across years and thereby determines if the emissions meet the overall, multi-year goals of the system.

Allowance Revenues

Any allowances that the regulatory authority sells in an auction will generate revenue. The regulatory authority must decide what to do with this revenue. Options include rebates to the participants, tax reductions, lowering national debt, direct reduction of GHG, investments in energy efficiency, investments in GHG reducing technologies, or any other purpose deemed beneficial.

ENERGY 2020 computes these revenues and then passes them to the macroeconomic model or the other ENERGY 2020 sectors.

Macroeconomic Feedback

The cap-and-trade system will have an impact on the economic growth, employment, and personal income of the area being regulated. These impacts will come from the requirement to purchase permits, the investments in new energy and emission reduction technologies, the increases in energy prices, and the method of utilization of the allowance revenues.

ENERGY 2020 passes the cost impacts to the macroeconomic model which processes the impact on the economy.

ENERGY 2020 and GHG Cap-and-Trade Systems

Since 1998, ENERGY 2020 has been used to analyze GHG reduction policies and programs, including the establishment of GHG cap-and-trade systems, in the following studies:

1. Environment Canada—Canada; Canada and US; industrial and power sectors in Canada; Alberta industrial and power sectors; and all sectors in Quebec
2. Western Climate Initiative (WCI)—all sectors for eight states and British Columbia; all sectors for eight states plus British Columbia, Ontario, Quebec, and Manitoba; all sectors for California and Quebec
3. California Air Resources Board—all sectors for California only; California and other Western states; California, Western states, and British Columbia, Ontario, Quebec, and Manitoba
4. Illinois Climate Change Advisory Group—US Midwestern states
5. Wisconsin's Governor's Task Force on Global Warming—US Midwestern states
6. BPA—US Western states

Within this group of projects, ENERGY 2020 simulated a myriad of levels of reduction, sectors covered both fully or partially, pollutants covered in different types of allocation for allowances, price constraints, economic impacts, and GHG policies, that support the cap-and-trade system.

Lessons Learned

"Emission Reductions Require Higher than Expected GHG Allowance Prices"

Most sectors have significant barriers to reducing GHG emissions. So, inducing the change through only the price mechanism requires prices much higher than policy makers expect. GHG prices of USD $25–50/tonne (which seems like a reasonable range) only translate into a USD $0.25–0.50/gal increase in gasoline prices which appears to be a relatively minimal impact given that prices often change by USD $0.25 in a week and certainly by USD $0.50 over a year over a year. In 2002, gasoline prices were less than USD $1.00/gal and the conventional wisdom was that everyone would drastically reduce driving if the prices doubled to USD $2.00/gal. Now, at nearly USD $4.00/gal, we are seeing only modest reductions (a 6 % drop from their 2007 high) in gasoline demands (EIA 2012). At this point a USD $0.25 increase in gasoline prices should not be expected to have a significant reduction in gasoline demands.

The reason is that in the US people have few options in traveling to work except to drive their automobiles. Most alternatives require more travel time which offsets the impact of an increase in gasoline prices. For example, assuming a 20 miles per gallon vehicle, USD $3.50/gal gasoline, a 20 mile (30 minutes) commute, and a free bus which requires 75 minutes of commute time, taking the bus saves the person USD $4.67/hour. This is less than the minimum wage, so the person would be better off driving and working more hours. If you add an unthinkable USD $200/tonne (USD $2.00/gal) GHG allowance cost, then the bus saving is still just USD $7.33/hour or roughly equal to minimum wage. Obviously, there are many additional factors on the automobile and bus side of the equation, but the imposition of a significant GHG allowance cost has only a minimal impact on the decision to ride the bus.

"Complementary Energy Efficiency Policies Matter"

Since it is so difficult to drive emission reductions with price only, complementary policies need to be in place to promote GHG reductions. These policies are best directed at sectors of the economy which are not expected to be sensitive to price impacts anticipated from the cap-and-trade system. The most obvious sector is the residential sector. Home-owners are generally focused on other aspects of their lives rather than trying to save a few dollars a week by reducing their GHG production. The people in the residential sector do not have sophisticated tools or analysts to help them determine the most cost effective allocation of their income. Even if they did, they may decide that a night out on the town is more important than staying home to save USD $5.00 worth of gasoline.

Compared to price-based mechanisms, conservation in the residential sector can be more effectively introduced using energy efficiency programs which influence people's beliefs about energy use and energy efficiency. If it becomes "cool" or "sexy" to save energy, then energy efficiency can be increased and GHG emissions reduced.

The commercial and industrial sectors are more apt to respond to prices. However efficiency standards for devices and processes will help those businesses that would like to be energy-efficient successfully compete with businesses that prefer not to invest in energy efficiency.

Policies which encourage consumers to purchase from companies that actively reduce their GHG emissions will provide a stronger incentive to companies than modest cost savings. These policies could include labeling products, not just with how much energy they use, but also the energy (and emissions) required to build the products.

"GHG Offsets Matter"

Many types of offsets provide a relatively inexpensive way to meet GHG emission goals. Assuming an offset protocol can be developed to ensure the GHG emission reductions are incremental and permanent, offsets provide the flexibility of meeting goals without excessive costs.

"Allocated Allowances May Generate Significant Winners and Losers"

Allowances allocated freely to participants in the cap-and-trade system have the potential to increase the wealth of the "winners" in the allocation scheme. In developing an allocation scheme, it is important to analyze all the potential winners and losers in the scheme. Developers of an allocation mechanism can be tempted to provide extra benefits to the sectors that "need" the benefit or sectors whose buy-in is needed by the cap-and-trade system to function. This leads to complex allocation schemes which contain loopholes and discontinuities which is true in historical, forecasted, intensity, and combination allocation schemes.

"Is There Competition or Manipulation?"

Cap-and-trade systems are beneficial under the simple theory of allowing the participants to trade allowances at a price determined by the market so that the lowest cost GHG reductions options in the economy will be found and developed. Ideally, the entrepreneurial competitive spirit will ensure that this is the result at the lowest possible cost to the system.

The competitive spirit, however, is not directly driven to lower costs for society, but to increase profits for itself. The participants will use every legal option to put themselves in position to increase their profits and reduce their exposure to competition from others. Some participants may be in a position to develop and exercise market power.

The regulatory authority can generate thousands of pages (ARB 2012) of regulations designed to prevent the development and exercising of market power and the manipulation of GHG prices. These rules (which of necessity change slowly), however, will be up against a myriad of large companies with well-funded analysts trying to develop ways to neutralize regulation and thus reap excess profits. In fact, the success of the system depends on competitive, aggressive participants.

The impact of GHG emissions are global, so any cap-and-trade market will face pressure to include more and more sectors and more and more countries, thus, making this a potentially very, very large market (billions, trillions?). Even a small, short-term manipulation of the price of GHG allowances could result in significant fortunes being made.

Of course, the regulatory authorities are aware of this and are attempting to plan for it with various minimum and maximum prices, allowance reserves to stabilize prices, and rules to control market power, but the benefits of a cap-and-trade system are based on the efficiency of a freely traded competitive market. As the market is constrained, the benefits are reduced.

Conclusion

ENERGY 2020 has been used extensively since 1998 to simulate cap-and-trade systems. Whereas there are many challenges to cap-and-trade modeling, through experience we have learned the value of building a model that incorporates the details needed to analyze complex designs. Modeling efforts must include simulation of a wide array of cap-and-trade constructs, such as emissions coverage and goals, allocated allowances, offsets, allowance price limits, allowance reserves, banking and borrowing allowances, allowance revenues, and macroeconomic feedback. Using ENERGY 2020 to analyze various cap-and-trade designs in both US and Canada has taught us some key lessons about cap-and-trade that are consistent across studies.

References

Amlin, J., Backus, G., Nabors, O.: A history of making energy policy. System Dynamics Conference, Albuquerque, NM, 2009. http://www.systemdynamics.org/conferences/2009/proceed/papers/P1377.pdf (2009). Accessed 21 May 2013

Air Resources Board: Climate change scoping plan, California Air Resources Board, December, 2008. http://www.arb.ca.gov/cc/scopingplan/document/adopted_scoping_plan.pdf (2008). Accessed 21 May 2013

Air Resources Board: Agreement 11-415 Between Air Resources Board and Western Climate Initiative, Incorporated, California Air Resources Board, January, 2012, Mary D. Nichols, Chairman, Sacramento, California 95812. http://www.arb.ca.gov/cc/capandtrade/wci/agreement.htm (2012). Accessed 21 May 2013

Energy Information Agency: Annual energy review, Table 5.13c Petroleum Consumption Estimates: Transportation Sector, 1949–2011, Energy Information Agency, US Department of Energy, September 27, 2012. http://www.eia.gov/totalenergy/data/annual/showtext.cfm?t=ptb0513c (2012). Accessed 21 May 2013

Granger, C.W.J., Engle, R.F.: Co-integration and error-correction: Representation, estimation and testing. Econometrica **55**(2), 251–276 (1987)

Granger, C.W.J., Newbold, P.: Forecasting Economic Time Series. Academic Press (1977) (second edition: 1986)

McFadden, D.L., Domencich, T.: Urban Travel Demand: A Behavioral Analysis. North-Holland Publishing Co. (1975) (Reprinted 1996)

State of Illinois: Illinois Climate Change Advisory Group. Report of the Illinois Climate Change Advisory Group, July 2008. http://www.epa.state.il.us/air/climatechange/ (2008). Accessed 21 May 2013

State of Wisconsin: Wisconsin's Strategy for Reducing Global Warming, Governor's Task Force on Global Warming Final Report to Governor Jim Doyle, July 2008. http://www.wicci.wisc.edu/download.php?fl=wisconsin_gov_taskforce_final_2008.pdf (2008). Accessed 21 May 2013

Systematic Solutions, Incorporated: ENERGY 2020 Model Documentation. Systematic Solutions, Inc. (1996)

Western Climate Initiative: Design-Recommendations-for-the-WCI-Regional-Cap-and-Trade-Program. Western Climate Initiative, March, 2009. http://www.westernclimateinitiative.org/document-archives/general/design-recommendations/Design-Recommendations-for-the-WCI-Regional-Cap-and-Trade-Program/ (2009). Accessed 21 May 2013

Wikipedia: Carbon Offset, May, 2013. http://en.wikipedia.org/wiki/Carbon_offset (2013). Accessed 21 May 2013

Part III
System Dynamics and Agent-Based Models in Action

Energy Policy Planning for Climate-Resilient Low-Carbon Development

Andrea M. Bassi, Prakash (Sanju) Deenapanray and Pål Davidsen

Introduction

Climate change has emerged as arguably the biggest threat facing human development in the twenty-first century. The current stock of atmospheric greenhouse gas (GHG) is large enough to cause climate change and climate variability. International efforts have been undertaken to stabilize atmospheric GHGs and to limit average global temperature rise to 2 °C (Randall 2010). If current emissions continue unabated, it is expected that the temperature rise will be between 4 °C and 6 °C, that can be reached towards the end of this century. Under this "do nothing" scenario, all nations would be losers. It is, therefore, in humanity's interest to do something about the current state of affairs. Although adapting to climate change and climate variability is important, the safest adaptation would be large-scale reduction in atmospheric GHG emissions. It has been shown recently that limiting global temperature increase to 2 °C above pre-industrial levels could be achieved through the "wedging the gap" approach consisting of 21 coherent major initiatives that together would trigger greenhouse gas emission reductions of around 10 Gt CO_2e by 2020, plus the benefits of enhanced reductions in air-pollutant emissions (Blok et al. 2012). Emissions reductions can be achieved broadly through a combination of: (1) policy measures that provide for financial and economic incentives (e.g., feed-in tariffs for renewable energies) or disincentives (e.g., carbon tax), and (2) market-based

A. M. Bassi (✉)
Founder and CEO, KnowlEdge Srl (KE) and Director, Centre for Systemic Planning (CSP) via Col Di Lana 35, 21053 Castellanza (VA) Italy
e-mail: andrea.bassi@ke-srl.com

P. (S.) Deenapanray
Director, Ecological Living in Action (ELIA) and Director, Centre for Systemic Planning (CSP) 74, Societe la Fleche, La Gaulette, Mauritius
e-mail: sanju@ecolivinginaction.com

P. Davidsen
System Dynamics Group, University of Bergen, Bergen, Norway
e-mail: Pal.Davidsen@geog.uib.no

H. Qudrat-Ullah (ed.), *Energy Policy Modeling in the 21st Century,*
Understanding Complex Systems, DOI 10.1007/978-1-4614-8606-0_8,

mechanisms such as carbon trading, both of which would be required to implement the "wedging the gap" approach. Further, this novel approach would require unprecedented global scale coordination and cooperation.

Global coordination for GHG emission reductions is typically carried out under the aegis of the United Framework Convention on Climate Change (UNFCCC). The use of new market mechanisms (NMMs) for achieving global reductions in GHG emissions was adopted at 16th session of the Conference of Parties (COP16) in Cancun (2010), and further referenced at COP17 in Durban (2011). According to Intergovernmental Panel on Climate Change (IPCC (AR4)), global energy use and supply (26 %), industry (19 %), and transport (13 %) are major drivers of GHG emissions; estimates of CO_2 emissions from agriculture and forestry have a higher level of uncertainty. Since the sectoral scope of NMMs depends, among others, on data availability and low degree of uncertainty in emission estimates the sectors that are recommended for NMMs are: (1) energy supply; (2) industry (e.g., oil refineries, natural gas facilities, iron and steel production, cement production); and (3) transport. Consequently, the case studies presented here place the focus of this chapter squarely on these key sectors.

System dynamics modeling (SDM) provides a useful approach to better understand the multi-dimensional socio-economic and environmental impacts of current climate variability and projected climate change that would be necessary to inform adaptation policies and strategies. Similarly, the system's approach allows the impacts of policies and emission reduction through market-based mechanisms to be investigated. Importantly, it allows energy policy to be tied with emission reduction across all economic sectors, and it also offers a way to investigate novel approaches to NMMs. In this chapter, we will demonstrate the practical use of SDM for policy planning to achieve climate-resilient, low-carbon development pathways, in the context of national development planning. In particular, we will use examples from developing (Mauritius and Kenya) and developed (USA) countries to make the case for the use of SDM for climate proofing of the energy sector and to develop nationally appropriate mitigation actions (NAMAs) as one type of NMM.

The case studies form a part of the work carried out for the United Nations and national governmental agencies, including the creation of cross-sectoral climate mitigation and/or adaptation simulation models for national policy formulation and evaluation. The main objectives of the modeling work presented in this chapter are to: (1) create an innovative simulation model, and to (2) improve policy formulation and evaluation analysis for the elaboration of coherent and comprehensive climate change mitigation and/or adaptation strategies. The models are built up on social, economic, and environmental sectors and integrate the best sectoral knowledge in one single model framework representing a full incorporation of economic and bio-physical variables. In particular, they capture: (a) feedbacks within and across sectors, aiming at identifying both synergies and potential bottlenecks (unexpected side effects); (b) time delays, whereby policies and investment allocations may lead to a "worse before better" situation; and (c) nonlinearity, leading to the identification of potential thresholds and tipping points.

Methodological Approach

Energy is a meta-technology that pervades all aspects of modern societies. Further, the global energy system is so pervasive that it now has a direct impact on the climate system, which in turn impacts all socio-economic sectors including the energy sector. Therefore, energy systems are better understood in terms of complex systems, which are characterized by nonlinear relationships that cause feedback loops to vary in strength, depending on the state of the system (Meadows 1980). In systems built on a variety of feedback loops, nonlinearity creates shifts in dominance of such loops, which become very important in determining how structure defines behavior, even at different times and with different states of the system.

Nonlinearity allows for a clearer interpretation and understanding of the context of analysis. A wide range of scenarios with different assumptions on nonlinear relations existing within the system can be simulated to test and evaluate the impact of various policy choices, and system responses to their implementation. Nonlinear relations highlight the creation of raptures as well as stronger or weaker approaches in response to unprecedented issues. Though this approach may not be perfectly accurate, it provides insights on the potential medium to longer-term impact of policies that cannot be discerned from linear tools. Both dynamic and detailed complexity should be represented to reach improved understanding of the context in which issues manifest themselves and have to be faced. Combining feedback loops, nonlinearity, and delays contribute to the creation of a consistent and coherent framework for the analysis of the properties and structure of complex systems.

System Thinking and System Dynamics

In order to design and evaluate national development policies the structure of the system analyzed (e.g., social, economic, and environmental) should be properly understood. Economic volatility, as well as natural disasters and other unexpected events can have a considerable impact on the effectiveness of policies over time. For these reasons scenarios have to be defined, to reduce the uncertainty coupled with the analysis carried out. Policies would then be evaluated based on the structure of the system analyzed as well as on a variety of possible scenarios. Policies are "shocks" to the system, which in turn responds to these changes. Hence, the system itself should be analyzed focusing on feedbacks and causal relations, with a specific interest on medium to longer-term impacts (which go beyond the implementation delays of policies—i.e., inertia of the system).

The understanding of the functioning mechanism of the system allows for the identification of medium to longer-term sectoral and cross-sectoral implications of policy implementation. These impacts have to be analyzed with the understanding that different sectors are influenced by different key causes defining the success (or failure) of policies. In other words, a policy can have very positive impacts for certain sectors and create issues for others. Furthermore, successful policies in the longer term may have negative short-term impacts, for which mitigating actions may be designed and implemented.

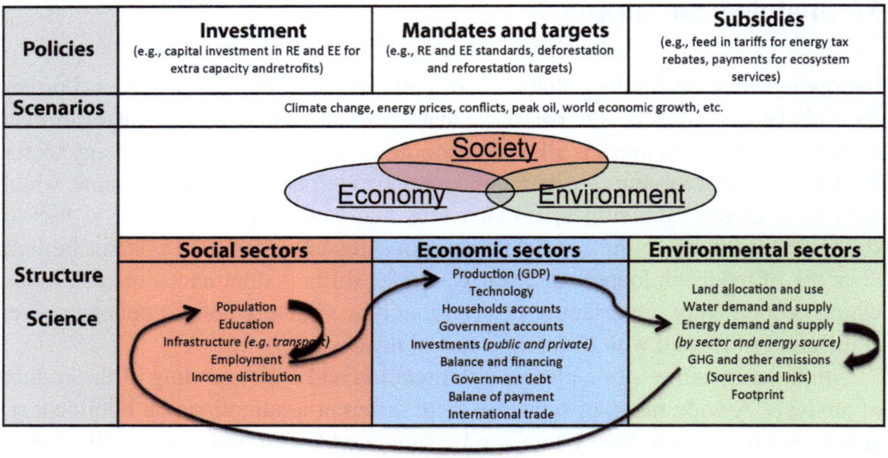

Fig. 1 The three main layers for carrying out integrated policy formulation and evaluation: structure, scenarios, and policies

Simulation models exist which aim at understanding what the main drivers for the behavior of the system are. In the case of system dynamics, this implies identifying properties of real systems, such as feedback loops, nonlinearity, and delays, via the selection and representation of causal relations existing within the system analyzed. This is advantageous for integrated policymaking because, while optimization models are prescriptive and econometric models are heavily relying on the history of the system analyzed, simulation models are descriptive and focus on the identification of causal relations influencing the creation and evolution of the issues being investigated.

Three Layers for Effective Policy Analysis

In order to make progress towards a low-carbon, climate-resilient development in the twenty-first century, an integrated approach that incorporates environmental, social, and economic (ESE) implications of policy implementation is needed. Underlying this approach is the recognition that the algebra among the social, economic, and environmental pillars of sustainable development (how the variables relate and affect one another in context, how they combine towards the equation of sustainable development) is more important than the arithmetic among them (added or subtracted as convenient) (United Nations Environment Management Group 2011).

Figure 1 indicates that policy formulation and evaluation need to be carried out in the context of scenarios (e.g., technological development, natural disasters), and policies (e.g., subsidies, incentives, and/or mandates) have to be evaluated across a variety of indicators (social, economic, and environmental) simultaneously. How these three levels are supported with solid and coherent information and interact with each other will greatly determine the success of any national development plan over the medium to longer term. This information is also crucial to truly understand the drivers of change and to effectively design policies that have the desired impacts.

More specifically, firstly, in order to design and evaluate national development policies the structure of the system analyzed (e.g., social, economic, and environmental) should be properly analyzed and understood. Using the example of the energy sector, this includes the investigation of the main drivers of demand, and how supply can respond to its needs. This is a broad investigation heavily relying on soft and hard data analysis, as we are in rapidly changing times and various cross-sectoral interdependencies are emerging.

Secondly, economic volatility, as well as climate impacts, natural disasters, and other unexpected events can have a considerable impact on the effectiveness of energy and environmental policies over time. For these reasons scenarios have to be defined to reduce the uncertainty coupled with the analysis carried out. Policies would then be evaluated based on the structure of the system analyzed as well as on a variety of possible scenarios.

Thirdly, the implementation of policies for climate change mitigation and adaptation should be tested in the context of longer-term national development, while possibly also taking into account broader issues such as globalization. The issue of globalization may have severe ramifications. For instance, the delocalization of energy intensive industries from developed to developing countries that do not have binding emission reduction commitments under the UNFCCC would have social impacts through job loss (or creation). In order to investigate whether they create synergies, bottlenecks, or unexpected side effects across sectors, the impacts of policies have to be evaluated for a variety of social, economic, and environmental indicators.

Model Development and Validation

The development of a SDM (conceptualization, customization, and validation) proceeds through a variety of tasks, and the generic steps are discussed below:

STEP 1: *Identification of key issues and opportunities*

Definition of key issues in the energy climate change nexus: As every model application is unique, the issues to be analyzed have to be carefully designed and agreed upon. A multistakeholder process (MSP) is adopted to obtain the widest possible stakeholder views. The steps of the MSP adopted are shown in the Appendix. The MSP is used to engage stakeholders during all the steps of model conceptualization, customization, and validation.

Definition of key opportunities and policy options in the energy and trade sectors: the options and opportunities, together with the issues, serve to define the boundaries of the model and always keep in mind the end goal of the project.

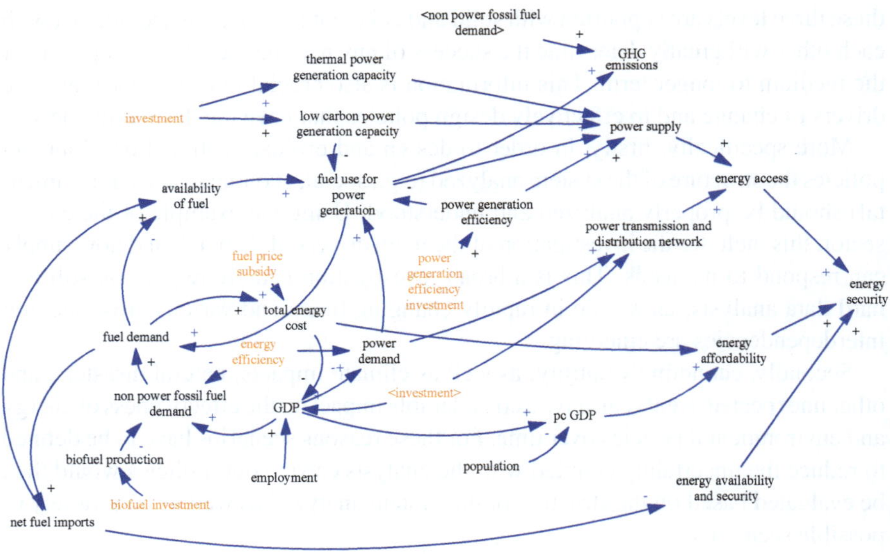

Fig. 2 Causal relations between energy and the environment-economy-society system

STEP 2: *Data collection and consistency check*

This is a time consuming task, and, on top of data mining, cross-sectoral data consistency checks are an essential step.

STEP 3: *Causal mapping and identification of feedback loops*

Causal mapping of system drivers: This step constitutes of creating causal loop diagrams (CLDs) of the issues or sectors that were identified in Step 1. The CLDs provide the high-level system view of issues that would be addressed by the modeling, and its development also draws from the availability of data identified in Step 2. Creating a map of the system analyzed has several purposes. First, it brings ideas, knowledge, and opinions of the core team of modelers together. Secondly, it highlights the boundaries of the model and analysis. Thirdly, it allows all participants to reach a basic to advanced knowledge of the energy sector and how it relates to society, the economy, and the environment. Finally, it serves as a starting point in the development of the mathematical (stock and flow) model.

Identification of key feedback loops in the causal map: Identifying the key drivers and feedback loops in the system allows considering the reinforcing and balancing nature of our complex environment. Also, feedback loops highlight potential side effects, synergies across variables and sectors, to make the best of the available investment and maximize returns. An example of a CLD that was developed to build the energy sector model of Mauritius (see Sect. 3.3) is shown in Fig. 2.

The parameters shown in orange in Fig. 2 are policy interventions. As shown by this CLD, energy security is a combination of energy availability, access, and affordability. Energy affordability is a function of energy costs which in turn depends on energy supply and demand. Energy supply is from a mix of renewable and nonrenewable sources. In the case of power, supply is constrained by generation capacity that depends on investment in infrastructure (e.g., power plants). In the CLD, energy availability and security is dictated by net imports that gives an indication of energy dependence. Once power is generated, its access depends on the network of electricity transmission and distribution. It is worthy to note that energy is a parameter in "total factor productivity" (not shown here) that is used in combination with a standard Cobb-Douglas production with constant elasticities to calculate the gross domestic product (GDP).

STEP 4: *Creation of customized mathematical models*

This step consists of a sequence of iteration involving key stakeholders, and it consists in the translation of CLDs into mathematical models, with numerical inputs and equations. This step adds the quantitative layer to the analysis. At this stage the model is built up on social, economic, and environmental sectors, practically integrating the best sectoral knowledge in one single model framework representing a full incorporation of economic and bio-physical variables, capturing (a) feedbacks within and across sectors, aiming at identifying both synergies and potential bottlenecks (unexpected side effects); (b) time delays, whereby policies and investment allocations may lead to a "worse before better" situation; (c) nonlinearity, leading to the identification of potential thresholds and tipping points. Further, the model is created making use of existing expertise (through the Causal Diagram and participatory modeling), acting as a knowledge integrator of successful—technically valid and already effectively utilized—models for policy analysis, and being fully customized to the national context (with an extensive cross-sectoral dataset and structure, for a more holistic approach to planning in the energy sector).

STEP 5: *Validation and analysis*

- Validation of model (structure): Variables and equations have to be validated to ensure that all experts feel comfortable with the overall structure of the model, reflecting reality. This is done primarily by simulating the base case. This is done by testing the outputs of simulations against historical data (Step 3), and this is done for a multitude of socio-economic and environmental indicators. The confidence that the causal relationships in the model are well established emerges from the ability of the model to replicate historical data. Where necessary the model can be calibrated to obtain a consistent and reliable baseline simulation—i.e., the business-as-usual (BAU) case.
- Simulation of alternative scenarios: Once the BAU is confirmed, scenarios can be simulated to test the impacts of alternative policy options that were identified in Step 1.
- Validation of model (behavior): Simulations (BAU and policy interventions) have to be validated to ensure that all stakeholders feel comfortable with the overall behavior of the model. Here again, the multistakeholder process described in the Appendix is used.

Practical Applications: Three Case Studies

Various energy contexts are unique in different geographical areas. A wide range of properties ranging from political environment to richness of natural resources characterizes these contexts. When reducing them to a simulation model, boundaries are set. These apply to the geographical area analyzed, including the socio-economical dimensions of the society scrutinized. In order to represent such diverse properties of the system, customization is needed. In addition, given the numerous interrelations existing among society, economy, and environment, complexity has to be simplified to account for the key mechanisms influencing the course of events (historical, present, and future).

Different geographical areas can have similar characteristics and show similar behavior while being structurally different. The approach proposed by the authors aims at decoupling the properties of the real systems analyzed, in order to better understand how the underlying structure of the system generates its behavior. Reality is complex, for two reasons: there is a very high level of detail in every real system (i.e., every major process is built up on smaller ones, that contribute to the formation of the aggregated behavior of the system), and there are dynamic relationships existing among both the elements forming the system analyzed and the ones surrounding it. While conventional modeling tools can extensively represent the details of each linear process involved in a real system (e.g., energy transformation from crude oil to refined fuels), a closer investigation of the dynamic relationships contributing to the growth and progress of the system itself is needed.

The studies presented in the next sections provide a diversity of countries covering widely differing geographical scales and levels of socio-economic development. Together, they form a good combination of examples to demonstrate the versatility of SDM for energy policy planning, including the integration of climate change.

The validation of such models takes place in different stages, and the most peculiar tests, when compared to optimization and econometrics, are the direct comparison of projections with historical data, which simulation models can backtrack, and the analysis of structural soundness with respect to reality (Central Intelligence Agency 2011; Barlas 1996). Potential limitations of simulation models include the correct definition of boundaries and a realistic identification of the causal relations characterizing the functioning of systems being analyzed.

United States of America

Under the current state of negotiations under the UNFCCC, developing (nonAnnex 1) countries do not have any binding responsibility for curbing emission reductions. As a consequence, a variety of policy interventions are being evaluated in several countries, including the USA. If this situation were to prevail in the future and the onus would remain on developed countries to carry out significant emissions reductions, the door opens up for the possibility of an exodus of energy intensive industries

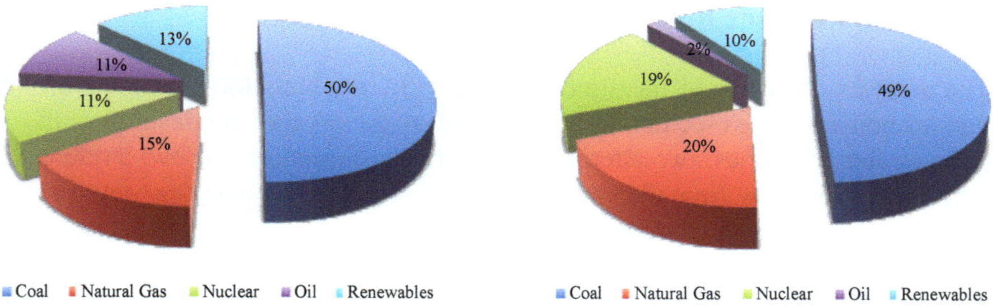

Fig. 3 Power generation shares for *coal, oil, natural gas, nuclear,* and *renewables,* 1980 and 2009. (Source: EIA)

from developed to developing countries, in search for a low-cost solution to their competitiveness problems. The impacts of possible several intervention options to reduce the GHG in developed countries are analyzed, also investigating the potential delocalization of energy intensive industries to developing countries.

Context

The USA is the largest energy consumer as well as the largest economy in the world (Central Intelligence Agency 2011; US Department of Commerce 2008). America's economic growth, on the other hand, is being challenged by domestic and international events. This is a unique opportunity, where the world's largest economy can serve as example for other countries that intend to move toward a low-carbon path.

The USA experienced the fastest economic development in North America over the last few decades. GDP grew by 63 % between 1990 and 2007 (US Department of Commerce 2008), whereas population has increased by 18 %, reaching 300 million in 2005 (United Nations Population Division 2007). Total energy demand increased by 20 % in the same period, while supply has remained just about flat, leading imports to increase by 56 % (US Department of Energy 2010b). As a consequence of increasing energy consumption, emissions are now 15 % above the 1990 level (see also Fig. 3 for the composition of power supply in 1980 and 2009).

The model used for the US analysis focuses on the impact of various policies and investments allocated to different sectors, such as energy, transport, industry, buildings, and waste management, all aiming at reducing GHG emissions. These include fuel efficiency standards for passenger vehicles (corporate average fuel economy, CAFE), the electrification of the intercity rail system and the expansion of urban rail, the enactment of renewable energy standards (RES) and exploitation of unconventional fossil fuel reserves with carbon capture and sequestration, the introduction of a cap-and-trade mechanism, energy conservation, and waste-to-energy (and reuse)mechanism.

Analysis of Modeling Results

The analysis of increased investments in the USA accounts for the simulation of various policies across sectors. In order to analyze the broader implications of these policies and provisions, two alternative scenarios are examined below: a reference case, with a recovery from the current economic crisis in the next few years, and a peak oil scenario, with world conventional oil production reaching a plateau phase shortly after 2020—or about 10 years earlier than in the reference case.

The key investments and supporting policies simulated in the green economy scenario are as follows:

- Nuclear energy: We assume that 380 GW of capacity will be added between 2015 and 2050. Capital and maintenance costs use International Energy Agency (IEA) assumptions, and are set to US$ 5,500 and US$ 165 per MW respectively (International Energy Agency 2010).
- Renewable energy: We simulate investments in renewable energy to reach a 20 % market penetration in the power sector by 2020. Investments are mostly directed to wind (80 %) and solar (20 %), which are assumed to cost, according to the IEA, between US$ 1,800 and US$ 1,750 per MW and US$ 3,900 and US$ 2,750 per MW respectively in 2015 and 2030 (International Energy Agency 2010).
- Electrification of rail (urban and freight): With this provision we assume the electrification of over 100,000 miles of existing inter-city railroads,[1] with the average cost per mile estimated at US$ 3.5 million and a 70 % shift of long-range freight to rail by 2050. The urban rail investment includes building new urban rail lines, reaching higher density on existing urban rail lines and electrifying current diesel commuter lines for a total of 500 miles per year over the next 40 years. The average cost per mile is estimated at US$ 54.5 million.[2]
- Industrial energy efficiency: Investments are assumed to be allocated to support energy efficiency interventions in energy intensive industries, especially in light of the possible enactment of a cap-and-trade mechanism with rebates declining before 2025. As a result, the energy efficiency of the industrial sector is projected to be 5 % higher than BAU by 2020. Investment estimations are based on IEA's world economic outlook (WEO) 2010 (International Energy Agency 2010), resulting in emissions mitigation costs that average about US$ 68/t of CO_2 over the simulation period, increasing from US$ 64/t on average between 2010 and 2030 to US$ 72/t on average between 2030 and 2050.

[1] These include 32,421 railroad miles that the Department of Defense has classified as being "strategic" (Military Traffic Management Command 1998), 14,000 miles of grade separated three or four track services (comparable to CSX plans from Washington DC to Miami), with one or two tracks devoted to 100–110 mph passenger and express freight service, and electrification of upgrade of additional 60,000 miles.

[2] It is assumed that, out of the 500 miles built or upgraded each year, 350 will be light rail (costing on average US$ 35 million per mile) and 150 will be metro—or subway—systems (costing on average US$ 100 million per mile). See John Schumann, "Status of North American Light Rail Transit Systems", "8th Joint Conference on Light Rail Transit, Dallas, Nov. 2000; Portland Tribune, 18 June 2002."

- Carbon capture and sequestration: It is assumed that all new plants (coal and gas) starting operations after 2020 will be equipped with carbon capture and sequestration (CCS) capabilities. We account for an efficiency loss of 12 %, assumed to be replaced with new capital and used cost assumptions from the IEA, averaging about US$ 600 per MW (International Energy Agency 2010).

The total investment simulated averages of 1 % of GDP throughout the simulation period. Administrative expenses are assumed to account on average for 30 % of this investment and the remainder is productive, or capital, expenditure. Capital investments amount to US$ 120 billion (Bn), US$ 125 Bn and US$ 238 Bn in 2015, 2030, and 2050 respectively. The capital investment equals on average 3.5 % of total public and private investment throughout the simulation and is divided into the different sectors as follows: renewable energy, 46 %; electrification of rail, 24 %; carbon capture and sequestration, 20 %; nuclear energy, 7 %; energy efficiency, 3 %.

With a baseline scenario (BAU) closely replicating economic and energy projections from the US Department of Energy's Energy Information Administration (EIA), the allocation of the green investments listed above is projected to produce higher GDP growth rates relative to BAU after the recovery, averaging 2.91 % instead of 2.68 % between 2010 and 2050. While each investment will individually impact GDP and emissions, among others, their timing and strength differs. By implementing all of the policies, growth rates will remain around 2.7 % pa in the 2020s and 2030s and rise to over 4 % pa in the 2040s. This development is due to the growing cumulative impacts of investments and the projected increase in fossil fuel prices, especially after the peak of conventional oil. Worth noting, GDP growth could be enhanced or curbed by the financing strategy implemented. Predominant public financing could increase the national debt, while excessive reliance on private financing may not be effective, due to already low savings and very high consumption.

As in the case of GDP, total employment would also be higher than the base case over the whole period, with 1 million and 10 million new jobs created by 2030 and 2050 respectively. Of these, 300,000 and 1 million net additions by 2030 and 2050 are attributable to the energy sector, with an average of 14,600 and 240,000 jobs gained in nuclear and renewable power generation, 50,000 lost in thermal generation and 131,000 gained through CCS.

GHG emissions on the other hand, despite the projected increase in GDP, will be significantly reduced by the policies simulated. The combination of policies would generate a reduction of about 0.5, 0.65, and 1.8 billion t of GHG by 2020, 2030, and 2050 relative to the BAU case. This reduction would be worth US$ 14 Bn, US$ 38 Bn and US$ 287 Bn in the same years using the American Clean Energy and Security Act of 2009 (ACES) and corresponding EIA study (US Department of Energy 2010) as references for emission allowance prices. The total avoided energy cost would reach about US$ 220 Bn, US$ 376 Bn, and US$ 821 Bn in 2020, 2030 and 2050, respectively. These achievements are reached after investing US$ 110 Bn in the energy sector (including energy efficiency, renewable energy, nuclear, and CCS) each year on average between 2010 and 2050.

It is worth noting that the timing of the impacts of the different policies simulated differs. Implementing the higher CAFE standard, a known case and embedded in the BAU case, and CCS have a relatively slow impact. Both policies apply only to newly manufactured cars and thermal power plants in our scenarios, with the average fuel economy of the whole auto fleet and the carbon intensity of the power sector improving gradually as the old is retired and new one built. In the same way as in the CAFE case, where it takes over 15 years for the whole fleet to reach the new standards—and the impact on GHG emissions and oil demand rises slowly over time—, the capital life time of power plants (30 years) has the consequence that the impacts of CCS on US fossil fuel emissions from the power sectors grow slowly but steadily.[3] Increased renewable energy production under a RES has a more immediate impact, being regulated by sharply increasing mandates on renewable energy generation, on jobs and income for the production and installation of new and additional wind turbines, solar panels, and biofuels plants among others. Also, the effect on GHG emissions is more immediate, as renewable energy under a RES would initially both fill the gap between growing demand and supply and gain market share replacing older capital.[4] The electrification of rail would also quickly reduce demand for oil as commuting and freight is shifted to electrified rail, and it would involve infrastructure construction spread over time, which helps job creation and GDP growth. Reduced transport costs will also increase funds available for other consumption. However, this rail transport will increase the demand for electricity, which at the margin would be coal-based, so the impacts on GHG emissions will be more modest.

It is also worth noting that with this combination of policies, but excluding CCS, the total amount of CO_2 emissions from fossil fuels will still increase by 0.5 % per year on average between 2010 and 2050, with a short-term decline and stronger longer-term increase relative to BAU. With the introduction of CCS for all new plants coming on stream after 2020 instead, emissions are projected to decline to 5.4 Bn t/year in 2050, reaching 1995 levels and scoring a 35 % reduction when compared to BAU (see Fig. 4a and Table 1). Results of the simulation show that the share of plants with CCS will steadily grow to 54 % of gas turbines and 50 % of coal-fired plants by 2050. With no CCS investment, per capita CO_2 and GHG emissions instead will stabilize after 2020, with GHG emissions reaching 17.5 t/person/year, to slowly pick up after 2040 due to higher GDP and energy consumption and climb up to 18.2 t/person/year. With the CCS investment though, the decline of emissions will be prolonged throughout the simulation horizon, reaching 13.95 t/person/year in 2050. Considering emission allowance price projection published by the EIA and used by the US Congressional Budget Office (CBO) (US Congressional Budget Office 2009; Yudken and Bassi 2009) and a 5 % yearly increase after 2030 (see Fig. 4b), CCS

[3] Worth noting, on top of requiring more water (NETL 2009), thermal power plants with CCS capabilities suffer a 12 % efficiency loss (IEA 2008), which increases capital investment, fuel input, and emissions, but also employment.

[4] It is assumed that the grid would be able to support a 20 % RES by 2020 and that thermal or nuclear plants would still be preferred for base load supply.

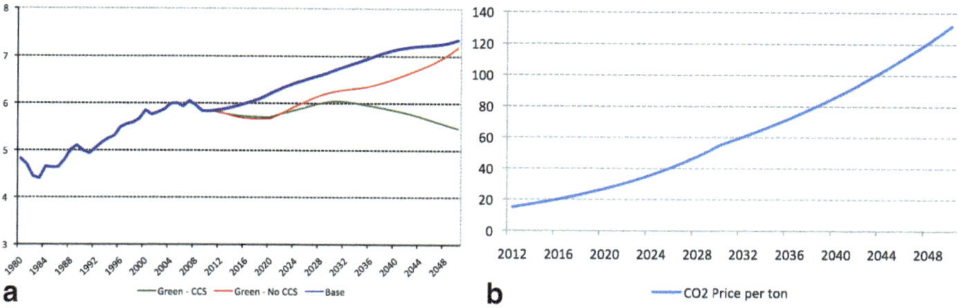

Fig. 4 Fossil fuel CO_2 emissions in Gt for the BAU, green case with and without CCS (**a**); emission price, US\$ 00/t of CO_2 (**b**)

Table 1 Key indicators for comparative analysis across countries: USA

	Unit	2010	2020		2030		2040		2050	
		BAU	BAU	Low GHG	BAU	Low GHG	BAU	Low GHG	BAU	Low GHG
Real GDP per capita	US\$/person	37,198	44,089	44,323	53,731	54564	64,144	66,663	81,571	89,461
Emission per capita	ton/person/year	18.62	18.05	16.59	18.06	16.30	18.23	14.92	17.85	13.27
Emission per GDP	kg/year/US\$	0.50	0.41	0.37	0.34	0.30	0.28	0.22	0.22	0.15
Energy intensity	Btu/US\$	8,311	6,770	6,685	5,562	5616	4,839	4,818	4,200	4,188

costs are going to be fully offset by avoided emission charges starting from 2035, while positive employment impacts remain and grow over time.

When analyzing the impact of a cap-and-trade mechanism— of a similar impact of other interventions on energy costs—on energy intensive industries (primary and secondary aluminum, steel, paper and paperboard, petrochemicals, and alkalies and chlorine), a highly debated topic in the US Congress, results of the simulation indicate that the manufacturing sector will likely have to face higher—policy-driven—costs, and that these could be fully offset, at least until 2020/2025, by the free emission allowanced once considered by the Government (see ACES bill, US Congress 2009). In this scenario, the emissions cost will start impacting the profitability of these industries as soon as the free allowances will decline, to be only partially offset by a proposed border adjustment fee (Yudken and Bassi 2009). The efficiency gains required to offset losses related to the implementation of moderate and high-CO_2 prices are in the range of 5 % by 2020 and 10 % by 2030 with the planned allowance allocation, assuming that no major energy efficiency investments will be made by the industries by then (Yudken and Bassi 2009). It is worth considering that, in the absence of free allowances, the energy efficiency improvement required to offset increasing energy prices could reach up to 30 % for iron and steel production, and over 20 % for paper and paperboard. These are sectors with historically low profit margins, implying that the former may relocate to less stringent countries (a growing trend, not only for the USA, in recent years), and the latter (being traditionally related to a localized production system) may slowly vanish and disappear from the American soil (Bassi et al. 2009; Bassi and Yudken 2011).

Summary

The case study of the USA indicates that the allocation of investments aiming at reducing GHG emissions may well result in higher GDP growth and employment relative to BAU, generating double and triple (social, economic, and environmental) dividends. This indicates that the investment required to implement the interventions simulated will have a positive socio-economic return on investment, especially when taking into account avoided energy costs and the potential to generate employment and reduce the economic vulnerability to the volatility of fossil fuel prices. GHG emissions, the primary objective of the interventions analyzed, are projected to decline and reach the 1995 level by 2050 thanks to the implementation of both demand and supply-side interventions.

It is worth noting that the timing of the impact of the green investments simulated differs. Indeed, implementing CCS has a relatively slow impact on the creation of jobs and reduction of emissions, as opposed to the enactment of a federal Renewable Portfolio Standard (RPS). Further, the implementation of a cap-and-trade mechanism, such as the ACES bill, would allow CCS costs to be fully repaid by 2035 and, with a provision to allocate free allowances to energy intensive sectors in early years, would greatly limit potential negative consequences on the profitability of manufacturing sectors.

Concluding, the use of a system dynamics model customized to the USA has allowed the simulation and analysis of the impact of simultaneous policies and investments at the source (such as the expansion of renewable energy and the introduction of CCS), as well as interventions to improve energy efficiency on the demand side. The results of the analysis, carried out across social, economic, and environmental indicators, indicate the potential generation of social and economic medium and long-term benefits, while considerably improving energy and national security.

Kenya

Kenya is a developing country member of the East African Community (EAC). Development in the EAC is constrained by power supply, and historical extreme events have shown that the large hydro-electric component of the Kenyan power sector is highly vulnerable to variability in precipitation. Energy security is therefore a concern for the economic growth and overall development of the country. Using the integrated SDM that has been developed for Kenya, scenarios to build climate resilience into the power sector and the development of Kenya according to Vision 2030 are investigated.

Context

Kenya is already prone to cyclical droughts and flooding because of its geographic location and it is likely to see an increase in the intensity and frequency of these events as global climate continues to change. In Kenya, where about 75 % of the population depends directly on land and natural resources for their livelihoods, the impact of climate change and related disasters on land and natural resources have the potential to severely affect the lives and livelihoods of most Kenyans. This expectation was expressed in the First National Communication of Kenya to the Conference of the Parties to the United Nations Framework Convention on Climate Change, and the State of the Environment Report 2006/2007, which stated that adverse environmental, economic, and social repercussions are anticipated as the impacts of climate change become increasingly manifested. Some of the adverse impacts include water and food shortages, famine, energy shortages, desertification, forced mass migration, diseases, and overall economic, environmental, and human degradation. Existing climate variability has significant economic costs in Kenya. With increasing climate variability in the future, aggregate models indicate additional (on top of existing climate variability) net economic costs which could be equivalent to a loss of almost 3 % of GDP each year by 2030 in Kenya (Stockholm Environment Institute 2009).

This case study forms part of a larger adaptation project to provide the Government of Kenya (GOV) with a dynamic, quantitative, and transparent planning tool for climate adaptation defined here as strategies, policies, programs, projects, or operations aimed at enhancing resilience or reducing vulnerability to observed or plausible changes in climate. It includes activities implemented to create changes in decision environments as well as actual adjustments to address climate risks (Adger et al. 2007). Since there is convergence between climate change and human development, the dynamic planning tool also serves to carry out integrated or multisectoral development planning over the multidecade time horizon, while offering the capacity to carry out scenario analyses of strategies and actions under uncertainty. A system dynamics model has been developed for Kenya that integrates the analysis of the risks and impacts of climate change across the major sectors in the economy, society, and environment, in order to inform coherent national development policies that encourage sustainable development, poverty eradication, and increased wellbeing of vulnerable groups, especially women and children, within the context of *Vision 2030* (Government of Kenya 2007). Four priority sectors—i.e., energy, agriculture, water, and health—have been identified for detailed analysis.

The Kenya model is composed of 50 modules (see Table 2). These modules are regrouped under 18 sectors (6 social sectors, 6 economic sectors, and 6 environmental sectors) based on their functional scope. The strength of system dynamics model is its

Table 2 Modules, sectors and spheres of the Kenya system dynamics model

Society	Economy	Environment
Population sector	**Production sector**	**Land sector**
1. Population	15. Production and income	34. Land
2. Fertility	16. Agriculture	
3. Mortality	17. Husbandry-fishery-forestry	**Water sector**
	18. Livestock	35. Water demand
Education sector	*19. Fisheries*	36. Water supply
4. Primary education	*20. Forestry*	
5. Secondary education	21. Industry	**Energy sector**
	22. Services	37. Energy demand
Health sector	*23. Tourism*	38. Energy supply
6. Access to basic health care		
7. HIV/AIDS	**Households sector**	**Emissions sector**
8. HIV children and orphans	24. Households accounts	39. CO_2 and GHG emission
9. Nutrition		
	Government sector	**Sustainability sector**
Infrastructure sector	25. Government revenue	40. Ecological footprint
10. Roads	26. Government expenditure	
11. Irrigation	27. Public inv. and consumption	**Extra modules**
	28. Gov. balance and financing	*41. MDGs*
Labor sector	29. Government debt	*42. HDI and GDI*
12. Employment		*43. Indicators*
13. Labor avail. and cost	**ROW sector**	*44. Climate impacts*
	30. International trade	*45. Climate interventions*
Poverty sector	31. Balance of payments	*46. Climate investments*
14. Income distribution		*47. Malaria transmission*
	Investment sector	*48. IVM interventions*
	32. Relative prices	*49. Malaria treatment*
	33. Investment	*50. Malaria cost accounting*

flexibility to accommodate additional modules or sectors depending on new issues to be analyzed, and also in its structural nature, being able to integrate economic sectors with biophysical variables for the environment and society. Using the MSP discussed in "Model Development and Validation", over 15 specific climate impacts were modeled across sectors. While climate adaptation can reduce the economic costs of climate change, it has a cost as well. The costs of adaptation are still emerging. Over 18 categories, accounting for more than 25 specific interventions, have been identified and included in the Kenya country model that relate to the balance between development and climate change. To highlight the potential results of the analysis and the value addition of the projects, several scenarios were simulated. The main results of the analysis are discussed below for the energy sector, while results of the full analysis can be found elsewhere (Bassi et al. 2011).

Analysis of Modeling Results

The following climate impacts were studied for Kenya: (1) reduced hydropower generation during droughts and floods; (2) damage to power infrastructure (e.g.,

Table 3 Summary of adaptation measures, including timeframe for implementation and annual investments

Policy interventions	Timeframe (years)	Investment per year (billion KSh)
Accelerated development of geothermal power by the government and its development partners	10	20.3
Accelerated development of geothermal power by the private sector (GDC will take up if there are no suitable investors)	10	12.1
Accelerated development of green energy (solar, wind, renewable biomass, etc.) by the govt. and its development partners	5	15
Accelerated development of green energy (solar, wind, renewable biomass, etc.) by the private sector	5	22.5
Provision of efficient (fluorescent) bulbs to domestic consumers	10	0.36
Water catchments protection programmes e.g., afforestation	10	0.375
Provision of improved jikos	10	0.075
Promotion of low-end solar devices including solar drip irrigation, solar water heating, etc.	10	3

power cables during floods); and (3) increased demand for electricity for services like refrigeration, air conditioning, and irrigation (due to increase in evapo-transpiration). The policy interventions that were simulated to climate-proof the energy sector against these impacts are summarized in Table 3.

The BAU assumes that no climate change adaptation measures would be implemented, including green energy measures. However, the reality is that climate variability and climate change will have an impact on power generation. In the BAU, electricity generation is assumed to be solely from hydro, geothermal, and thermal power generation. In the climate change adaptation scenario, the energy sector includes both promotion of renewable energy on the supply side and use of efficient bulbs on the demand side with a total investment of 73.71 billion KSh per year (see Table 3).

With the use of efficient bulbs, the actual residential lighting will be 2,441.88 GWh by 2030 compared to 3,860.96 GWh if efficient bulbs were not used, translating to 36.75 % of energy saved. By 2050, this will have reduced to 2,186.37 GWh compared to 5,975.07 GWh if the old bulbs are used (see Fig. 5). This translates to a saving of 63.4 % of energy with the use of efficient bulbs. Reduction in electricity demand puts less pressure on the need to increase electricity generation, and constitutes an adaptation strategy in the face of climate vulnerability.

On the supply side, construction of geothermal generations by the government and the private sector would lead to geothermal power capacity to increase from 0.1 GW to 1.33 GW, compared to 0.66 by the year 2031 without the interventions. Further, the government and private sector interventions on green energy development would boost the green energy power generation capacity from 0.01 GW to 0.7 GW and thus have a net increase of 0.7 GW by the year 2017, since the investment is proposed to take five years (2012–2017).

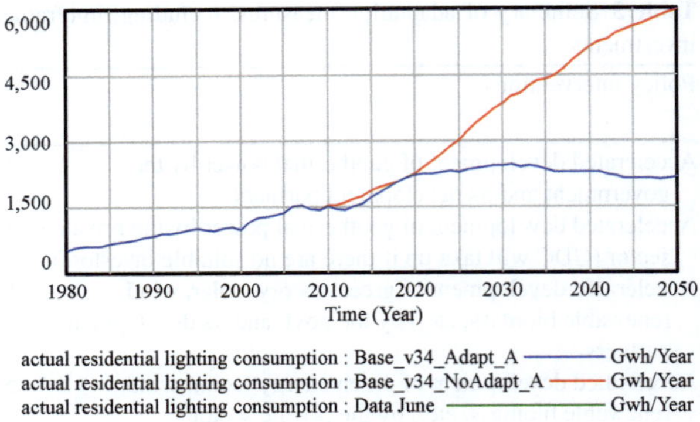

Fig. 5 Residential lighting energy consumption

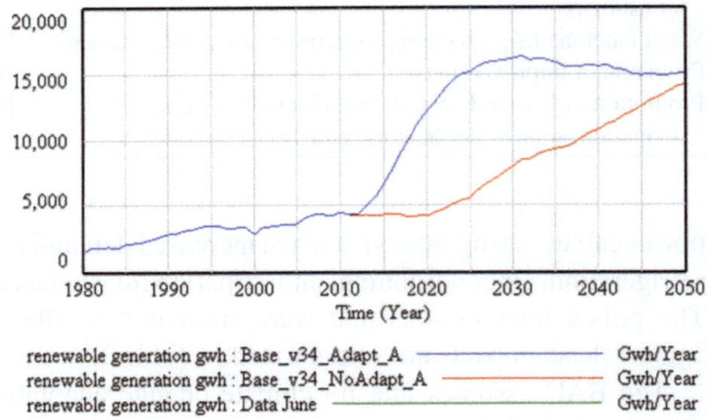

Fig. 6 Total renewable (geothermal, wind, and PV) power generation

Total renewable power generation increases with the addition of solar and wind power generation by 104 % in 2030, by 44.2 % in 2040 and by 6.8 % in 2050 (Fig. 6). The decline in percentage increase in renewable power generation is due to installed plant for wind and solar generation reaching its maximum lifetime.

As a result of these combined mitigation measures, the climate change adaptation scenario (Adaptation) is projected to reduce total CO_2 emissions to 33.6 Mt per year in 2050, compared to 36 Mt in the BAU case.

Summary

This case study has demonstrated the use of system dynamics modeling to climate-proof the power sector in Kenya that is highly vulnerable to current climate variability and projected climate change. Once climate vulnerabilities have been established using a MSP process, policy interventions have been identified to reduce the vulnerability of the power sector. In the climate adaptation scenario, a total investment of US$ 2.7 Bn per year between 2011 and 2030 in climate adaptation and mitigation measures among a number of sectors in Kenya including agriculture (crop cultivation, livestock, fishery, and forestry), energy, and tourism have been simulated. Results of this study show that adaptation has potentially very large benefits in reducing present and future damages.

In the energy sector, CC investment to the tune of 73.71 Bn KSh per year leads to energy savings through efficient bulbs and expanding energy production from renewable sources, both also reducing power cuts. The intervention of using efficient bulbs from the energy demand side would result to a net total energy saving of 1,843 GWh from the current level of 8.54 GWh. On the supply side, public and private sector interventions would increase geothermal power capacity from 0.1 GW now to 1.34 GW by 2031 (compared to 0.66 in the BAU), and green energy power generation capacity from 0.01 GW to 0.7 GW by the year 2017. Despite the fact that wind and solar generation would reach their maximum lifetime, total renewable power generation will increase with the addition of solar and wind power generation by 104 % in 2030, by 44.2 % in 2040, and by 6.8 % by 2050. As a result of these combined mitigation measures, the CC scenario is projected to reduce total CO_2 emissions to 33.6 Mt per year in 2050, 7 % lower than the BAU case (36 Mt). Correspondingly, the total ecological footprint will decline to 1.25 by then relative to 1.4 in BAU, and the ratio of footprint to biocapacity will be 8.3 to 9.3 in the CC and BAU scenarios, respectively.

Mauritius

Mauritius is an upper-middle-income country and a Small Islands Developing State (SIDS). A SDM has been developed for the power and transport sectors, and this section will demonstrate how the tool can be used to develop NAMAs for the power sector. In addition, we will show that SDM is a versatile tool for sectoral carbon crediting under dynamic baselines. Since development of NAMAs will require stringent measurement reporting verification (MRV), countries under the UNFCCC would be required to submit their GHG inventories every second year. We will show how relevant indicators can be built into the SDM so that the model can also be used for monitoring and evaluation of policy implementation. This can be achieved dynamically while being in line with IPCC guidelines for national GHG inventories.

Context

Mauritius is classified as a small island developing state (SIDS) within the UN system. A commonality of SIDS is their inherent vulnerabilities to external shocks including extreme weather events that would be made worse by climate variability and projected climate change, and energy security due to the volatility in the price of oil and lack of indigenous supply of hydrocarbons. It is an upper-middle-income country that has witnessed substantial economic growth over the past 3 decades. The rapid economic growth has been accompanied by an increase in its dependence on imported fossil fuels, and correspondingly on GHG emissions (see Fig. 7).

The main GHG in Mauritius is CO_2, and arises from the burning of fossil fuels (coal, fuel oil, diesel oil, gasoline, and kerosene). Although the national emission for Mauritius is small by global standards, it, nevertheless, has a relatively significant

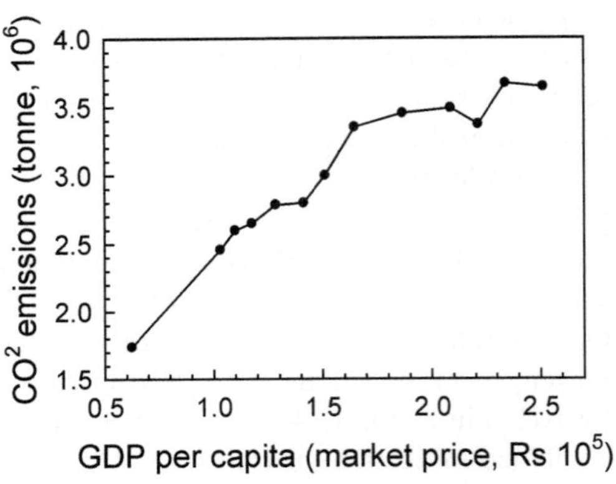

Fig. 7 Variation of GDP per capita and CO_2 emissions in Mauritius, 1995–2011

per capita CO_2 emission. For instance, emissions amounted to 3957.4 kt CO_2 in 2010, while removal by sinks was 250 kt CO_2. The energy industries contributed around 54.6 % of this emission, followed by the transport sector which contributed 31.9 % of the total emissions and the manufacturing industries with 9.1 %. Probably of more importance here is the relatively high per capita CO_2 emission of 2.8 t CO_2 for Mauritius in 2011 (Central Statistics Office 2012). Figure 7 shows that the per capita emission is 2.5 times higher than the per capita emission of Africa (~ 1.1 t CO_2/person), and 1.8 times higher than that of India (~ 1.5 t CO_2/person) (Flavin 2008). If the equivalent emissions of methane and nitrous oxide are taken into account, the net per capita GHG emission in 2010 was 3.8 t CO_2e. Figure 7 does not take land use, land use change, and forestry (LULUCF) into account. Including LULUCF the increase of per capita emission is expected to be higher. As an example, a comparison of the national annual emissions statistics is compared with the GHG inventory carried out under the second national communications (SNC) under the UNFCCC. After carrying out detailed emissions from LULUCF, the SNC reports a per capita emission equal to 3.6 t CO_2e in 2006 (Government of Mauritius 2010). Using the annual emission statistics that does not include emissions from LULUCF, the per capita emission in 2006 is found to be just over 3 t CO_2e.

Mauritius is categorized as a nonAnnex 1 country under the UNFCCC and does not have any responsibility or duty to reduce its GHG emissions. However, there are several reasons why emission reduction would be justified in the case of Mauritius:

1. *Energy security*: Development cannot be carried out without the access to reliable sources of energy. From a strategic perspective, reducing dependence on fossil fuel sources over which Mauritius has little control is desirable, especially in a context of rising fuel prices (over a long term), price volatility, and issues related to the geopolitics of fossil fuels (Kuik et al. 2011).
2. *Ethical considerations*: The UNFCCC calls for the "common but differentiated" approach to stabilizing GHG emissions to levels that would limit the long-term rise in temperature to 2 °C. The "common but differentiated" approach recognizes that developed countries are primarily responsible for the bulk of atmospheric GHGs (i.e., the historical perspective) and that countries do not all have the same capacity

to carry out emission reductions or adapt to the consequences of climate change. However, as emissions emanating from developing countries exceed those from developed countries,[5] and considering that: (1) the atmosphere has a fixed budget as a sink for GHGs, and (2) all sentient beings share the only one and same atmosphere, a case can be made on ethical considerations for an equal emissions entitlement—i.e., equal per capita emissions (Moellendorf 2012). Under future contraction and convergence (Global Commons Institute 2010) or common but differentiated convergence (Höhne et al. 2006) scenarios, Mauritius will have to reduce its per capita GHG emissions.

3. *Access to carbon finance*: As a nonAnnex 1 country, Mauritius can benefit from carbon credits by implementing projects that reduce GHG emissions. Further, nonAnnex 1 countries are encouraged to formulate low-carbon development strategies. Investigating the potential of leverage carbon finance as a supplementary flow of foreign direct investment to support the sustainable development of Mauritius is desirable, especially considering the increasing importance of NMMs and NAMAs (Hinostroza et al. 2012).

Analysis of Modeling Results

In the case of Mauritius, the two scenarios that were analyzed are the BAU and government's official long-term energy strategy (LTES) (Ministry of Renewable Energy and Public Utilities 2009). The BAU assumes that current trends will continue and that current policies and decisions will take their course (e.g., planned investments in the construction of new energy supply), while the LTES scenario simulates the policy interventions of the LTES. Table 4 summarizes the main assumptions of the scenario analyses.

This case study being limited to economy-wide GHG emissions calculated in terms of emission of CO_2 does not provide any discussions of the social and economic impacts of scenario analyses. Such results can be found in another recent study that has used system dynamics modeling to investigate the impacts of green investments in several sectors (energy, water, and agriculture) of Mauritius, especially on the creation of green jobs (Bassi and Deenapanray 2012). The total energy consumption in the BAU scenario is shown in Fig. 8, and it reveals that the model simulation (red) can very well replicate historical data (blue). This was the case for a host of socio-economic and environmental indicators that are not shown here.

Figure 9 shows the simulations of GHG emissions for the BAU (red) and LTES (blue) scenarios. Since the modeling can capture dynamic complexity, the BAU simulation represents a dynamic baseline of what would happen in the absence of any policy interventions. Hence, the net emission reductions arising from the

[5] Total carbon dioxide emissions by developing countries are expected to surpass that of developed countries by 2015. Please see http://www.epa.gov/climatechange/emissions/globalghg.html—accessed 30 July 2011.

Table 4 Main assumptions used in scenario analyses

BAU scenario
GDP growth is calculated endogenously and it follows the declining trend of GDP over the last 18 years (1.8 % average between 2009 and 2025, with 2009 and 2025 at 2.2 % and 1.1 %, respectively). Simulations to 2040 have assumed constant GDP growth rate at 1.1 % after 2025
Maintenance of existing electricity production capacity through 2025, with a net increase in 70 MW between 2010 and 2011 in thermal power plants burning heavy fuel oil (HFO); construction of a 100 MW pulverized coal plant in 2013; in the baseline case only, the construction of additional 100 MW of HFO capacity after 2020 to supply peak power, demand is assumed (this investment is not necessary in the policy case, which projects lower energy demand)
Domestic retail prices of imported primary energy sources (fossil fuels) are exogenous in the model. Historical data use those published by the Central Statistics Office (CSO), Mauritius and projections are calculated applying a future growth rate to the anchor price of 2008 in Mauritius. Projections on the yearly price change for fossil fuels and end-use petroleum-derived fuels are taken from the EIA of the US Department of Energy
Long-term energy strategy (LTES) scenario
Energy efficiency
Applied to electricity consumption, for the domestic, commercial, industrial, irrigation sectors and other uses. Energy efficiency is projected to increase by 2 % in 2010, 4 % in 2015, 6 % in 2020 and 10 % in 2025, relative to 2008
Renewable energy, power sector, and fossil fuels
Efficiency increase in the use of bagasse for electricity generation, to reach 600 GWh of output by 2013
Construction of three wind farms for a combined production capacity of 70 MW (20 + 20 + 30 MW). The project is assumed to start in 2011, with full operational status being reached in 2012
Construction of about 4 MW of small solar energy (2016) and 1 MW of small hydro generating units (2010), for a total of 5 MW installed by the SIPP—an amount considered safe for the grid according to Central Electricity Board (CEB)
Construction of a waste-to-energy plant of 20 MW capacity at La Chaumière. This plant will start producing electricity in 2013
Construction of a landfill gas plant, for 3 MW of capacity (as indicated by CEB) at Mare Chicose. This plant will start producing electricity in 2013 and will be operational for 6 years
Cost-reflective electricity prices
A feed-in tariff for wind is set at Rs 5/kWh, while the one for large scale solar is set at Rs 8/kWh; small scale solar is assumed to receive instead an Rs 15/kwh feed in tariff
The maximum load factor of the various units is assumed as follows: wind 33 %, solar 30 %, waste 80 %, hydro 16 %, landfill gas 76 %, geothermal 70 %, gas turbine 20 %, HFO 56 %, coal 75 %, cogeneration 80 %
The capital lifetime of the various plants is assumed to range between 20 and 30 years
The base load share of electricity demand is set at 60 %
Transport
Reduction of the vehicle age between 2009 and 2025, reaching a 5 % yearly depreciation of the vehicle stock by the end of the period of analysis
Reduction of the age of buses to 10 years, by 2020
Increase in the efficiency of all road vehicles, 20 % between 2008 and 2025, due to the lowered age of vehicles and improvements in fuel efficiency and tuning of engines and tires
Construction of the bus way (for which costs and capacity are not defined yet)
A subsidy to public transportation, of Rs 5/vehicle/day
A congestion charge, of Rs 5/vehicle/day
Capacity of the bus way, assumed at 25 buses
Effectiveness of public transportation subsidy, congestion charge and bus way, each, 5 % in 2012 and 7.5 % in 2015

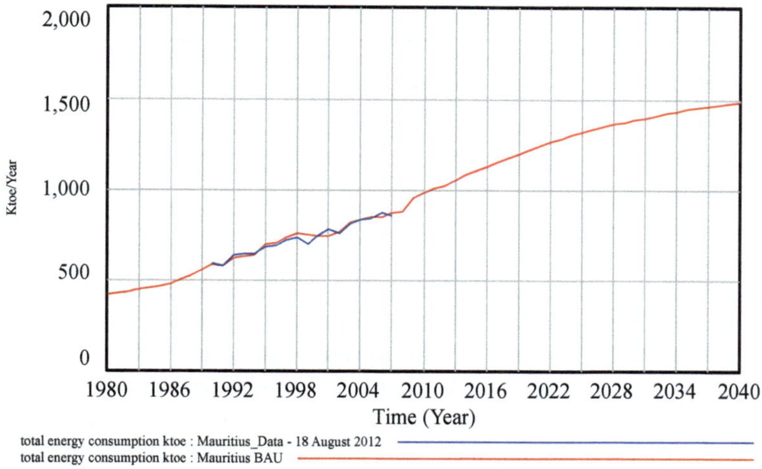

Fig. 8 Total energy consumption—historical data (*blue*) and BAU (*red*) simulation

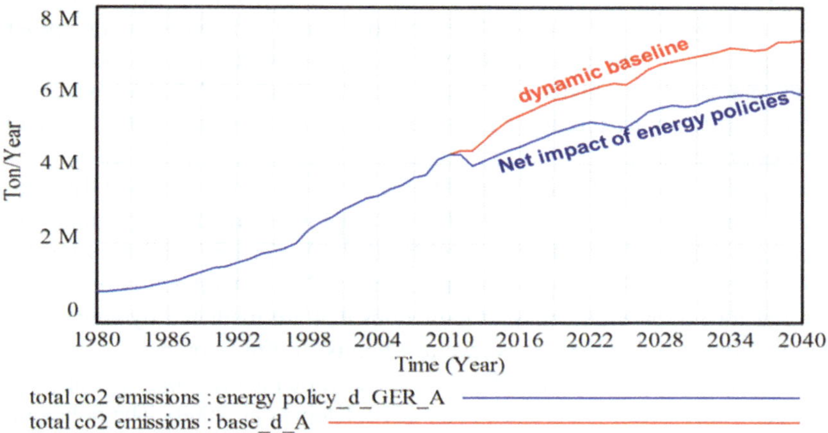

Fig. 9 Emissions reduction through policy implementation against a dynamic baseline

implementation of LTES, which is represented by the curve shown in blue in Fig. 9, can be obtained from the difference between the two curves. Existing baseline-and-credit approaches typically use static baselines that are determined predominantly *ex-ante* (i.e., before a project has been implemented), and are generally applicable to stand-alone projects or program of activities where the activities are of the same type (Beaurain and Schmidt-Traub 2010). There are also cases where the baseline can be updated periodically based on *ex-post* (i.e., after a project has been implemented) observations and emission reductions are calculated based on the most current baseline (Kollmus et al. 2008). For all practical purposes the baseline is considered as counterfactual or hypothetical—i.e., the real future baseline cannot be known once the low-carbon project is implemented. In contrast, system dynamics modeling

Table 5 Emission reductions from policy interventions are calculated against a dynamic baseline, 2015–2040

Year	2015	2020	2025	2030	2035	2040
Emission reduction (million ton CO_2)	0.758	0.786	1.100	1.205	1.190	1.417

allows emission reductions, and hence carbon credits,[6] to be calculated using a dynamic baseline, which would represent a significant advantage of calculating real emission reductions, albeit at the added difficulty of model development. The simulated reductions in emission of CO_2 are summarized in Table 5 as would happen if all elements of the LTES depicted in Table 4 were implemented. The results shown in Table 5 are related to the absolute level of GHG emission reduction. A MRV framework would still need to be put in place for assurance purposes which is discussed later. Further, an assumption is made here that any absolute emission reduction relative to the dynamic baseline would be additional to what would have happened in the absence of the implementation of energy policies.

The simulations shown in Fig. 9 offer two additional advantages that would support NAMAs and NMMs, with the intention to catalyze larger-scale emission reductions beyond what can be obtained from existing project-based or programmatic approaches. The usefulness of system dynamics modeling as a tool to develop: (1) sectoral crediting (under a dynamic baseline), and (2) NAMAs are discussed broadly, while noting that there are several outstanding issues that go beyond the scope of this chapter.

It was discussed in the introduction that stabilizing atmospheric GHGs to levels that would prevent irreversible climate change would require profound emission reductions that would need to go well beyond what can be realistically achieved using the project-based or programmatic approaches. This juncture has led to the proposal for sector-wide emission reductions that are mediated by economy-wide energy policies and strategies as leverage points. In this approach, called sectoral NAMAs, multiple policy-induced emission reduction interventions are implemented across an entire sector (van Asselt et al. 2010; Klein et al. 2009). The policy instruments would be of the sort listed in Table 4, and any sectoral carbon credits generated by emission reductions may be used to support any combination of the sector-specific policies or policy instruments. The decomposition of the total CO_2 emission reduction (Fig. 9) into sectoral emission reductions is given in Fig. 10. In this case, most of the emission reductions are generated in the power sector that contributes around 55 % of the total national CO_2 emissions (see "Context"). The difference between the power sector (green) and dynamic baseline (blue) curves represents the emission reduction from interventions in the power sector, while the difference between the red and green curves give the emission reductions in the transport sector. Thus far, the analysis has shown how system dynamics energy modeling can be used to potentially establish sectoral NAMAs under a dynamic baseline, which may or may not be credited.

[6] One carbon credit is equal to the reduction in the emission of 1 t of carbon dioxide equivalent.

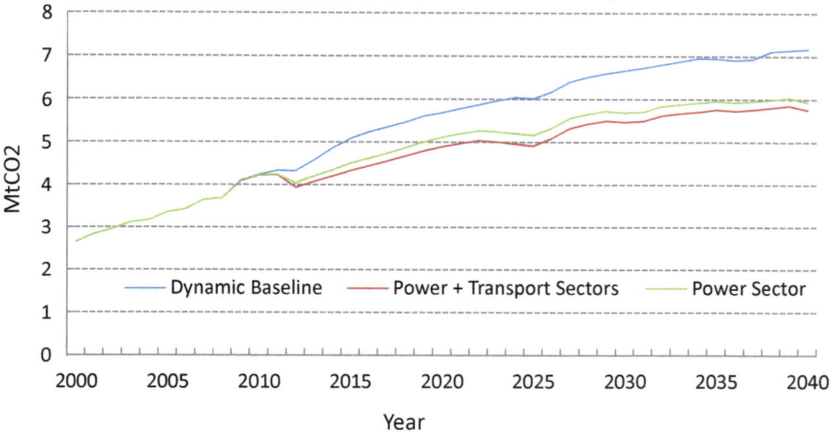

Fig. 10 Sectoral decomposition of total emission reductions from LTES scenario

The next step would be to determine which actions at the sectoral level could be credited. For this, a categorization of sectoral activities will have to be carried out, involving the necessary government agencies, based on the typology provided under the UNFCCC for NAMAs (Hinostroza et al. 2012; Klein et al. 2009):

- ***Unilateral NAMAs***—NAMAs that a country intends to implement completely on its own but for which recognition of this effort is desired. Here the host country sustainable development may not be necessarily driven by the need for emission reductions. Instead, emission reductions could be the result of improving energy security to increase the resilience of the country against externally driven energy shocks or for the generation of green jobs through green investments (Bassi and Deenapanray 2012).
- ***Conditional (or supported) NAMAs***—NAMAs that will only be implemented with the help of international assistance, in the form of financing, technology transfer, and/or capacity building. These NAMAs would typically go beyond unilateral efforts and represent greater emission reduction ambitions for which assistance would be needed for the incremental effort. Actions that contribute to the "no lose or no regrets" goals of the host country would be classified as conditional NAMAs. In this case, the necessary support would have to be specified.
- ***Credited NAMAs***—NAMAs that are eligible for support through full or discounted crediting in the carbon market for activities beyond the BAU scenario.

The categorization of the sectoral interventions as NAMAs is beyond the scope of the work presented here. The process by which a developing country would officially declare its NAMAs has yet to be determined, but it has been proposed that each developing country put forward a climate plan or low-carbon development strategy, such as the LTES, that would also describe the NAMAs that it intends to implement. Generic steps have been developed (GIZ 2011) and best practices (Center for Clean

Table 6 Cumulative investments and O&M costs, and emission reductions for actions in the power sector

Sectoral action	Cumulative investment (capital and O&M)—million Rs 2011–2025	2011–2040	Cumulative emission reduction (Ton CO_2) 2011–2025	2011–2040	In-country capacity
Hydro	0	0	0	0	Exists
Wind	12,094	31,216	3,095,318	10,420,031	Does not exist
Waste energy	6,340	8,905	1,604,613	3,391,653	Exists
Cogeneration	0	0	1,554,004	3,232,159	Exists
Solar	615	669	116,925	284,460	Does not exist
Energy efficiency	12,045	29,757	1,762,900	6,932,600	Exists to varying degrees
Total	*31,095*	*70,548*	*8,133,760*	*24,260,903*	

Air Policy 2011) identified that can nevertheless assist countries to develop NAMAs. Sectoral programs could be a part of the developing country's plan or strategy, since they allow any country to grow their economic sectors in a more climate-friendly manner without compromising the country's sustainable development (Klein et al. 2009). The added benefit is that the sectoral approach takes a more systemic view of low-carbon development, and hence offers the opportunity for accelerated energy transformation. Although there are no clear guidelines for categorizing NAMAs, it is expected that cost and technical capacity would be key considerations. Table 6 shows the cumulative investment and operation and maintenance (O&M) costs of various interventions in the power sector, as well as the corresponding emission reduction potentials. The status of available capacity is also given for each intervention. So, the system dynamics modeling could also be used as a useful tool to guide the categorization of NAMAs once the criteria and indicators for that categorization are defined either at the international or national level.

Finally, the MRV of these mitigation actions is important to generate transparency on their effectiveness and facilitate decision making, especially in the case of credited NAMAs. MRV can be thought of as a knowledge management system for tracking GHG emissions, actions to reduce GHG emissions, and climate change mitigation support (GIZ 2011). The system dynamics model can also be used as a monitoring and evaluation tool since, as shown in Fig. 8, the simulation of scenarios can be compared to historical or measured data. Hence, by defining what to measure (i.e., indicators such as direct and indirect emissions, electricity generated by renewable sources, or electricity saved by energy efficiency, etc.) and how to measure the necessary indicators (i.e., what methodology needs to be adopted), the modeling tool can calculate the indicators using the predefined methodologies. In fact, the definition of indicators to monitor and evaluate policy interventions is an integral part of Step 1 in the model development process (see "Model Development and Validation"). For instance, the model used here is fully compliant with the guidelines used for preparing national GHG inventories using the guidelines of the Intergovernmental Panel on Climate Change (IPCC) that are used for national communications under the UNFCC. Not all requirements of a MRV system can be captured by the tool, such as how often, and who should be responsible for MRV.

Summary

The main objective of this case study was to demonstrate yet another novel application of system dynamics modeling. Using the long-term energy strategy of Mauritius as example, it has shown how scenario analysis of energy policy could potentially be used for the following:

1. Quantify GHG emission reduction under a dynamic baseline. The dynamic baseline is generated under the BAU scenario (i.e., absence of policy interventions) and captures the complex interactions between key social (e.g., population dynamics and disposable income), economic (e.g., GDP growth), and environmental spheres (e.g., GHG emissions) of sustainable development.
2. Develop sectoral GHG emission reduction or sectoral NAMAs based on the implementation of sector-wide strategies. The case study was used to show how total GHG emission reductions could be decomposed into sector-wide emission reductions in the power sector and transport sector separately.
3. Although the modeling tool alone cannot be used to categorize sectoral mitigation actions into unilateral-, supported-, or credited-NAMAs, it can provide some of the information such as investment and O&M costs and action-specific GHG emission reductions to assist in this categorization. Further, it has been shown elsewhere that the modeling tool can also be used to calculate other key indicators like green job creation, reduction in the bill of imported fossil fuels and contribution to economic growth, among others, that may also be relevant to categorize NAMAs (Bassi and Deenapanray 2012).
4. Predefined indicators and methodologies to calculate the indicators based on agreed international benchmarks can be embedded in the system dynamics model to provide key elements of a stringent MRV system. The use of the model as a monitoring and evaluation tool has been discussed, especially in the context that the scenario analyses allow the simulations of the model to be compared directly with historical or measured data.

Conclusions

Climate change has emerged as arguably the biggest threat facing human development in the twenty-first century. It is, therefore, in humanity's interest to do something about the current state of affairs.

Global coordination for GHG emission reductions is carried out under the aegis of the UNFCCC. The use of NMMs for achieving global reductions in GHG emissions was adopted at COP16 in Cancun (2010), and further referenced at COP17 in Durban (2011). Since the sectoral scope of NMMs depends, among others, on data availability and low degree of uncertainty in emission estimates, the sectors that are recommended for NMMs are: (1) energy supply; (2) industry (e.g., oil refineries, natural gas facilities, iron and steel production, cement production); and (3) transport.

In this chapter, we have demonstrated the practical use of system dynamics modeling for policy planning to achieve climate-resilient, low-carbon development pathways, in the context of national development planning. In particular, we have used real examples from developing (Mauritius and Kenya) and developed (USA) countries to make the case for the use of SDM for climate proofing the energy sector and to develop (NAMAs) as one type NMM.

In all cases, the impacts of policy interventions was discussed for several cross-sectoral indicators, highlighting the strengths of system dynamics models in analyzing broader impacts of policy implementation and in identifying potential double and triple dividend interventions. This advantage of SDM also makes these models relevant to analyze energy issues in the context of national development, be it climate adaptation and/or mitigation, or a broader framework such as the green economy.

Appendix: Multistakeholder Process for Model Development

It is widely acknowledged now that the knowledge required to articulate what would constitute sustainable development in any given context (i.e., country or subregions therein) is often dispersed within the system boundary (i.e., country and its subregions), which is why a multistakeholder approach is necessary for successful outcomes. In other words, the complex system of socio-economic conditions existing within the natural ecosystems characterising any given territory can only be seen collectively for the adequate response to the increasing demands for policy-relevant interventions. MSPs can also help ensure better coordination between different institutions and agencies, in addition to ensuring that knowledge is combined and properly utilized by sharing common mental models. MSP is also an appropriate means to achieve consensus and ownership of the modeling tool for planning and decision-making purposes.

The five sequential steps of a generic MSP are illustrated in Fig. 11 (Hemmati 2002). Each step involves specific actions to ensure maximum ownership of the process by the beneficiary stakeholders and ensuring them that climate change related actions are discussed through dialogue and consequently integrated in the national and local agenda. Briefly, the steps of the MSP are defined in generic terms while noting that the central issues are related to climate change:

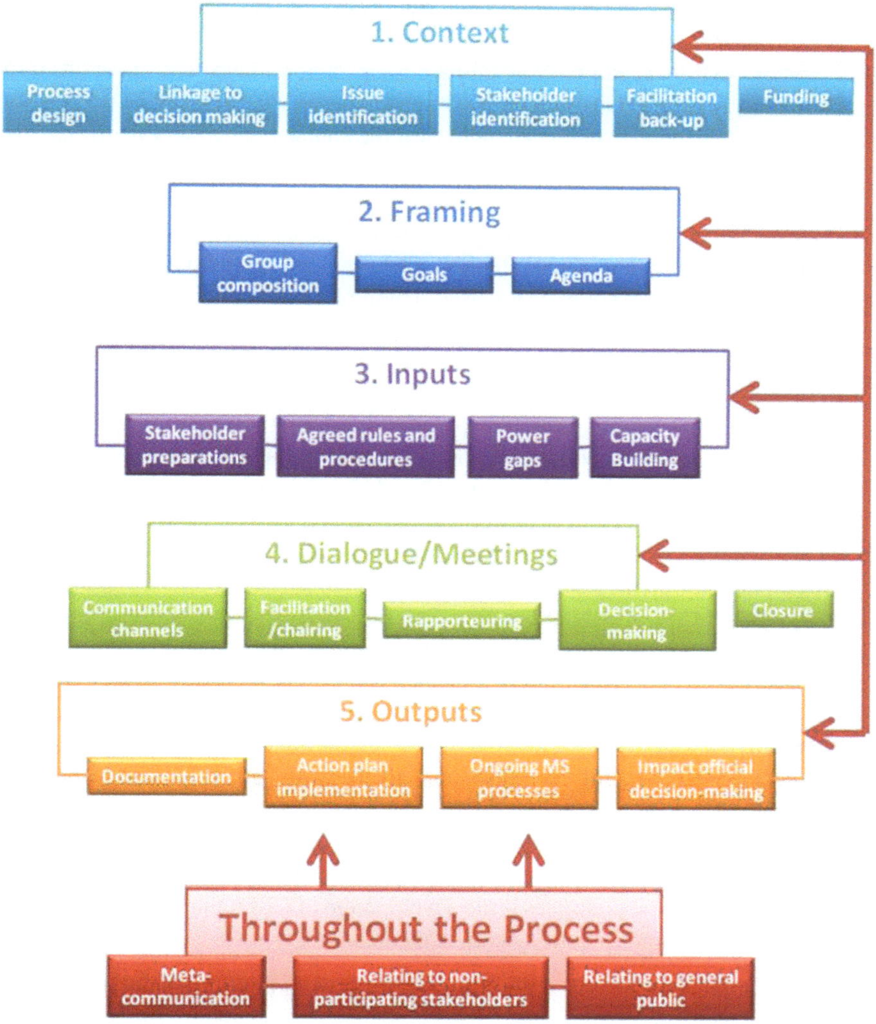

Fig. 11 The five sequential steps of the multistakeholder process used in model development

1. *Context*—Setting the context is probably the most critical step in the process, and the "one-size-fits-all" cannot be applied. After the key stakeholders have been identified based on principles of inclusiveness, diversity, and size, they should be involved in every aspect of the design process to generate legitimacy, credibility, and trust. This does not mean that conflicts will not arise, but that any conflicts may be better dealt with later on. In designing the process, it should be made very clear how the output of the dialogues will permeate the policy decision-making process. Productive dialogue can only take place when all participants share a common understanding of the agenda of the MSP. This requires a clear definition

of what issues the MSP will address. Successful MSPs require facilitation and organizational back up, also implying the need for adequate financial resources.
2. *Framing*—MSPs need precisely defined issues before them. The questions to be addressed and the goals of the process need to be very clear to all the participants and agreed by them. Possible changes over the course of an on-going process also need to be agreed on by the group, allowing for consultations within constituencies if necessary.
3. *Inputs*—In order to facilitate dialogues, several inputs must be in place or be made available to participants. First, all participants must have equitable access to all information, and they should be given sufficient preparation time. The ground rules for the purpose of dialogue must be agreed on within the group, while noting that no one has all the answers but that the output required will be the collective wisdom and knowledge of the participants. Fundamental differences exist between stakeholders in such things as knowledge and information, communication skills, size, nature, and the amount of resources that define significant power gaps and unfair distribution of bargaining and negotiating power. Care must therefore be taken to identify and address power gaps, and this is also a reason why facilitation of dialogues is critical. Bilateral meetings can be used where necessary to prepare participants for plenary sessions.
4. *Dialogue*—MSPs are about creating a space where dialogue can take place. An atmosphere that cultivates equity, respect, dignity, humility, and hope will create a space where people can interact in such a way that their differences and their commonalities become clear so that they can begin to explore possible ways forward.
5. *Outputs*—MSPs should be transparent all the way. So, they should not only publish and communicate their deliberations and outcomes but also keep record of their design. A critical aspect is to be able to demonstrate to stakeholders how the outcomes of their dialogues impacted policy decision-making. One of the key initial outputs of the dialogues is the development of CLDs that are then translated mathematically into system dynamics models.

MSP was adopted for the conceptualisation, customization, and validation of system dynamics models developed in the three case studies presented in this chapter.

References

Adger, W.N., Agrawala, S., Mirza, M.M., Conde, C., O'Brien, K., Pulhin, J., et al.: Assessment of adaptation practices, options, constraints and capacity. In: Parry, M.L., Canziani, O.F., Palutikof, J.P., Linden, P.J.v.d., Hanson, C.E. (eds.) Climate Change 2007: Impacts, Adaptation and Vulnerability. Contribution of Working Group II to the Fourth Assessment Report of the Intergovernmental Panel on Climate Change, pp. 717–743. Cambridge University Press, Cambridge (2007)

Barlas, Y.: Formal aspects of model validity and validation in system dynamics. Syst. Dyn. Rev. **12**(3), 183–210 (1996)

Bassi, A.M., Deenapanray, P.N.: A green investment analysis using system dynamics modelling—the case study of Mauritius. Small States: Econ. Rev. Stat. **16**, 256 (2012)
Bassi, A., Yudken, J.: Climate policy and energy-intensive manufacturing: a comprehensive analysis of the effectiveness of cost mitigation provisions in the American Energy and Security Act of 2009. Energy Policy. **39**, 4920–4931 (2011)
Bassi, A., Yudken, J., Ruth, M.: Climate policy impacts on the competitiveness of energy-intensive manufacturing sectors. Energy Policy. **37**, 3052–3060 (2009)
Bassi, A.M., Wang'ombe, C., Kirui, G., Njaramba, L.N., Barasa, D.O., Muema, J.N., et al.: Strengthening Institutional Capacity for Integrated Climate Change Adaptation & Comprehensive National Development Planning in Kenya. Government of Kenya, Nairobi (2011)
Beaurain, F., Schmidt-Traub, G.: Developing CDM programmes of activities: a guidebook. South Pole Carbon Asset Management (2010)
Blok, K., Höhne, N., van der Leun, K., Harrison, N.: Bridging the greenhouse-gas emission gap. Nat. Clim. Chang. **2**, 471–474 (2012)
Center for Clean Air Policy: The Road to NAMAs—Global Stories of Successful Climate Actions. Center for Clean Air Policy, Washington (2011)
Central Intelligence Agency: The World Factbook 2011. Retrieved from The World Factbook 2011. www.cia.gov/library/publications/the-world-factbook/index.html (2011). Accessed October 2012
Central Statistics Office: Environment Statistics—2011. Central Statistics Office, Port Louis (2012)
Flavin, C.: Building a low-carbon economy. In: Flavin, C. (ed.) State of the World. WorldWatch Institute, Washington (2008)
GIZ: Nationally appropriate mitigation actions (NAMAs): steps for moving a NAMA from idea towards implementation, draft 8.2. GIZ (2011)
Global Commons Institute: Contraction and convergence—an international conceptual framework for preventing dangerous climate change (2010)
Government of Kenya: Kenya Vision 2030. Government of Kenya, Nairobi (2007)
Government of Mauritius: Second National Communication Under the UNFCCC. Government of Mauritius, Port Louis (2010)
Hemmati, M.: Multi-Stakeholder Processes for Governance and Sustainability: Beyond Deadlock and Conflict. Earthscan, London (2002)
Hinostroza, M., Lütken, S., Aalders, E., Pretlove, B., Peters, N., Holm Olsen, K.: Measuring Reporting Verifying—A Primer on MRV for Nationally Appropriate Mitigation Actions. UNEP Risø Centre, Technical University of Denmark, Copenhagen (2012)
Höhne, N., den Elzen, M., Weiss, M.: Common but differentiated convergence (CDC): a new conceptual approach to long-term climate policy. Clim. Policy. **6**, 181–199 (2006)
International Energy Agency: World Energy Outlook 2010. Power generation cost assumptions, Paris (2010)
Klein, D.E., Ma, H., Helme, N., Wang, C.: Technology-based sectoral NAMAs: a preliminary case study of China's cement and iron and steel sectors. Center for Clean Air Policy (2009)
Kollmus, A., Zink, H., Polycarp, C.: Making Sense of the Voluntary Carbon Market: A Comparison of Carbon Offset Standards. WWF, Germany (2008)
Kuik, O.J., Lima, M.B., Gupta, J.: Energy security in a developing world. WIREs Clim. Chang. **2**, 627–634 (2011)
Meadows, D.: The Unavoidable A Priori, excerpt from Randers, Elements of the System Dynamics Method (1980)
Ministry of Renewable Energy and Public Utilities: Long-Term Energy Strategy 2009–2025. Government of Mauritius, Port Louis (2009)
Moellendorf, D.: Climate change and global justice. WIREs Clim. Chang. **3**, 131–143 (2012)
Randall, S.: History of the 2 °C climate target. WIREs Clim. Chang. **1**, 598–605 (2010)
Stockholm Environment Institute: The economics of climate change in Kenya (2009)
United Nations Environment Management Group: Working towards a balanced and inclusive green economy, a United Nations system-wide perspective (2011)

United Nations Population Division: World population prospects: the 2006 revision, New York (2007)
US Congress: H.R. 2454, the American Clean Energy and Security Act of 2009, Washington (2009)
US Congressional Budget Office: The estimated costs to households from the cap-and-trade provisions of H.R. 2454, Washington (2009)
US Department of Commerce: Bureau of Economic Analysis. National economic accounts, national income and product accounts table, Washington (2008)
US Department of Energy: Energy market and economic impacts of the American Power Act of 2010. Energy Information Administration, Washington (2010a)
US Department of Energy: International Energy Annual 2010. Energy Information Administration, Washington (2010b)
van Asselt, H., Berseus, J., Gupta, J., Haug, C.: Nationally appropriate mitigation actions (NAMAs) in developing countries: challenges and opportunities, Free University Amsterdam, Amsterdam (2010)
Yudken, J., Bassi, A.M.: Climate policy impacts on the competitiveness of energy-intensive manufacturing sectors. National Commission on Energy Policy, Bipartisan Policy Center, Washington (2009)

Understanding the Dynamics of Electricity Supply and Demand in Canada

Hassan Qudrat-Ullah

Introduction

Currently, the electricity production industry in Canada consists of four major nonrenewable sectors, and two major renewable sectors. The majority of nonrenewable electricity comes from crude oil, coal, natural gas, and uranium. On the other hand, the majority of electricity from renewable sources comes from hydroelectric and wind production. The relationship between supply and demand of electricity has changed over the past few economic cycles (Qudrat-Ullah, 2013). It is important to note that conventionally, as the demand for electricity increased, the production of electricity also increased. A notable change came with the 1989–1993 recession, when the demand growth for electricity stalled. With the stall of demand came the stall of supply (IFC Consulting 2006). However, as the economy recovered from the recession, demand growth resumed, however supply did not follow. Instead, the focus on maintaining alignment between supply and demand was on productivity. Demand is driven by increase in electricity using economic activities, and efficiency gains. Productivity may be further divided into mechanical efficiency and conservational efficiency or electricity spent for value addition. The two may be further divided into current machinery efficiency improvements, the invention of more efficient machinery, and the devising of new techniques that improve the value adding capabilities of processes. The driver behind such productivity improvements is research and development, which in turn is driven by investment (Park et al. 2007; Kilanc and Or 2008). For renewable energy sources, technological efficiency does not depend on the demand side dynamics directly. However, the economics and cost competitiveness of the technologies do (IFC Consulting 2006).

Despite considerable improvements in the productivity area, Canada's electricity supply and demand system has experienced significant imbalance in recent history (IFC Consulting 2006; Canada's sector council program powering up the

H. Qudrat-Ullah (✉)
School of Administrative Studies, York University, 4700 Keele Street,
Toronto, ON M3J 1P3, Canada
e-mail: hassanq@yorku.ca

H. Qudrat-Ullah (ed.), *Energy Policy Modeling in the 21st Century,*
Understanding Complex Systems, DOI 10.1007/978-1-4614-8606-0_9,

future 2008). In fact, complexity of the system makes sustainable policy decision making a difficult task. Complexity of this system primarily comes from the existence and interactions of nonlinear and dynamic variables including various stocks of electricity generation capacity, restricting and regulatory regimes, fuel supply and price dynamics, and advances and challenges in technologies for electricity generation, transmission, and consumption.

Understanding of such complex policy issues and decisions necessitates the use of system simulation (Ford and Bull 1989; Olaya and Dyner 2005; Qudrat-Ullah and BaekSeo 2010). Specifically, researchers from the system dynamics community have found system dynamics simulation models capable of modeling and analyzing complex energy systems. For instance, system dynamics models have successfully been applied to various complex energy issues including (i) national energy policy design and evaluation (Ford 1983; Naill 1992; Qudrat-Ullah and Karakul 2007; Ochoa 2007), (ii) energy conservation analysis (Ford and Bull 1989), (iii) privatization of electricity industry (IFC Consulting 2006; Bun and Larsen 1992; Ford 1997; Dyner and Bunn 1997; Qudrat-Ullah and Davidsen 2001; Assili et al. 2008), (iv) generation expansion planning (Kilanc and Or 2008; Adelino and João 2011), and (v) assessment and mitigation of CO_2 emissions (Qudrat-Ullah and Davidsen 2001; Anand et al. 2005; Ansari and Seifi 2012). Therefore, to better understand the demand and supply dynamics of the electricity industry of Canada, we develop, validate, and utilize a system dynamics based simulation model.

The remainder of this paper is organized as follows: in "Sectorial Overview," sectorial overview of various electricity consumption sectors in Canada is presented. Key dynamics of the electricity sector are described in "Key Dynamics in the Electricity Sector." "Development of the Dynamic Model" details the model structure and model validation. Results with status quo scenarios as well as additional investments-based scenarios are discussed in "Results." "Concluding Remarks" concludes this paper.

Sectorial Overview

Energy Consumption by Sector

Residential Sector

This sector accounts for electricity consumed in Canadian households, and includes energy for space and water heating, air conditioning, appliances and other end use energy devices. In the year 2007, residential energy consumption accounted for 18.4 % of the total electricity end use. The forecasted rate of change for this sector over the next 9 years is + 0.5 %/year. Changes in usage rate are most heavily influenced by government policies and changing consumer preferences (NEB 2011). Government programs aimed at energy use reduction include policies for stricter building codes in Canada's most populous provinces. New furnace and boiler efficiency standards improve energy intensity of all new homes nationally (NEB 2011; Statistics Canada 2007). Policies for electricity usage in lighting are also becoming strict. Furthermore, much of the common appliances found in households that were formerly unregulated, now have minimum energy performance guidelines (Qudrat-Ullah, 2013).

Natural gas and electricity make up the majority of demand in this sector. Though impressive improvements have been achieved in energy efficiency of space heating and major appliances, aggregate demand still experiences growth as a result of increasing house sizes, preference for air conditioning, and the increase in adoption for electronic goods (NEB 2011; Working document of the NPC global oil & gas supply 2007).

The methods by which electricity demand for the residential sector has been met depend heavily on the regional availability of fuel, energy prices, and end use demand. In Atlantic Canada for example, hydroelectricity is the dominant source of electricity. In the prairies, natural gas holds the majority of market share (NEB 2011).

Commercial Sector

This sector includes offices, retail, warehousing, government and institutional buildings, utilities, communications and other service industries. It also includes energy consumed by oil and gas pipelines and street lighting (NEB 2011). Electricity demanded by this sector is generally used for similar functions as that of the residential sector, namely, space heating and cooling, water heating, lighting, and electrical plug load. In the year 2007, commercial energy demand, like residential demand, was also 18.4 %. However, growth rate in this sector is much more significant, averaging at 1.4 %/year for the next 9 years (Government of Canada 2011; Electricity generation, by utilities, by source 2005). It is important to note that demand growth rate for both the residential and commercial sectors is at a historical low due to aggressive improvements in energy efficiencies. Energy related policies also severely impact energy consumption in this sector. Building codes, for example require more efficient insulation, heating/ventilation and air conditioning. Such requirements aim at reducing the energy demand by 25 % relative to the Model National Energy Code for Buildings of 1997 (Qudrat-Ullah, 2013). Demand is also being reduced by equipment standards, including minimum boiler efficiency and packaged heating/cooling units as well as improvements in lighting efficiency.

The Industrial Sector

This sector includes manufacturing, forestry, fisheries, agriculture, construction, and mining. Much of the energy demand for this sector comes from a select few energy intensive industries, namely iron and steel, aluminum manufacturing, cement manufacturing, chemicals and fertilizers, pulp and paper, petroleum refining, and oil and gas extraction (NEB 2011). This sector is by far the greatest consumer of electricity in Canada, accounting for 63.2 % of electricity demand in 2007. Market share however is projected to drop to 60.5 % by 2020, reflecting slower economic growth in the Canadian goods producing sector (NEB 2011). This sector is by far the largest contributor to environmental pollutants, and is the target of many regulatory policies, including the cap and trade program.

Key Dynamics in the Electricity Sector

The Shift in Production Mix

Conventionally, much weight has been given to nonrenewable, highly contaminating sources of power (Cappers et al. 2010). As of late however, environmental consciousness, as well as greater acceptance of the finite life of nonrenewable energy, has led to an emergence of new environmentally friendly and high-yield strategies. According to a study by Canada's National Energy Board, hydroelectric, nuclear, natural gas, and wind capacity are projected to increase in the future (NEB 2011; Statistics Canada 2007). Wind power is expected to achieve the greatest relative growth, reaching 10 % of installed capacity by 2020 (NEB 2011). Though some hazardous sources, such as biomass, landfill gas, and waste heat are experiencing growth, technologies such as carbon capture and storage are expected to experience parallel growth, as a method of mitigation of environmental pollution (Qudrat-Ullah, 2013).

Macroeconomic Influences

The stocks and flows of the Canadian energy market are greatly influenced by global trends in energy pricing, technology development, as well as government regulation.

Crude oil supply and pricing for example greatly influence Canada's ability to produce electricity domestically. Over the past decade, many emerging economies including India and China averaged yearly economic growth rates of approximately 7 % each. The sustainability of such growth was largely attributed to the ability of global crude oil supplies to meet growing demand. Such supply did not keep up with demand, leading to an increase in crude oil pricing. In early 2008, oil prices were at a record high US$ 100/barrel. They continued to climb throughout the year to as high as US$ 147/barrel. For most countries, the increase in the commodity's price reduced demand for crude oil. However, demand in some countries including Canada increased as a result of government subsidies to the industry (NEB 2011; NPC 2007). Growth in crude oil prices however stalled and declined throughout the 2008 recession as a result of decreased demand. Crude oil prices and demand, like that of most other energy sources, parallel the health of the economy. The higher the economic growth, the higher the demand is which then leads to higher prices for such resources.

Similar to the demand for crude oil, natural gas demand also fell as a result of the economic slowdown. This slowdown, however, was synchronous with the increased supply of natural gas due to emerging technologies especially with regards to tight gas and shale gas production (NEB 2011). Such an imbalance in the supply and demand for natural gas caused prices to fall by nearly 75 % from peak prices of US$ 13.32/MMBTU in July 2008. Likewise, gas-drilling activity also fell by 50 % in the year 2009. However, such a reduction in prices and supply combined with the slow but sure recovery in the economy drove the demand, prices, and production up once more in a balancing act (Qudrat-Ullah, 2013).

Notable in terms of policy and regulation is the increasing trend promoted by the world governing agencies towards environmental protection. The Western Climate Initiative is an example of a policy which aims to make emission production much more costly and difficult. The initiative has developed a carbon market cap-and-trade program (NEB 2011; IEEE Xplore 2011). This program allows participants to emit only as much pollutants as specified in their permits. Should a company wish to produce more, and furthermore emit more, it would have to acquire more permits from other participants—a costly venture. Four Canadian provinces, and seven US states have already been inducted into this program (CIA 2011; NRC 2009; Statistic Canada 2011). Canada also has several provincial level policy directives, including the BC Energy Plan, Alberta's Climate Change and Emissions Management Amendment Act, and Manitoba's Beyond Kyoto (NEB 2011). Such programs, as well as many others, impose emission restrictions and mandate energy efficiency.

Microeconomic Influences

On a micro level, much influence on the price and supply of electricity comes from generation ability, transmission, and distribution costs. Notably, electricity prices are lowest in provinces with a high proportion of supply coming from hydro production (NEB 2011). This suggests high yield and efficiencies in the hydroelectricity industry. Hydroelectric plants have substantial start-up costs, however because of the longevity of such assets, hydro-generating stations that have been installed, and paid off many years ago, are now still in use at low operational costs. At a provincial level, regulators aim at finding a balance between low cost heritage assets, and high start-up cost new assets. Though new assets and technologies have high start-up costs, their long-term profitability is much higher than that of their predecessors. Prices in most jurisdictions are highly dependent on the cost of service provision and the regulated rates of return. Cost of service varies greatly, with large-scale consumers enjoying lower prices due to economies of scale, and small-scale, usually residential, consumers incurring higher costs. It is worth noting that the short-term trend is towards higher electricity prices due in a large part to the development of higher cost generation resources and planned improvements to transmission systems (NEB 2011).

Though the use of coal for power generation is on the decline, it is still Canada's second largest source of electricity (Statistics Canada 2007). The use of coal is more dominant in western provinces where supply is high, and the cost of distribution is relatively low. Coal prices are expected to remain relatively stagnant due to increasing competitive pressures and productivity increases in mining and rail transportation (Qudrat-Ullah, 2013).

Electricity Restructuring

Traditionally, the Canadian industry was composed of integrated companies that performed all electricity logistics, from generation, all the way to customer distribution (Qudrat-Ullah, 2013). Restructuring aims at shifting from monopolistic production to separate generation, transmission, and distribution service companies. The purpose of such restructuring is to promote competition among generators and new entrants to the market, and to provide more open access to the transmission incumbent systems, a system also known as wholesale access (Centre for Energy 2009). Such unbundling also increases competition with regards to marketing of electricity, providing more choices to consumers. In such a system, consumers would have a choice of supplier, expanded metering services, and options with respect to green power. As a result of restructuring, trade is likely to increase. Areas with high electricity prices are likely to begin adopting suppliers they previously did not have access to (Stone 2008).

The Shift to Productivity

A shift from increases in capacity to increases in productivity is made evident by the decrease in energy usage relative to Canada's GDP of 1.3 % per year. Such a decline is heavily attributable to efficiency improvements in electricity and natural gas end use devices, as well as declining heavy industry sectors (NEB 2011).

To exemplify the notion of increased productivity, we will examine the "Smart Grid" initiative being considered by the Ontario Energy Board. Such an initiative is the result of the Green Energy Act that took effect in 2009, requiring an increase in electricity conservation and demand management efficiency. It also requires implementation of a Smart Grid, and promotes the increased use of renewable energy sources (U.S.D. of Energy 2009). Contrary to the conventional response to increased demand, the Green Energy Act facilitates productivity improvements rather than supply increases. The implementation of a Smart Grid does not increase the aggregate supply of electricity, but rather improves flow, turning one-way flow of electricity into a bilateral exchange between households and utility stations. As more and more entities external to the utilities are implementing their own energy producing initiatives, supply no longer comes solely from the government generation stations. Other entities are increasingly producing wind and solar energy, and surplus of such energy may be sold back to the utilities (U.S.D. of Energy 2009). Such efficiencies remove redundant provisions of power and reduce the costs associated with such inefficient electricity transportation. A Smart Grid also greatly improves informational flows through networks, greatly improving demand management potential, and reducing costs associated with information aggregation. Further benefits of implementing a Smart Grid include (Qudrat-Ullah, 2013):

- Self-healing from power disturbance events
- Enabling active participation by consumers in demand response
- Operating resiliently against physical and cyber attack
- Providing power quality for twenty-first-century needs
- Accommodating all generation and storage options
- Enabling new products, services, and markets
- Optimizing assets and operating efficiently

Another macro level initiative that aims to improve productivity within the power generation and distribution system is the increase in interoperability standards. Collaboration among all entities within the network ensures alignment between operating standards, and compatibility among operating and delivery systems. The unification of the power system leads to a flexible, uniform, and technology neutral environment that improves customer choice, and yields economies of scale (U.S.D. of Energy 2009). Closely related to the alignment of standards is interconnection planning and analysis. Collaboration in industry analysis and forecasting reduces volatility with respect to future generation. Such collaboration also encourages the development of uniform industry-wide strategies for dealing with the supply and demand of power in Canada. It is important to note however, that such initiatives as indicated above are impeded by a shortage of workers knowledgeable in emerging technologies such as Smart Grids. For this reason, many workforce development programs are under development in attempts to update the practical knowledge of industry employees. The Consortium for Electric Reliability Technology Solutions for example is a consortium of national laboratories, universities, and industries that performs research and develops and disseminates new methods, tools, and techniques to protect and enhance the reliability of the electric power system (U.S.D. of Energy 2009). The consortium works with energy boards in developing employee development programs. A final notable industry-wide initiative aimed at productivity enhancement is the increasing use of stakeholder engagement and outreach activities. Such activities disseminate information regarding changes in industry practices, cost performance data, environmental considerations, etc., in attempts to encourage investment primarily for research and development purposes.

Grid Energy Storage

Another benefit of a Smart Grid is its ability to store electricity produced in excess of demand. Conversely, when electricity demand exceeds supply, the electricity stored within the grid is released to various destinations with an electricity deficit. This system of electricity storage allows generation plants to run more efficiently, as production shifts both up and down need not be extreme, as electricity stored within the grid will aid in balancing supply and demand, even if the amount being produced differs from the amount demanded (Centre for Energy 2009). Below we will present a circumstance, which greatly benefits from the storage capacities of the grid.

Peak Demand

Currently, peak demand in Canada is growing much faster than average demand. Peak demand in the country occurs in the summertime, and results from the high-usage rate and high-adoption rate of air conditioning systems. Much of this increase comes from the adoption of large-scale industrial air conditioners. However, it is important to note that this growth in peak usage, which contributes to an increasing gap between supply and demand, will not continue on the same trend for much longer. This is because the air conditioning market will become saturated, preventing adoption from continuing indefinitely.

Traditionally, such a deviation from regular production as described above would require large swings in electricity production. However as grid storage prevalence increases, such swings in production need be less and less severe, as electricity stored in times of electricity surpluses will be used to balance the supply and demand gap.

Demand Side Factors

In this section, we will discuss demand side changes, which hold stake in the discourse regarding the gap. According to forecasts, there are sufficient "demand side" resources available to close the gap or at least to delay its appearance beyond 2020 (IFC Consulting 2006). In fact, research shows that even if the increasing productivity trends within the last 15 years were to slow down or reverse, a gap would still not materialize if about 50 % of potential for fuel substitution, demand management, and energy efficiency were realized, along with a modest growth in cogeneration (IFC Consulting 2006). Subsequently, we will explore the dynamics of such demand side developments:

Fuel Substitution

Currently, electric space and water heating account for 37 % of total residential electricity usage (IFC Consulting 2006). However, alternative methods of electric space and water heating are becoming cheaper and more attractive. High efficiency gas heating for example, is now 40 % cheaper than electric heating (IFC Consulting 2006). Such a wide variance in cost is projected to lead to significant electricity savings, through substitution.

Presently, baseboard heaters provide over half of electric space heating. Traditionally switching to more energy efficient alternatives has been made difficult by the costs of retrofitting the infrastructure of the dwelling. Recent advances however, in small diameter, flexible piping, hydronic heat distribution systems that allow conversions from electric baseboard heating with relatively little disruption to the household and at a much lower capital cost than regularly, have made such substitutions much more attractive. Fuel substitution for the above uses of electricity is projected to reduce the potential gap in 2020 by 400 MW.

Demand Side Management Potential

This notion involves changing the level or pattern of demand for energy. Examples of tactics in demand side management include improving efficiency with which a service is provided, decreasing the underlying demand for the service, and influencing or controlling the timing of the service demanded (IFC Consulting 2006). The Smart Grid discussed above is a key factor in managing demand through efficient informational flow. Studies show that the gap may be reduced by 1,500 MW by 2020 through demand management efficiencies (IFC Consulting 2006).

Cogeneration

Cogeneration or "combined heat and power" is a method of simultaneously producing both electricity as well as heat. Traditionally, heat is produced as a byproduct of electricity production, and such heat is released into the environment. Through cogeneration however, such heat can be captured in steam or water, and reused to produce more electricity. A study conducted by the Ministry of Energy suggested that cogeneration could reduce the supply and demand gap by as much as 8,250 MW (IFC Consulting 2006). Cogeneration was also found to be much more cost competitive than regular generation, having prices for electricity delivered at about 40 % below the average market price.

Energy Efficiency

According to an ICF Consulting study, technical potential for energy efficiency improvements could reduce electricity use by 36.6 TWh, and cut peak demand by 8.2 GW. This is equivalent to 26 % of Ontario's current electricity use and 33 % of system peak. However, improvements only as far as economically feasible yielded lower savings, at 29.6 TWh and 5.2 GWh for peak times (NPC 2007). Though significantly lower than total potential, such improvements still represent 21 % of total energy sales. According to the same study, such improvements may help reduce the gap between supply and demand by 2,150 MW by 2020 (Qudrat-Ullah, 2013).

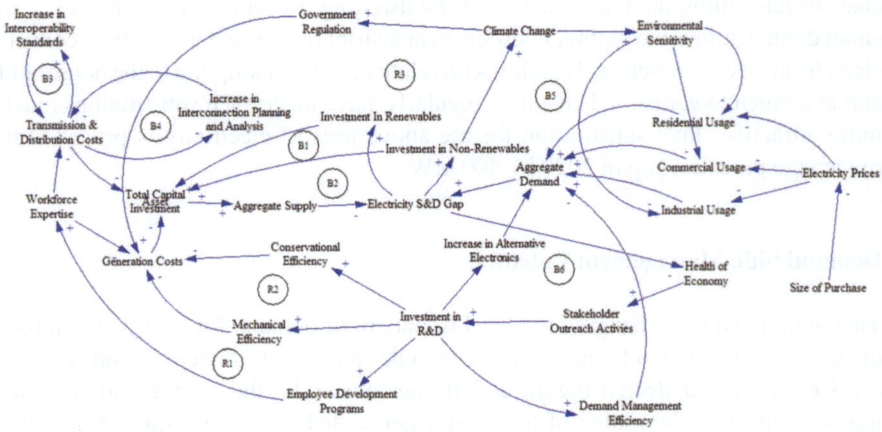

Fig. 1 Dynamic hypothesis. *Ri* represents reenforcing or positive feedback loops and *Bi* represents balancing or negative feedback loops (for details, please see in Sterman (2000))

Dynamic Hypothesis

Based on the comprehensive review of factors and policies on both the demand side and the supply side of electricity generation sector of Canada, we postulate a dynamic hypothesis, given in Fig. 1.

Figure 1 aims to describe the cause and effect relationships within our system. Below, we will identify and describe each loop in our diagram.

Investment in Renewables Loop Beginning with a large supply and demand gap (S&D gap), we notice that a large investment in renewable energy takes place. Furthermore, the higher the investment in renewable energy is, the higher the total investment in energy capital assets. The higher the investment in capital assets, the higher our aggregate supply will be, and the smaller our gap between supply and demand will be.

Investment in Nonrenewables Loop Beginning with a large S&D gap, we notice that a large investment in nonrenewable energy takes place. Furthermore, the higher the investment in nonrenewable energy is, the higher the total investment in energy capital assets. The higher the investment in capital assets, the higher our supply will be, and the smaller our gap between supply and demand will be.

Interoperability Standards Loop The higher our costs, the higher our increase in interoperability standards will be. This will result in cost reductions. As our costs decrease, the marginal benefit of increasing interoperability standards will decrease, and the rate of increase in interoperability standards will decrease as well, as its opportunity cost increases.

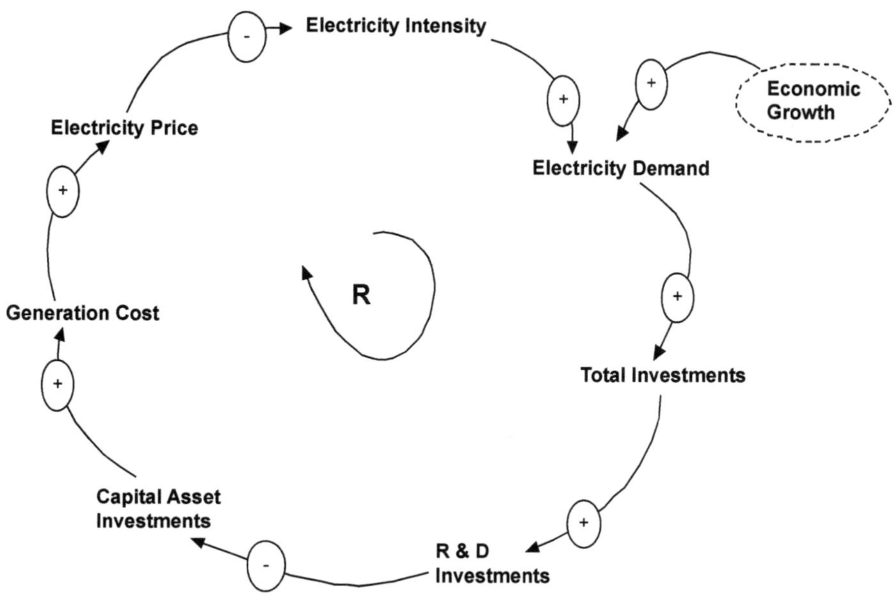

Fig. 2 Causal loop diagram of electricity pricing feedback loop

Interconnection Planning and Analysis (IP&A) Loop The higher our costs, the higher an increase in IP&A will be. This will result in cost reductions. As our costs decrease, the marginal benefit of increasing IP&A will decrease, and the rate of increase in interoperability standards will decrease as well as its opportunity cost increases.

Environmental Sensitivity Loop The higher our S&D gap, the higher our investment in nonrenewables will be. This will result in an increase in climate change, and as climate change becomes more noticeable, environmental sensitivity will increase. Environmental sensitivity will decrease our propensity to use electricity throughout the industry as a whole. Aggregate demand will decrease as a result, and our gap will narrow down.

Alternative Electronics Loop The smaller our S&D gap, the higher the health of our economy will be. Economic prosperity will foster an environment for stakeholder outreach activities, which in turn will increase investment in research and development. As a result of the increased research and development, our rate of increase in alternative electronics and equipment will increase, effectively reducing our aggregate demand through efficiency improvement. This in turn will narrow the S&D gap.

Employee Development Loop The higher our investment in R&D, the more employee development programs we will have. As a result, workforce expertise will increase, decreasing our generation costs, and increasing our capital asset investment. As a result, aggregate supply will increase, decreasing our gap, eventually leading to further increases in R&D, and more employee development programs. Figure 2 presents the electricity pricing loop explicitly.

Efficiency Loop The higher our investment in R&D, the higher both our conservational and mechanical efficiencies will be, decreasing our generation costs and increasing our total capital asset investment as a result. Aggregate supply will increase, decreasing our S&D gap, eventually leading to further investments in R&D, perpetuating an increasing cycle.

Government Regulation Loop The higher the climate change, the more government regulation we will have. This will increase our generation costs and decrease our capital asset investments, and furthermore our supply, increasing our gap. An increase in our gap results in further investment in substitute nonrenewables, which will continue climate change, increasing government regulation further later on.

Development of the Dynamic Model

Model Assumptions

In Ontario alone, the gap between supply and demand for electricity is expected to reach 15,000 MW by 2020. In order to get the total gap between supply and demand in Canada as a whole, we will use a relative value calculation. Ontario generates 26 % of total capacity in Canada. Assuming the gap in capacity will be a constant percentage throughout the country, we face a total gap of approximately 57,700 MW. This represents a 46 % requirement for capacity increase. We will also make the assumption that the % deficit in total capacity may be applied to each source of generation in the same manner. Therefore, the gap in hydro production will be 33,780 MW. The gap attributed to nuclear capacity will be 8,915 MW. The gap attributed to coal will be 10,401 MW. The gap attributed to natural gas will be 2,800 MW. Finally, the gap attributed to crude oil will be 1,314 MW (percent of total capacity by individual source provided by Statistics Canada).

In 2000, Canada had a total installed capacity of 111,000 MW. Today, capacity is 12 % higher, at 124,240 MW (Statistic Canada 2011). Because throughout the first decade of the new millennium capacity has grown in similar proportions, we will attribute this 12 % gain to each source in the same manner, to isolate for individual capacity growth. Between 2000 and now, hydroelectricity has therefore experienced an increase in capacity of 12,216 MW. Nuclear capacity experienced an increase of 2,077 MW (WNA 1996). Coal capacity experienced a growth of 2,422 MW. Natural gas experienced a growth of 652 MW. Finally, crude oil capacity experienced a growth of 306 MW. Accordingly, the growth rates per year are 2,036 MW/year for hydroelectricity, 346 MW/year for nuclear energy, 404 MW/year for coal energy, 109 MW/year for natural gas energy, and 51 MW/year for crude oil energy (WNA 2011).

As noted above, our forecasted deficit in electricity capacity will be 57,700 MW by 2025. For the purpose of this paper, we used Canada's 2006 capacity of 124,240 MW. Accordingly, in order to avoid a deficit in the year 2025, Canada's aggregate capacity needs to be 181,940 MW.

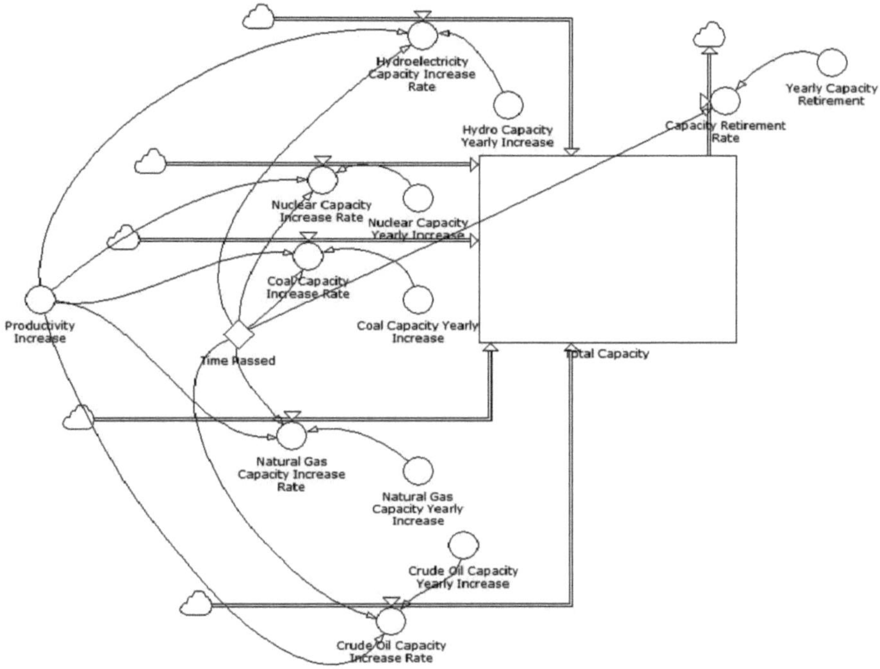

Fig. 3 The stock and flow structure of the dynamic model

Stock and Flow Structure of the Dynamic Model

The stock and flow structure of our dynamic model is presented in Fig. 3.[1] This model aims to assess various outcomes according to changes in rates of change in capital assets within the system.

Total capacity is a stock that has an initial value of 124,240 MW, Canada's total capacity to produce electricity in 2000. It is the aggregate capacity of the above 5 major sources for electricity production. This stock is increased by the flows representing rates of change of capacity by each individual source. Hydroelectricity capacity increase for example, is a function of the historical yearly change in capacity, productivity increases, and the time elapsed, which is set to one year, as all rates of change are assessed on a yearly basis. The same formula is applied to each of the five sources of energy. The auxiliary "productivity increase" represents a 0.3 % increase in usage efficiency, decreasing the amount that is required to be supplied (Canada's sector council program powering up the future 2008). Thus, all rates are multiplied by 99.7 %. Draining the stock is the decrease in capacity caused by asset retirement rate. This rate was attained by multiplying the retirement rate of 5,750

[1] Interested reader can contact the author for mathematical equations of this dynamic model.

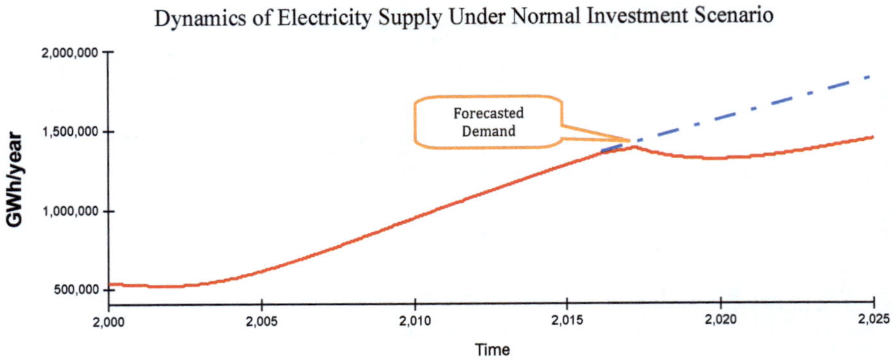

Fig. 4 Dynamics of electricity supply under status quo scenario

GWh worth of production per year by the average utility efficiency rate of 35.6 (NPC 2007; WNA 1996).

Validation of the Dynamic Model

System dynamics models are causal models (Barlas 1989). The essence of system dynamics modeling lies in identifying how the structure and decision policies help generate the observable patterns of behaviors of a system. Therefore, both the structural and behavioral validity procedures constitute the core of validation process for any system dynamics model. Our developed model was successfully exposed to both the structural and behavior validity procedures (for details on these tests, please see (Qudrat-Ullah 2012)).

Results

Capacity Under Status Quo Scenario

Our first simulation attempts to model Canada's electricity generation capacity if it continues to change at the current rate, as shown in Fig. 4. Involved in this simulation are the growth rates of capacity by each industry according to historical trends. We have only included major sources of electricity in this model, accounting for over 97 % of total capacity. The omitted sources are negligible at this time. According to this initial simulation, capacity will reach a total of 149,396 MW by 2025. This is 32,544 MW short of covering the forecasted gap. Therefore, maintaining current policy will not help in achieving the goal of having a balanced and sustainable supply and demand system of electricity in Canada.

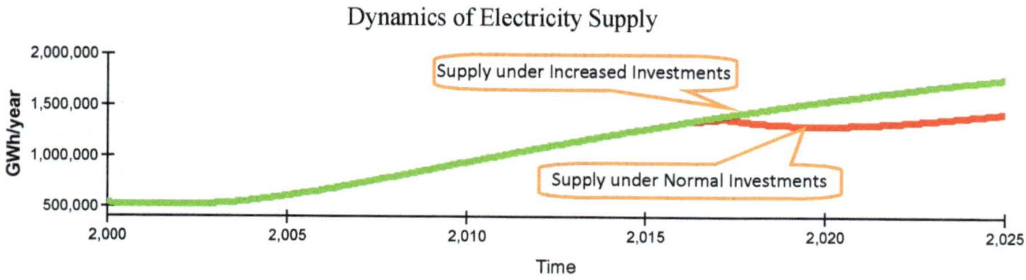

Fig. 5 Electricity Supply Dynamics

New Investments-Based Scenario

Our second simulation, however, involves the recommended US$ 95 billion investment in generation, US$ 27 billion investment in transmission, and US$ 63 billion investment in distribution (Canada's sector council program powering up the future 2008), as shown in Fig. 5. With this investment, capacity growth rate is expected to rise further by 58 % per year (IFC Consulting 2006). The adjusted rates of growth would therefore be 3,217 MW/year for hydro, 547 MW/year for nuclear, 638 MW/year for coal, 172 MW/year for natural gas, and 81 MW/year for crude oil. This policy is forecasted to close the projected gap between supply and demand for electricity. Accordingly, our model displays that according to this growth rate, capacity by 2025 will be 181,867 MW. Therefore, under this scenario, we can see the possibility of achieving the goal of having a balanced and sustainable electricity supply and demand system in Canada.

Concluding Remarks

With our theoretical review, dynamic hypothesis, and simulation model-based scenarios, we have attempted to explain the dynamics of variables acting within the electricity supply and demand system of Canada. Specifically, we have looked at variables within our generation capacity system. The key to the avoidance of a gap between electricity supply and demand, as well as sustainable, safe, and cost-competitive production, is to take advantage of the identified factors and potential policy decisions. In addressing our current supply and demand gap issue, we must not only continue to invest in capital assets for electricity production, but also continue our increased investments in R&D and productivity initiatives. Demand management and reduction, as well as production and end use machinery efficiency, play prominent roles in maintaining stability throughout the system. Canada must be prepared to diverge from traditional adjustment methods and adopt new strategies focused on *capital assets, productivity, and efficiency* in order to avoid a downward spiral of electricity industry deficiency.

As per our model-based analysis, an additional investment of about US$ 10 billion over a decade (2015–2025) will not only allow Canada to effectively close the supply and demand gap but also in a relatively greener way. With these additional investments, Canadian economy can also expect better energy intensity (0.21 versus 0.25 toe/million US$). This will result in wider recognition of Canada as a green economy (Conference Board of Canada 2010; Qudrat-Ullah, 2013).

By utilizing our developed simulation model, future research can investigate other related issues in the context of alternative policy design for Canadian electricity sector. For instance, in the identified capacity-mix, which capital asset should be preferred the most to support low-carbon economic regime. Our developed model is flexible enough to be adapted to model and analyze such issues. Therefore, besides providing useful policy insights on electricity generation capacity dynamics in Canada, this research contributes with an effective policy analysis and design tool in the form of a unique system dynamics-based simulation model. Researchers can calibrate the developed model to their case-specific data and can perform desired scenario-based analysis.

Acknowledgments The author acknowledges the contribution of his students Vald, Karim, and Yogesh. Especially in the earlier version of literature review, a part of this paper is based on their group project under the author's supervision.

References

Adelino, P., João, S.: Generation expansion planning (GEP)—A long-term approach using system dynamics and genetic algorithms (GAs). Energy **36**, 5180–5199 (2011)

Anand, S., Vrat, P., Dahiya, R.P.: Application of a system dynamics approach for assessment and mitigation of CO_2 emissions from the cement industry. J. Environ. Manag. **79**, 83–398 (2005)

Ansari, N., Seifi, A.: A system dynamics analysis of energy consumption and corrective policies in Iranian iron and steel industry. Energy **43**, 334–343 (2012)

Assili, M., Javidi, H., Ghazi, R.: An improved mechanism for capacity payment based on system dynamics modeling for investment planning in competitive electricity environment. Energy Policy **36**(10), 3703–3713 (2008)

Barlas, Y.: Multiple tests for validation of system dynamics type of simulation models. Eur. J. Oper. Res. **42**(1), 59–87 (1989)

Bun, D., Larsen, E.: Sensitivity reserve margin to factors influencing investment behavior in the electricity market of England and Wales. Energy Policy **29**, 420–429 (1992)

Canada's sector council program powering up the future. Labour market information study. http://www.brightfutures.ca/lmi/etc/en/docs/ESC%20LMI%20Report%20FINAL%20HR.PDF (2008). Accessed 01 Dec 2011

Cappers, P., Goldman, C., Kathan, D.: Demand response in U.S. electricity markets: empirical evidence. Energy **35**(4), 1526–1535 (2010)

Centre for Energy: Electricity transmission in Canada. http://www.centreforenergy.com/AboutEnergy/Electricity/Transmission/Overview.asp?page=6 (2009). Accessed 13 Nov 2012

CIA: The world factbook. https://www.cia.gov/library/publications/the-world-factbook/geos/ca.html (2011). Accessed 22 Nov 2011

Conference Board of Canada: Energy intensity. http://www.conferenceboard.ca/hcp/details/environment/energy-intensity.aspx (2010). Accessed 02 May 2013

Dyner, I., Bunn, D. W.: A system simulation platform to support energy policy in Columbia. Syst. Model. Energy Policy. Chichester: Wiley (1997)

Electricity generation, by utilities, by source, 2005. Table 5. http://www.statcan.gc.ca/pub/16-002-x/2007002/t/4129477-eng.htm (2005). Accessed 02 Dec 2011

Ford, A.: Using simulation for policy evaluation in the electric utility industry. Simulation **40**(3), 85–92 (1983)

Ford, A.: System dynamics and the electric power industry. Syst. Dyn. Rev. **13**, 57–85 (1997)

Ford, A., Bull, M.: Using system dynamics for conservation policy analysis in the Pacific Northwest. Syst. Dyn. Rev. **15**(1), 1–16 (1989)

Government of Canada: Canada—U.S. energy relations. http://www.canadainternational.gc.ca/washington/bilat_can/energy-energie.aspx?lang=eng (2011). Accessed 22 Nov 2011

IEEE Xplore: The United States of storage (electric energy storage). http://ieeexplore.ieee.org/xpl/freeabs_all.jsp?arnumber=1405868 (2011). Accessed 01 Dec 2011

IFC Consulting: The electricity supply/demand gap and the role of efficiency and renewables in Ontario. http://www.pollutionprobe.org/old_files/Reports/elec_supplydemandICF.pdf (2006). Accessed 01 Dec 2012

Kilanc, P., Or, I.: A decision support tool for the analysis of pricing, investment and regulatory processes in a decentralized electricity market. Energy Policy **36**(8), 3036–3044 (2008)

Naill, R.: A system dynamics model for natural energy policy planning. Syst. Dyn. Rev. **8**(1), 1–19 (1992)

NEB: Statistics—Estimated production of Canadian crude oil and equivalent. http://www.neb.gc.ca/clf-nsi/rnrgynfmtn/sttstc/crdlndptrlmprdct/stmtdprdctn-eng.html (2011). Accessed 02 Dec 2011

NPC (National petroleum council): Electricity generation efficiency, power generation efficiency subgroup of the demand task group of the NPC committee on global oil and gas (2007). Accessed 06 Dec 2011

NRC (Natural resources Canada): The atlas of Canada—Crude oil and natural gas resources. http://atlas.nrcan.gc.ca/site/english/maps/economic/energy/oilgas/1 (2009). Accessed 21 Nov 2011

Ochoa, P.: Policy changes in the Swiss electricity market: a system dynamics analysis of likely market responses. Socio-Econ. Plan. Sci. **41**(4), 336–349 (2007)

Olaya, Y., Dyner, I.: Modeling for policy assessment in the natural gas industry. J. Oper. Res. Soc. **56**(11), 22–31 (2005)

Park, J.Y., et al.: Investment incentives in the Korean electricity market. Energy Policy **35**(11), 5819–5828 (2007)

Qudrat-Ullah, H.: On the validation of system dynamics type simulation models. Telecommun. Syst. **51**(2–3), 159–166 (2012)

Qudrat-Ullah, H., BaekSeo, S.: How to do structural validity of a system dynamics type simulation model: the case of an energy policy model. Energy Policy **38**(5), 2216–2224 (2010)

Qudrat-Ullah, H., Davidsen, P.: Understanding the dynamics of electricity supply, resources and pollution: Pakistan's case. Energy **26**(6), 595–606 (2001)

Qudrat-Ullah, H., Karakul, M.: Modeling for policy assessment in the electricity supply sector of Pakistan. Int. J. Energy Sect. Manag. **3**(1), 240–256 (2007)

Qudrat-Ullah H.: Understanding the dynamics of electricity generation capacity in Canada: A system dynamics approach. Energy **59**, 285-294 (2013)

Statistics Canada: Electric power generation, transmission and distribution. http://www.statcan.gc.ca/pub/57-202-x/57-202-x2007000-eng.pdf (2007). Accessed 01 Dec 2011

Statistics Canada: Canadian coal consumption. http://www.coal.ca/content/attachments/article/62/Coal%20Consumption.pdf (2011). Accessed 29 Nov 2011

Sterman, J.: Business Dynamics: Systems Thinking and Modeling for a Complex World. McGraw-Hill, Boston (2000)

Stone, K.: Overview of Canada's coal sector, natural resources Canada. http://www.asiapacificpartnership.org/pdf/Coalmining/expo_vegas/Overview_Canada_Coal_Sector.pdf (2008). Accessed 13 Nov 2012

U.S.D. of Energy: Smart grid. http://energy.gov/oe/technology-development/smart-grid (2009). Accessed 12 Nov 2012

WNA (World Nuclear Association): Nuclear power reactors. http://www.world-nuclear.org/info/inf32.html (1996). Accessed 15 Nov 2011

WNA (World Nuclear Association): Uranium in Canada | Canadian uranium production. http://world-nuclear.org/info/inf49.html (2011). Accessed 01 Dec 2011

Working document of the NPC global oil & gas supply. Electricity generation efficiency. http://www.npc.org/Study_Topic_Papers/4-DTG-ElectricEfficiency.pdf (2007). Accessed 26 Nov 2011

Adoption of Renewable Energy Technologies: A Fuzzy System Dynamics Perspective

Michael Mutingi

Introduction

In the presence of fuzzy or linguistic and dynamic variables, system dynamics modeling of the adoption of renewable energy technologies (RETs) is inherently complex. In the past decade, academicians and practitioners have witnessed increased growth in systems dynamics-based and scenario-based analysis in the area of energy policy design and evaluation. Intriguingly, so much of research efforts are increasingly directed toward energy-economy interactions and impacts on a global level (Naill 1992), country level level (Roger et al. 1990; Naill 1992; John et al. 1998; Qudrat-Ullah and Davidsen 2001; Chandrasekara and Tara 2007; Han and Hayashi 2008; Qudrat-Ullah and Seong 2010), and at island level (Uemura et al. 2003; Chen et al. 2007; Demiroren and Yilmaz 2010; Trappey et al. 2012). However, the major challenge in modeling real-world systems is the existence of linguistic or fuzzy variables which often confront the decision maker (Levary 1990; Tessem and Davidsen 1994; Kikuchi 2005; Kosko 1995). For instance, key variables in the RET marketplace are oftentimes perceived and expressed as "low", or "high", or "somewhat high", among others. These tendencies are of common occurrence in human systems whose behavior is a result of informal models, called mental models, created by the human mind. Mental models influence the adoption or rejection of technology innovations, which ultimately affects how the policy maker will formulate energy policies. In such fuzzy environments with linguistic time-dependent variables, the right policies and the impacts of possible actions are not precisely known (Mutingi and Mbohwa 2012).

Further to the presence of uncertain variables in the marketplace, organizations and policy makers face serious problems in understanding the dynamics of the adoption of RETs, especially in a complex business arena consisting of RET adopters who promote technology adoption, inhibitors who oppose technology adoption, imitators whose decisions are largely influenced positively or negatively by adopters

M. Mutingi (✉)
School of Mechanical and Industrial Engineering, University of Johannesburg,
Bunting Road Campus, PO Box 524, Auckland Park, Johannesburg 2006, South Africa
e-mail: mmutingi@gmail.com

H. Qudrat-Ullah (ed.), *Energy Policy Modeling in the 21st Century*,
Understanding Complex Systems, DOI 10.1007/978-1-4614-8606-0_10,

and inhibitors, and various other conflicting technology policies (Chen 2011; Van den Bulte and Joshi 2007). Technology inhibitors, particularly in the renewable energy sector, often seek to increase the number of opponents to renewable energy innovation. In so doing, they hold back the spread of innovation, or even try to stop its diffusion. Word of mouth (WOM), a customer-to-customer communication at individual or firm level about the potential characteristics of a product plays a critical role in reducing the risk and uncertainty in technology purchase and adoption (Krushner and Leif 2008; Mutingi and Matope 2013; Wong and Sheng 2012). It is by word of mouth, through various media, that renewable energy information seekers obtain knowledge which eventually influences their adoption behavior. In such a complex market environment with multiple players, it is crucial to obtain in-depth understanding of the dynamics of the recursive relationship between the adopters' decision to adopt a product and how that adoption experience may trigger further contribution to word of mouth, which will again impact other potential adopters in a negative or positive manner (Cavusoglu et al. 2010; Mutingi and Matope 2013). In such environments, developing and evaluating appropriate RET policies is a difficult but important undertaking.

Not only is the renewable energy marketplace characterized with fuzzy linguistic variables and complex interactions of multiple players but also the dynamic complexities caused by nonlinear interactions, time delays, and information feedbacks. These complex dynamic complexities are difficult to model in a closed form. This suggests that a systems-based model that takes into account the existence of fuzzy variables is most appropriate, from a causal loop perspective.

Critical questions arising from the above issues include the following: How best can we represent fuzzy variables in the real world marketplace? How does the existence of multiple players in the renewable energy arena affect its diffusion and adoption? How best can we represent the interactive dynamics from a systems perspective? How can a policy maker utilize a systems model to design and evaluate renewable energy policies? This research work seeks to cover this void.

In light of the above mentioned complexities, the primary aim of this chapter is to disseminate the roles and application of fuzzy system dynamics approach aimed at improving the usefulness of energy policy system models in the real world where linguistic variables are commonly used. Fuzzy logic and system dynamics concepts are integrated from a systems perspective in order to model typical world energy-economy scenarios. The key focus of the fuzzy system dynamics paradigm is to enable decision makers to develop system models that can capture the imprecise fuzzy variables in human systems by hybridizing fuzzy logic and system dynamics paradigms. The paradigm is vital for RET policy design and assessment.

In pursuit of the current research purpose, the following objectives are adopted in this chapter:

1. To identify the complex dynamic factors and variables characterizing the RET adoption process;
2. To develop a causal loop analysis for the RET adoption process, taking into account the fuzzy variables necessary for RET policy formulation;

3. To develop a fuzzy system dynamics model that captures the fuzzy variables and relationships between the key factors;
4. To carry out simulation analyses, providing useful managerial insights based on "what-if" experiments.

The rest of this chapter is structured as follows: Section "Literature Review" provides a review of the related literature, covering renewable energy adoption, system dynamics simulation, and fuzzy logic systems. Section "System Dynamics Applications" provides a brief background to the system dynamics methodology. The next Section "A Fuzzy System Dynamics Model" describes the proposed fuzzy system dynamics model. Section "Simulation Experiments" presents a set of simulation experiments for the study. This is followed by simulation results and relevant discussions in Section "Simulation Results and Discussion". Finally, concluding remarks are presented in Section "Conclusions and Further Research", providing useful managerial insights and further research prospects.

Literature Review

Renewable energy is the energy that comes from natural resources, such as sunlight, wind, rain, tides, waves, and geothermal heat. The adoption of RETs, such as solar energy, wind, and biomass has significant positive impact on the environment (Peter 2010; Trappey et al. 2012). Because of their environmental friendliness, policy makers concerned with renewable energy policies are concerned with improving the momentum of diffusion and adoption of RETs (Huang 2009). Their application has been widespread across different disciplines, such as power generation, transportation fuel, and rural household energy supply (Qudrat-Ullar and Davidsen 2001). Further to that, RET is of critical importance for a sustainable ecological environment (Qudrat-Ullah and Seong 2010; Trappey et al. 2012).

Adoption of Renewable Energy Technology

Technology diffusion and adoption is an important concept in the environment and the modern day industry. In accordance with Rogers (2003, p. 177), adoption is defined as a decisive action taken by a customer to "make full use of an innovation as the best course of action available." In addition, rejection (or decline) is a decision "not to adopt an innovation". On the other hand, Rogers (2003, p. 5) defines diffusion as a process in which "an innovation is communicated thorough certain channels over time among the members of a social system". In this regard, participants in the market create and share information with one another so as to reach a mutual understanding before the actual adoption of the technology innovation. However, in actual fact, a number of barriers have been experienced in the RET adoption process due to the presence of barriers, inhibitors, and other adoption related factors. In this

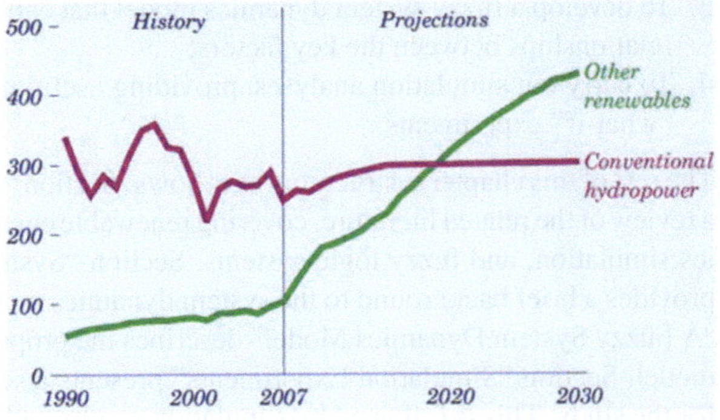

Fig. 1 Grid connected electricity generation from renewable energy sources, 1990–2030. (Billion kilowatts; EIA 2009)

vein, several researchers have investigated the possible inhibiting factors impacting the adoption of renewable energies (John and William 1998; Uemura et al. 2003; Raja et al. 2006; Chandraseka and Tara 2007; Vicki and Tomas 2008; Li et al. 2009; Huang 2009; Ernest and Mathew 2009; Mutingi and Matope 2013).

In real life, the dynamics of RET adoption comes as a result of complex interactions and feedbacks within a fuzzy dynamic environment. In other words, RET adoption is often associated with complex interactions and feedback mechanics between technology providers, individuals, organizations, policy makers, and other stakeholders. We envisage an adoption process in which a potential adopter goes through identifiable phases: preliminary knowledge dissemination, adopter attitude or perception, technology take up or rejection, technology practice, and finally confirmation of technology adoption confirming the adoption or usage. From this analysis, we see that the adoption process is composed of a series of dynamic decisions and actions: takeup, practice, adoption, and decline.

The adoption of RET generally follows the well-known Bass model (Bass 1969). In practice, technology adoption is expected to pass through four phases, namely; (i) introduction phase, in which the technology is launched into the market, (ii) growth phase, which is characterized by rapid exponential adoption of the technology, (iii) maturity phase, where technology reaches its maximum expected adoption, and (iv) decline phase, which refers to when the technology loses popularity and some customers discontinue using the technology. As such, the overall technology adoption process follows an S-shaped trend. This hypothesis is adopted in this chapter.

Figure 1 shows a real life example of an S-shaped graph of renewable electricity energy adoption as reported by the USA Energy Information Administration (EIA 2009). The graphical trends are based on the historical data from 1990 to 2007, and on projections thereafter. A closer look at the trend from 1990 to 2007 shows a prolonged delay in the adoption process. According to the report, the stagnation of the adoption of "other renewables" in the period was attributed to the need for robust renewable energy policies, state support, technology improvements, and other public environmental concerns. The presence of resistances from the market, inhibitors, and lack of supporting incentives discourage technology adoption in the real world marketplace, specifically power generation from wind and solar energy sources. To understand the complex dynamics behind the adoption of renewable energy, the adoption process should be modeled based on system models. System dynamics methodology offers a great potential in modeling the diffusion and adoption of renewable energy technologies.

System Dynamics (SD) Applications

System Dynamics is frequently used to analyze environmental policies and the ensuing environmental impact (Ford 1997; Trappey et al. 2011, 2012). The simulation methodology can be used to analyze complex systems with a focus on policy design and analysis (Forester 1961). One of the most famous applications of SD is the classical work on “The Limits to Growth” by Meadows et al. (2004). Following this work, Wang et al. (2008) developed a system dynamics simulation model for urban transportation systems based on cause-and-effect-analysis and feedback loop analysis technique. Furthermore, Jin et al. (2009) proposed a remarkable model for dynamic ecological footprint forecasting to support policy formulation and evaluation for improving urban sustainability. In the same vein, Han and Hayashi (2008) used system dynamics simulation to assess CO_2 policy mitigation for an intercity passenger transport network in China. Related studies also exist in Qudrat-Ullah and Seong (2010).

As can be seen from the above review of related literature, an appreciable number of system dynamics models have been applied to assess environmental impact and to formulate useful policies. System dynamics has also been applied to a number of practical problems, such as corporate planning, supply chain management, public management, economic behavior, healthcare modeling, and new product development (Morecroft 2007; Sterman 2004; Quadrat-Ullah 2005; Rodrigues and Dharmaraj 2006; Mutingi and Mbohwa 2012; Mutingi 2012; Reddi and Moon 2011). Considering its widespread application across various disciplines, SD is a potential tool for modeling and understanding the complex dynamics of the RET adoption process. However, to model the imprecise fuzzy variables involved in energy adoption, the use of fuzzy logic is imperative.

Fuzzy Logic System

A fuzzy logic system is a logical system that is based on the theory of fuzzy sets. A fuzzy set relates to classes of objects with imprecise boundaries in which membership is a matter of degree. The main part of a fuzzy logic system is a set of rules that converts the given inputs to outputs based on the theory of fuzzy sets (Zadeh 1978; Kosko 1992a, 1994). In practice, fuzzy approximation theorem is used to implement a fuzzy logic system (Kosko 1992b). Basically, the inputs to the system is the perceived information that relates to the state of the system, while the output is a specification of the decision or action to be taken. A fuzzy logic system incorporates a rule-base that contains a set of "if then else" rules of the form:

$$\text{IF } x \text{ is } A \text{ THEN } y \text{ is } B \tag{1}$$

where, A and B are linguistic values defined by fuzzy sets on the ranges (universes of discourse) X and Y, respectively.

In accordance with the fuzzy logic concepts, "x is A" is known as the antecedent, while "y is B" is called the consequent. This setting provides strong constructs for fuzzy inference processes. Fuzzy inference is the process of formulating the mapping from a given input to an output based on some fuzzy logic set of rules (Sugeno 1985; Mamdani 1975). This kind of mapping provides a basis from which decisions can be made based on a set of linguistic control rules obtained from expert policy makers. The process of fuzzy inference involves three main constructs, that is, membership functions, logical operations, as well as if–then rules. In addition, the fuzzy inference process involves crisp (non-fuzzy) inputs, linguistic (fuzzy) rules, the defuzzifier, and the crisp output.

A fuzzy logic system is built on top of the experience of experienced experts who already have prior understanding of the system under investigation. It is built on the structures of qualitative description used in everyday natural language, which makes it easy to use. This approach is necessary because most real life systems do not have enough precise data to allow statistical analysis which normally demand data collection over a long time. Being tolerant of imprecise data, a fuzzy logic system builds this understanding into the process rather than tacking it onto the end. Furthermore, fuzzy logic can model nonlinear functions of arbitrary complexity. In general, a fuzzy logic system can be defined in three steps: fuzzification, fuzzy rules, and defuzzification (Labibi et al. 1998).

A Fuzzy System Dynamics Model

This section presents a fuzzy system dynamics model developed for a typical adoption process of a renewable energy marketplace in a fuzzy environment. The renewable energy system is first conceptualized from a causal loop analysis perspective. A dynamic model is then developed as a control theoretic model in a Matlab environment.

Causal Loop Diagram

Figure 2 represents a high-level causal-loop diagram in a more aggregate form, indicating the dynamic interactions between adoption and word-of-mouth. Three major loops are realized. The loop on the extreme left, denoted by "B1," is a balancing loop that represents the impact of adopters on the adoption rate. In turn, the adoption rate depletes the potential adopters. In essence, this loop explains the concept of

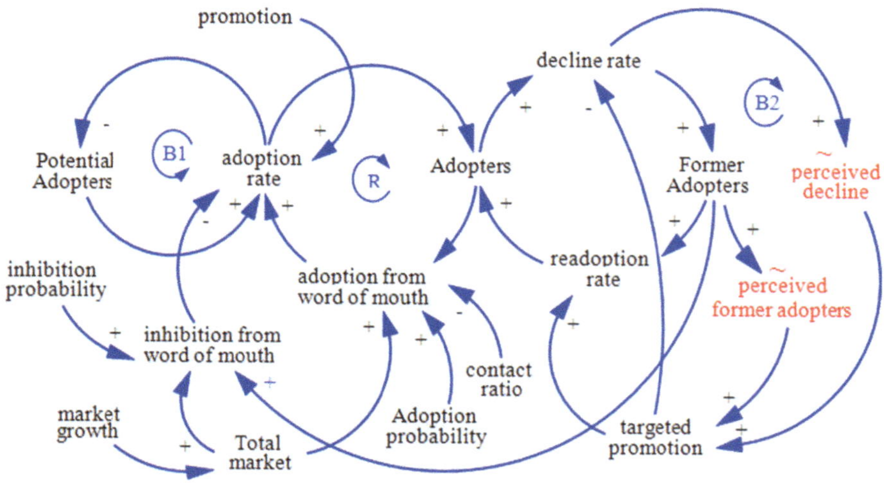

Fig. 2 Causal loop analysis for renewable energy adoption in a fuzzy environment

market saturation. On the other hand, the loop labeled "R" is a reinforcing loop that depicts the impact of the adoption rate on the current adopter population in the marketplace. The current adopter population, in turn, exerts an exponential impact on the adoption rate itself. In actual fact, the loop explains the word-of-mouth concept, a mechanism by which information is spread among potential adopters and adopters, thereby accelerating the adoption process. However, contrary to most extant adoption models, renewable energy adoption is influenced by technology inhibition which discourages or even tries to stop its adoption.

In real life, adopters who adopt the RET may ultimately decline the use of that renewable energy due to various reasons. These former adopters or technology decliners often fight further adoption of the technology by potential adopters, depending on the effectiveness of their influence. We represent this effectiveness in terms of inhibition probability. In the case of targeted promotion, rigorous promotion is focused at former adopters so that the former adopters may re-adopt the technology, thereby increasing the adopters' population. Otherwise, the adopters' population will continue to decrease over time due to continued technology decline.

As shown by the loop on the extreme right, labeled "B2," policy makers depend on perceptions on market behavior in terms of (a) perceived former adopters and (b) perceived decline rate. Unfortunately, these market variables are controlled by linguistic values such as "low", "fairly high", or "high". Faced by the difficulty in incorporating these complex fuzzy variables in a conventional system model, the policy maker has to utilize fuzzy logic systems so as to represent these complex variables in a more realistic manner. Thus, based on the two fuzzy variables, expert knowledge can be used to compute the appropriate targeted promotion so as to influence re-adoption in a cautious way. With a judicious use of the perceived values of the variables, the effective decline can be put under control. The momentum of the adoption process can then be maintained at an acceptable level. Therefore, we simulate these concepts from a fuzzy system dynamics view point, using fuzzy logic tools coupled with Simulink in Matlab. The next section presents the fuzzy system dynamics model the formulation and evaluation of renewable energy policy.

Fuzzy System Dynamics Model

In this study, we propose that RET adoption follows an s-shaped or sigmoid pattern. Thus our hypothesis generally agrees with the technology diffusion and adoption (Rogers, 1995); howbeit from a more complex fuzzy system dynamics view point. In developing our fuzzy system dynamics model, we realize that the RET adoption process passes through four identifiable stages. As such, we identify four stocks, namely: Potential_Adopters, Semi_Adopters, Adopters, and Former_Adopters. We briefly describe these four stocks as follows:

- *Potential_Adopters*: individual customers who may adopt the RETs available in the market;
- *Semi_Adopters*: individual customers who initially take up the RETs, but are yet to confirm their actual adoption of the RET;
- *Adopters*: individual customers who eventually confirm the adoption of the RET;
- *Former_Adopters*: individuals who discontinue using the RETs after some period of usage and may decide to re-adopt the same technology in the foreseeable future, subject to policy incentives and policy controls.

The above mentioned stocks are influenced by their respective flow rates along the ageing chain, beginning from RET take-up, to adoption, and finally to decline or termination rates. These rates are briefly described as follows:

- *takeup*: the rate at which potential adopters primarily take up the RET, before the actual adoption;
- *adoption*: the rate at which the semi adopters eventually confirm their adoption of the RET;
- *termination*: the rate at which adopters decisively quit using the energy technology due to various constraints and barriers to continued use of the RET that they previously adopted;
- *re-adoption*: the rate at which former adopters re-adopt the RET due to policy incentives, policy controls, and other initiatives.

In conventional system dynamics modeling, the simplest possible model is always used to capture the mental models of policy makers who are experienced in the field of interest. On the contrary, fuzzy system dynamics goes a step further to represent linguistic variables and intelligent decision rules of experts, which allows for human judgment to be included into the system dynamics model. The concepts of fuzzy set theory and system dynamics modeling are applied to capture fuzzy variables and their dynamic interactions in the renewable energy marketplace. Figure 3 shows the

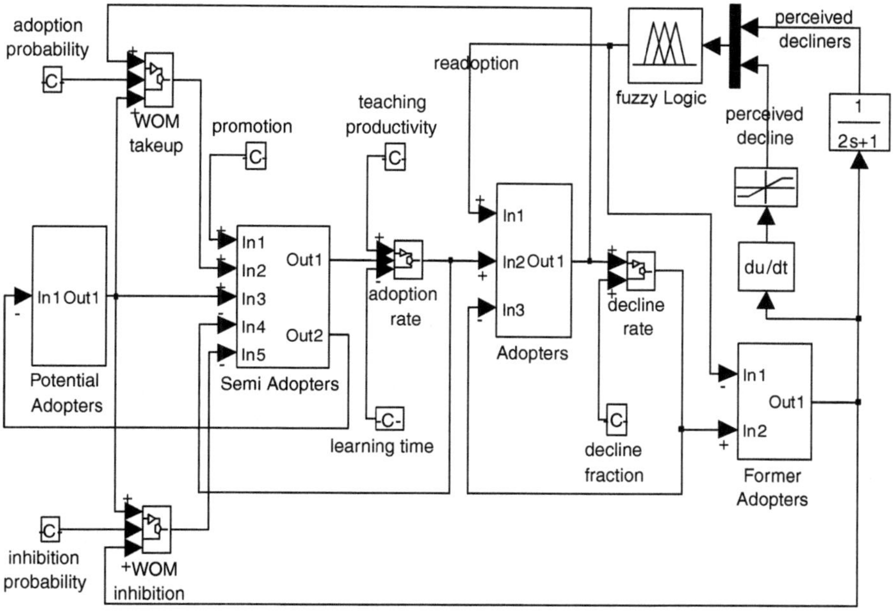

Fig. 3 Block diagram for renewable energy policy control in a fuzzy environment

overall block diagram for the system, derived from the causal feedback loop analysis diagram in Fig. 2.

To capture the mental models controlling the behavior of the adoption of RETs, two key input variables are defined. The first variable defines the magnitude of error between the preset acceptable number of former adopters and the perceived or observed number of former adopters. This implies that the observed number of former adopters should ideally be equal the acceptable number of former adopters. It follows that the preferred error should be as close to zero as possible. The second variable is defined by the current rate of change of the number of former adopters. In this connection, we define the error by the following equation,

$$error = \frac{Former\_Adopters}{Acceptable\_Former\_Adopters} - 1 \tag{2}$$

Since former adopters are supposed to be as close as possible to the acceptable former adopters, the range of values of the preferred error, error, is expected to be close to 1, and approximately in the range [−1, 1]. In this vein, error values close to zero are most preferable. The level of preference diminishes fast as the values of error approach 1 (or −1). In this connection, we represent this situation as a fuzzy set, or preferred error, as demonstrated in Fig. 4.

In addition to the *error* explained above, we define the rate of change of former adopters, *rate*, as follows:

$$rate = \frac{\mathrm{d}}{\mathrm{dt}}(Former\_Adopters) \tag{3}$$

Fig. 4 Fuzzy set (preferred error)

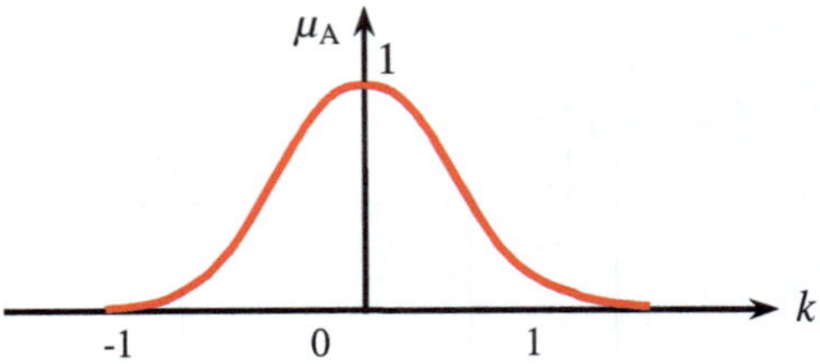

The *rate* variable, that is, perceived decline rate, defines whether the number of former adopters is increasing or decreasing. It follows that if the rate is increasing, then the corresponding energy policy initiatives should be amplified. On the other hand, if the rate is decreasing, then the desired policy efforts should be minimal. This then controls the investment, promotional, and other incentive-based policies that influence the re-adoption activities. Based on the fuzzy relationship depicted by the causal loop analysis in Section "A Fuzzy System Dynamics Model", a fuzzy rule base is constructed to represent the fuzzy policy design for renewable energy marketplace. As an illustration, let the variable *policy_amplification* denote the desired policy adjustment, and *error* represent the deviation from the acceptable quantity of former adopters. Then, a fuzzy rule base can be constructed in the format shown in Fig. 5.

With reference to Rule 1, the desired policy amplification is zero since the error is zero. As for Rule 2, the preferred error is low, which implies that the desired policy is to reduce the current policy amplification fast since the number of former adopters is much lower than the acceptable number. Conversely, when the preferred error is high, the implication is that the current policy amplification should be increased fast. Furthermore, if the error is "ok," that is, in the neighborhood of zero, then the consequent decision depends on whether the current trend (rate) of former adopters is increasing or decreasing. If rate is positive, then policy should ideally be reduced slowly. On the other hand, if rate is negative, then policy should be amplified slowly since the trend shows that former adopters are somewhat on the increase. All these fuzzy linguistic variables are coded using the fuzzy tool box and Simulink on a Matlab platform. As explained by Zadeh (1965, 1978), fuzzy set theory concepts are introduced to handle uncertain, fuzzy, or linguistic variables. A linguistic or fuzzy variable represents the ranges of values that the variable can take. For example, a fuzzy variable can take "high", "low", and "medium" values. The set of rules maps the input variable, or a combination of them, to a single output response variable. A membership function, in the form of a graphical representation, depicts the magnitude of participation of each input that associates a weighting with each of the inputs that are processed and defines the functional overlap between inputs (Retortillo et al. 2008). The advantage of using this approach is that the process of model building is fast, and fuzzy logic tools can represent real life scenarios.

In developing our fuzzy system dynamics model, we adopted and implemented the indirect structure test (Barlas 1996; Saysel and Barlas 2006). The indirect structure test involves carrying out simulation runs in order to provide information

Fuzzy rule base:

Rule 1: IF (error is ok) THEN (policy amplification is zero);

Rule 2: IF (error is low) THEN (policy amplification is negative fast);

Rule 3: IF (error is high) THEN (policy amplification is positive fast);

Rule 4: IF (error is ok) and (rate is positive) THEN (policy amplification is negative slow);

Rule 5: IF (error is ok) and (rate is negative) THEN (policy amplification is positive slow);

Fig. 5 Fuzzy rule base for renewable energy policy formulation

about possible flaws in the model structures. Thus, extreme condition and behavior sensitivity were applied.

- Structure verification: This tests whether the model structure is consistent with relevant descriptive knowledge of the system being modeled (Barlas 1996; Qudrat-Ullah and Davidsen 2001; Qudrat-Ullah and Seong 2010):
- Extreme conditions: This method tests whether the model exhibits a logical behavior when selected parameters are assigned extreme values (Barlas 1996; Sterman 2004).

The next section describes further simulation experiments. The results of the verification and further simulation experiments are illustrated in Section "Simulation Results and Discussion", deriving useful managerial insights.

Simulation Experiments

In this study, it is assumed that a total market size of 100 % is expected to adopt a RET, for instance, in power generation technologies. The simulation experiment is assumed to run over a period of 25 years. The modeler or policy maker can input policy parameter values according to his choices to represent the possible policy alternatives. The policy parameters are (i) promotion, and targeted promotion, which, in turn, is influenced by the modeler or policy maker's perceptions on former adopters' population and the decline rate. To run the model, the policy maker needs to input structural parameter values such as adoption fraction, inhibition fraction, decline fraction, and contact ratio. These parameters can be used for sensitivity analysis, to anticipate the energy system behavior in case of slight changes in the parameters. Initially, stocks of Potential Adopters and Former Adopters were set at 95 and 5 %, respectively. In addition, Semi Adopters and Adopters were both set at 0 %. Table 1 summarizes the input values that were used in the experiments.

In the simulation experiments, four scenarios were assumed for which four simulation experiments were undertaken, namely: (i) the base case, which is an ideal scenario, (ii) the real world scenario with technology inhibition due to presence of negative influence from former adopters, (iii) real world scenario with both technology

Table 1 Some input parameters and their assigned values

Model input	Assigned input value
Potential adopters	95 %
Semi adopters	0 %
Adopters	0 %
Former adopters	5 %
Usage period	5 years
Adoption fraction	0.1
Inhibition fraction	0.4
Decline fraction	0.05
Promotion	0.5
Contact ratio	0.6

inhibition and technology decline, and (iv) real world case with fuzzy policy control aimed at overcoming the effects of technology inhibition and decline. The respective detailed assumptions of these four scenarios are explained in the following sections.

Base Case

The base case scenario is borrowed from the well known Bass model that was first developed by Bass (1969). Deriving from the basic Bass model, the base scenario in this context reflects the most ideal case of the adoption process based on the following assumptions:

- Once potential adopters take up the available technology, they will not decline the use of the technology;
- There is no re-adoption of the technology, which follows from the previous assumption;
- There is no corrective policy control needed since the adoption process is assumed perfect.

However, in the real world, the presence of technology inhibitors should be taken into consideration.

Adoption With Decline

The scenario considers the presence of various barriers to continued use of the adopted technologies, which is a common occurrence in the RET marketplace. There is always a possibility that technology adopters will eventually decline the usage of the technology after a certain usage period. In this case, the following assumptions are considered:

- Potential adopters taking up the available technology will eventually decline the use of the technology;

- There is no re-adoption of the technology by the former adopters once they decline its usage;
- No corrective policy control is put into place as the adoption process is assumed perfect.
- The former adopters do not have any influence on the current potential technology adopters.

However, real world technology adoption is inundated with technology inhibitors who influence potential adopters or imitators whose behavior is largely influenced by word-of-mouth from other players in the RET marketplace. In the RET marketplace, former adopters influence the decisions of potential adopters who rely on the word-of-mouth from other players in the marketplace. Therefore, the influence of former adopters or technology inhibitors should be considered as well.

Adoption With Decline and Inhibition

This scenario assumes the real world case where the marketplace consists of potential adopters, adopters, and technology inhibitors who will always try to inhibit the adoption process. Therefore, in this scenario, we make the following assumptions:

- Potential adopters taking up the available technology will decline the use of the technology after some usage period;
- There is no re-adoption of the technology, once the adopter declines the use of the technology;
- Former technology adopters negatively influence the decisions of potential adopters or imitators;
- No corrective policy control is put into place to counter the effects of technology inhibition.

Nevertheless, in practice, policy makers desire to put in place robust dynamic policies that counter the negative impacts of technology inhibitor. At the same time, policy makers endeavor to put in place promotional policies that enhance RET adoption. The formulation of the dynamic policies depends on their perception of the variables that reflect the adoption process. Perceived decliners and perceived decline rate are the two fuzzy variables considered in this study.

Adoption With Fuzzy Policy Control

This is the most practical scenario where all the key influential factors are taken into account, that is, technology promotion, technology decline, technology inhibition, and policy control considering fuzzy dynamic factors: perceived decliners and perceived decline rate. In this chapter, we assume that the two fuzzy variables are defined by linguistic values in the domain "low", "ok", and "high." Effective

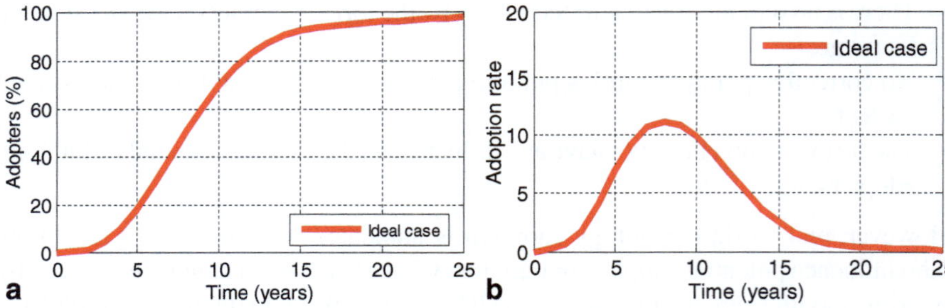

Fig. 6 RET adoption behavior for the base case scenario. **a** The adoption behavior in terms of quantity of RET adopters. **b** Behavior of the rate of adoption over the simulation period

dynamic policies can be formulated based on the dynamic linguistic values of the two fuzzy variables.

The next section presents the results of the four experiments outlined above; base case, adoption with decline, adoption with decline and inhibition, and adoption with policy control.

Simulation Results and Discussion

This section gives an analysis of the simulation results of the four scenarios described in the previous section, that is, the base case, adoption with decline, adoption with decline and inhibition, adoption with fuzzy policy control.

Base Case

Figure 6 illustrates the behavior of the RET adoption process for the base case scenario over a simulation period of 25 years. Figure 6a depicts the adoption behavior in terms of quantity of RET adopters, while Fig. 6b shows the behavior of the rate of adoption over the simulation period. As the RET is introduced at the beginning of the planning horizon, the quantity of adopters grows exponentially. However, as expected, the introduction phase begins with a short period of slow growth followed before exponential growth which characterizes the growth phase. The rapid growth is attributed to the dominant influence of the positive loop, as explained in the causal loop analysis in Section "A Fuzzy System Dynamics Model." However, the dominance of the positive loop diminishes over time as the time approaches 13 years, that is, when the adoption reaches the maturity phase. In this phase, the negative loop gains dominance over the positive loop and the exponential growth reduces to an asymptotic growth. The growth of RET adoption plateaus at the saturation point,

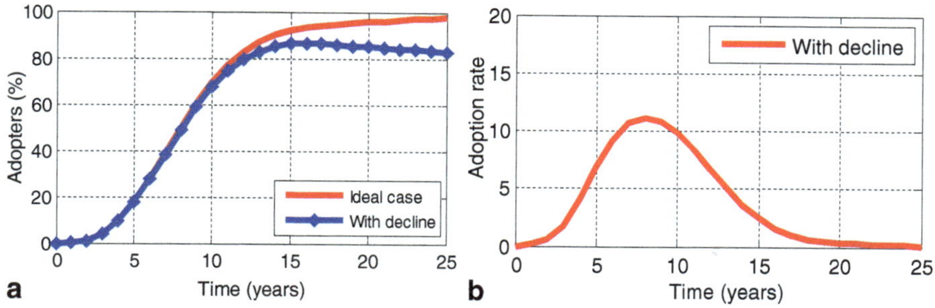

Fig. 7 RET adoption behavior with decline. **a** Variation of adopters. **b** Variation of the adoption rate over the simulation period

that is, at 100 %. As can be seen in Fig. 6b, the behavior of the adoption rate influences the behavior of the quantity of adopters. The rate of adoption grows slowly at the introduction phase, followed by rapid exponential growth, leading to the maturity phase. The peak of the adoption rate occurs at point (8, 11.26). In retrospect, the overall behavior of the adoption process is influenced by the change of dominance from the positive feedback to the balancing negative feedback.

However, in practice, the adoption process is characterized by decline as adopters stop the usage of the renewable energy for various reasons. The next section illustrates the simulation results taking into account the phenomenon of decline.

Adoption With Decline

Figure 7 demonstrates the adoption behavior considering the effects of decline. Figure 7a shows the variation of adopters, while Fig. 7(b) shows the variation of the adoption rate over the simulation period. The introduction phase in (a), bounded by the period from 0 to 5 years, is similar to the ideal (base) case in Fig. 6a. Howbeit, in the growth phase, the graph of adopters with decline gradually lags behind the ideal case over the simulation horizon. This can be explained in terms of the gradual increase of decline as adopters increasingly stop using the renewable energy due to barriers and constraints in the marketplace. This implies that in the absence of corrective policy controls, policy incentives, or incremental innovations, no re-adoption will take place, and the quantity of adopters will continue to decrease. As expected, the behavior of the adoption rate, as shown in Fig. 7b, is not different from the ideal case. This follows from the fact that the adoption rate is not affected by decline since decline can only take place after adoption.

In practice, the former adopters of renewable energy technologies tend to influence and inhibit the adoption behavior of potential adopters. Therefore, the overall behavior of adoption, defined in terms of adoption rate and quantity of adopters, is influenced by the presence of former adopters in the marketplace. The next section illustrates the most practical scenario, where decline and inhibition are taken into account, simultaneously.

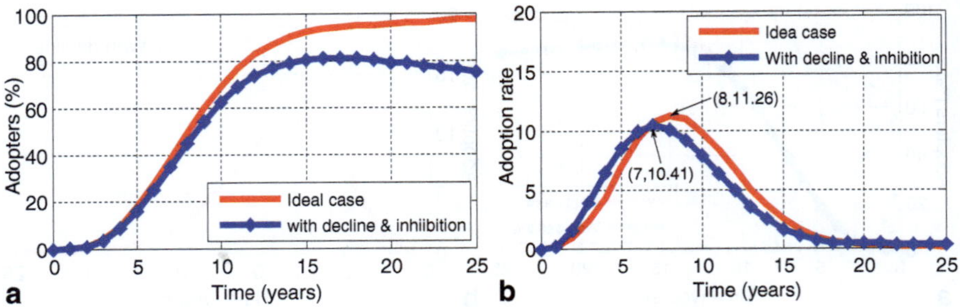

Fig. 8 RET adoption behavior with decline and inhibition. Behavior of the adoption process in terms of **a** quantity of adopters and **b** rate of adoption

Adoption With Decline and Inhibition

Figure 8 demonstrates the behavior of the adoption process in terms of quantity of adopters as in part (a) and rate of adoption as in part (b). According to Fig. 8a, the graph of adopters with decline and inhibition shows gradual lag in the introduction phase, contrary to the previous cases in which the graph of adopters showed no significant lag in the introduction phase. The presence of inhibitors influences the overall adoption process, right from the beginning. The impact of inhibition, by word-of-mouth, from former adopters continues to grow over the planning horizon as the number of former adopters increases. In comparison to the previous cases, that is, the ideal and the decline scenarios, the peak of the number of adopters is much lower. Therefore, the impact of decline and inhibition should be taken into account when designing and evaluating renewable energy policies, from a system dynamics point of view.

With regards to the adoption rate shown in Fig. 8b, the peak point for the ideal (base) case is at (8, 11, 26), while the peak point for the real world case is at (7, 10.41). It can be seen that the overall shape of the graph is much different from the ideal case, which demonstrates the impact of the presence of decline and inhibition in the renewable energy marketplace. Moreover, in comparison to the ideal case, the magnitude of peak adoption has reduced significantly, from 11.26 to 10.41. This essentially explains the apparent negative impact of decline-related factors and the influence of former adopters on the overall adoption process often observed in real life. Retrospectively, we note that decline and inhibition related factors affect the adoption process in three ways: (i) delaying the peak adoption time, (ii) reducing the magnitude of the peak adoption rate, and (iii) changing the entire shape of the adoption curve, which reflects the behavior of the adoption process. Further simulation analysis revealed the importance of the inhibition probability or inhibition fraction, which is a measure of the likely effect of the inhibitors on the adoption behavior of the imitators. It was noted from the study that when the inhibition probability is significant, opponents can delay the diffusion process significantly, though the inhibitors may be a small community. On the contrary, if the inhibition probability

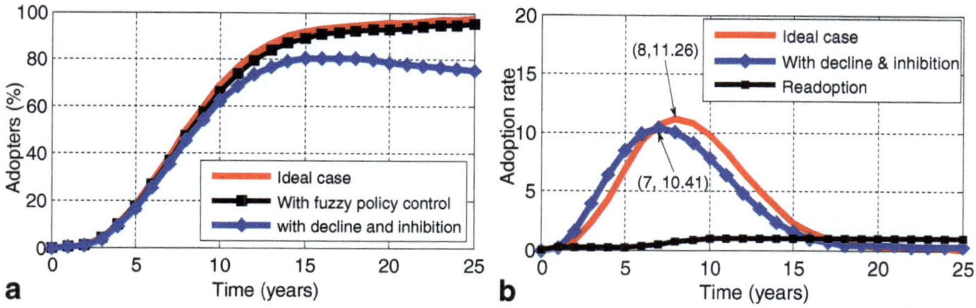

Fig. 9 RET adoption behavior with fuzzy dynamic policy control. **a** Compares the dynamic behavior of the ideal case, decline and inhibition case without control policy, and with fuzzy dynamic policy. **b** Demonstrates the simulation output for the real world case, with and without policy control

is insignificant, the negative influence of the inhibition process is quite marginal, though the inhibitors may be a large community. In summary, technology inhibitors and their likely negative influence should always be considered when analyzing RET management problems. The results emphasize the need for robust dynamic renewable energy policies that are centered on the information from the perceptions in marketplace. However, due to the imprecise and linguistic nature of the variables found in the marketplace, the use of fuzzy dynamic policy is the most viable and appropriate option. The next section illustrates the simulation results based on this phenomenon.

Adoption With Fuzzy Policy Control

Figure 9 illustrates the simulation results for the real world case characterized with decline and inhibition, and the impact of policy control. Figure 9a compares the dynamic behavior of the ideal case, decline and inhibition case without control policy, and with fuzzy dynamic policy. As can be seen from the graphical results, the graph under fuzzy policy control closely follows the ideal case throughout the planning period. By using fuzzy logic rules to capture the fuzzy variables in the marketplace, that is perceived decline rate and perceived number of former adopters, the policy maker can make informed dynamic policies that can effectively keep the adoption behavior under control. Fuzzy system dynamics employs the strengths of the system dynamics and the fuzzy logic paradigms to capture the dynamic and fuzzy variables from a systems simulation perspective.

Figure 9b demonstrates the simulation output for the real world case, with and without policy control. As far as the adoption rate is concerned, the peak points for the ideal (base) case and the real world case remain at (8, 11, 26) and (7, 10.41), respectively. However, due to the introduction of the dynamic fuzzy policy control policy, re-adoption increased gradually over the first 10 years, reaching a peak point of about 1.5 % per unit time and virtually remains constant over the simulation

Table 2 A summary of comparative simulation results

Sr. No.	Performance criteria	Ideal or base case	With decline	With decline and inhibition	With fuzzy policy control
1.	Peak adoption rate	8.0	8.0	7.0	8
2.	Peak adoption time	11.26	11.26	10.41	11.26
3.	Maximum adoption	100	85	80	97.5
4.	Maximum adoption time	24	24	16	25

period. This essentially explains the possible positive impact of policy incentives, policy controls, and other subsidy initiatives from governments, organizations, and other stakeholders.

It is important to note, from this study, that fuzzy-based approaches can be used to capture the imprecise market-based variables. As such, the suggested fuzzy system dynamics is a potential tool for renewable energy policy design and evaluation. Table 2 presents a summary of the simulation results, comparing the four simulated scenarios.

Conclusions and Further Research

The ultimate adoption of renewable energy technologies such as solar, wind, and hydro-electricity has a significant contribution towards energy-low carbon economies. As such, the dynamics of RET adoption is a key consideration in energy policy design and evaluation. This chapter highlighted the important elements affecting the adoption of renewable energy. Important among the key issues are (i) the RET marketplace is characterized with various technology barriers and enablers which affect the smooth adoption of the technologies, (ii) the marketplace consists of adopters who promote technology adoption, inhibitors who oppose technology adoption, and technology imitators whose adoption decisions are affected by word-of-mouth from both adopters and inhibitors, and (iii) policy design and evaluation is complex due to the presence of fuzzy linguistic variables that characterize the marketplace. To capture the impact of the fuzzy variables and their dynamic interactions, this chapter presents a fuzzy system dynamics approach that can handle the dynamics of fuzzy variables. The model is built from the system dynamics and fuzzy logic constructs on a Matlab® platform. The study provides important contributions to the academic community and to policy makers concerned with energy-low carbon economy interactions and impacts.

Contributions to Theory

This chapter highlighted important factors central to the dynamics of renewable energy adoption. It is realized in this study that there are three most critical players in the renewable energy marketplace, namely (i) adopters, (ii) inhibitors, and (iii) imitators. Contrary to most previous models, the current model captures the interactive dynamics of the three key players. In addition to this consideration, the model takes into account the presence of fuzzy linguistic variables to capture uncertainties often found in the adoption of renewable energy technologies. The inclusion of these two aspects, that is, the presence of multiple players and fuzzy variables, contributes to the body of knowledge in energy policy, providing further in-depth understanding of the interactive dynamics in renewable energy adoption in a fuzzy environment.

The use of the fuzzy system dynamics in energy policy modeling and evaluation is an invaluable extension of the possible areas of application of the system dynamics as well as the fuzzy logic paradigms. In particular, the fuzzy system dynamics paradigm fosters the integrative advantages of system dynamics and fuzzy logic which is otherwise impossible when the two are applied singly. Therefore the study is a useful addition to system models for policy analysis and evaluation for complex systems. Apart from its contribution to academician and researchers' community, the chapter also contributes to the decision makers in industry by providing useful managerial insights through dynamic simulation.

Managerial Implications

For effective energy policy formulation and evaluation, the policy maker requires robust tools that can handle fuzzy linguistic variables which characterize the renewable energy marketplace. The use of fuzzy logic system tools plays a crucial role in capturing fuzzy variables inherent in most human systems. Thus, by using the fuzzy tools, the policy maker is able to infuse more realism into the policy models. By so doing, robust energy policies can be designed, evaluated and possibly optimized in a more practical way. Furthermore, the infusion of system dynamics and fuzzy logic tools enables the policy maker to capture the complex interactive dynamics involving potential adopters, adopters, imitators, and inhibitors of renewable energy technologies. Ultimately, effective dynamic energy policies can be formulated for the medium to long-term decisions.

Oftentimes, decision makers need to make up their minds and make policies in a fuzzy environment within a given timeframe. As such, they need intelligent soft computing tools by which they can model energy policies in an uncertain environment. Fuzzy system dynamics provides a platform for building expert knowledge into energy policy models within a reasonable model-building time. These simulation models also provide a suitable environment for policy makers to discuss and make decisions collectively. Problem specific insights are easily derived from the simulation of possible policy alternatives, enabling decision makers to anticipate the impacts of their decisions, even in uncertain environments. Deriving from the simulation studies in this chapter, useful managerial insights, essential for renewable energy policy makers, can be summarized as follows:

- The RET adoption behavior is influenced by complex interactions between potential adopters, adopters, and former adopters, and the ensuing energy policies;

- Robust policy formulation can be done effectively by using perceived market information in terms of number of former adopters and rate of decline, and by incorporating expert knowledge into a fuzzy rule base;
- The use of expert knowledge for the design evaluation of renewable energy policies is essential in a fuzzy environment, especially during the decline phase;
- Policy makers should rely on the use of dynamic fuzzy policy control, so as to effectively capture the fuzzy linguistic variables in the RET marketplace;
- Targeted promotional efforts are especially crucial for encouraging re-adoption by former adopters, and should be planned according to the perceived market behavior in terms of number of former adopters and the rate of decline. Targeted promotion can be in the form of financial incentives and support services; and,
- Fuzzy rules can be constructed easily, to incorporate expert knowledge into fuzzy system dynamics so that policy alternatives can be tested and evaluated.

It is anticipated that, based on these managerial insights, policy makers can make their way forward in renewable energy policy formulation and evaluation in a fuzzy environment. Therefore, overall, fuzzy logic and system dynamics tools are viable tools for energy policy formulation and evaluation in complex fuzzy environments.

Further Research

Further research avenues in this area are twofold. First, the current research can be extended to include the impact of renewable energy adoption on the overall energy consumption and the overall energy-economy characteristics. This is achieved by integrating the renewable energy adoption model with the overall energy policy models at national level. In addition, the impacts of specific policy incentives, regulatory policies, and other policy controls such as carbon tax policies can be investigated further from a fuzzy system dynamics perspective. Second, it may be interesting to extend this study to the application of the fuzzy system dynamics model to real world cases, subject to the availability of data. It is anticipated that the application of fuzzy system dynamics models to real world energy-economy policy problems will bring model realism into energy-low carbon models. Ultimately, more practical and effective dynamic policies can be constructed and evaluated.

References

Bass, F.: A new product growth model for consumer durables. Manage. Sci. **15**(5), 215–227 (1969)
Barlas, Y.: Formal aspects of model validity and validation in system dynamics. Syst. Dynam. Rev. **12**(3), 183–210 (1996)
Cavusoglu, H., Hu, N., Li, Y., Ma, D.: Information technology diffusion with influentials, imitators, and opponents. **27**(2), 305–334 (2010)
Chandrasekara, B., Tara, C.K.: An opinion survey based assessment of renewable energy technology development in India. Renew. Sust. Energ. Rev. **11**, 688–701 (2007)

Chen, Y.: Understanding technology adoption through system dynamics approach: a case study of RFID technology. Ninth IEEE/IFIP International Conference on Embedded and Ubiquitous Computing, 366–371 (2011)
Chen, F., Duic, N., Alves, L.M., Carvalho, M.G.: Renew islands-renewable energy policies for islands. Renew. Sust. Energ. Rev. **11**(8), 1888–1902 (2007)
Demiroren, A., Yilmaz, U.: Analysis of change in electric energy cost with using renewable energy sources in Gokceada, Turkey: an island example. Renew. Sust. Ener. Rev. **14**, 323–333 (2010)
EIA.: Annual Energy Outlook. Government Printing Office, Energy Information Administration. (2009)
Ernest, F.B., Matthew, A.B.: Feasibility of solar technology (photovoltaic) adoption- a case study on Tennessee's poultry industry. Renew. Energ. **34**, 748–54 (2009)
Forrester, J.: Industrial dynamics. Pegasus Communications, Waltham (1961)
Ford, J.W.: System dynamics and the electric power industry. Syst. Dynam. Rev. **13**(1), 57–85 (1997)
Han, J., Hayashi, Y.: A system dynamics model of CO_2 mitigation in China's inter-city passenger transport. Transport. Res. D.: Tr. E. **13**(5), 298–305 (2008)
Huang, L.M.: Financing rural renewable energy: a comparison between China and India. Renew. Sust. Energ. Rev. **13**(5), 1096–1103 (2009)
Jin, W., Xu, L., Yang, Z.: Modeling a policy making framework for urban sustainability: incorporating system dynamics into the ecological footprint. Ecol. Econ. **68**(12), 2938–2949 (2009)
John, B., Bo, S., William, W.: The economics of sustainable energy for rural development: a study of renewable energy in rural China. Energ. Policy **26**, 45–54 (1998)
Kikuchi, S.: Study of transportation and uncertainty. Applied research in uncertainty modeling and analysis, vol. 20. In: Attoh-Okine, N.O., Ayyub, B.M. (eds.) pp. 303–319. Springer US. (2005)
Kosko, B.: Fuzzy systems as universal fuzzy approximation. In: Proceedings of the first IEEE International Conference on Fuzzy Systems, 1153–1162 (1992a)
Kosko, B.: Neural networks and fuzzy systems. Prentice-Hall, Englewood Cliffs (1992b)
Kosko, B.: Fuzzy thinking. Harpner Collins, London (1994)
Kosko, B.:Combining fuzzy systems. IEEE FUZZ-95, March, Yokohama, Japan (1995)
Krushna, M., Leif, G.: An adopter-centric approach to analyze the diffusion patterns of innovative residential heating systems in Sweden. Energ. Policy **36**, 577–9 (2008)
Labibi, A.W., Williams, G.B., O'Conor, R.F.: An intelligent maintenance model system: An application of the analytic hierarchy process and a fuzzy logic rule-based controller. J. Oper. Res. Soc. **49**, 745–757
Levary, R.: Systems dynamics with fuzzy-logic. Int. J. Sys. Sci. **21**(8), 1701–1707 (1990)
Li, J.F., Zhu, L., Hu, R.Q.: Policy analysis of the barriers to renewable energy development in the People's Republic of China. Energy Sus. Dev. **6**, 11–20 (2009)
Mamdani, E.H., Assilian, S.: An experiment in linguistic synthesis with a fuzzy logic controller. Int. J. Man-Machine Stud. **7**(1), 1–13 (1975)
Meadows, D.H.: Limits to growth: the 30-year update. Chelsea Green Publishing Company, USA (2004)
Morecroft, J.D.W.: Strategic modeling and business dynamics: a feedback systems approach. Wiley, Chichester (2007)
Mutingi, M. (2012). Dynamic simulation for effective workforce management in new product development. Manag. Sci. Lett. **2**, 2571–2580 (2012)
Mutingi, M., Mbohwa, C.: Fuzzy system dynamics simulation for manufacturing supply chain systems with uncertain demand CIE42 Proceedings of the International Conference on Computers and Industrial Engineering, South Africa, 1–12 (2012)
Mutingi, M., Matope, S.: System dynamics of renewable energy technology adoption. IEEE International Conference on Industrial Technology, South Africa (forthcoming) (2013)
Naill, R.: A system dynamics model for national energy policy planning. Sys. Dynam. Rev. **8**(1), 1–19 (1992)

Peter, W.: Biogas production: current state and perspectives. Appl. Microbiol. Biot. **85**, 849–60 (2010)
Qudrat-Ullah, H.: MDESRAP: a model for understanding the dynamics of electricity supply, resources, and pollution. Int. J. Glob. Energy. Issue **23**(1), 1–4 (2005)
Qudrat-Ullah, H., Davidsen, P.I.: Understanding the dynamics of electricity supply, resources and pollution: Pakistan's case. Energy **26**, 595–606 (2001)
Qudrat-Ullah, H., Seong, B.S.: How to do structural validity of a system dynamics type simulation model: The case of an energy policy model. Energ. Policy **38**, 2216–2224 (2010)
Raja, P., Laurence, D., Vasanthi, M.P.: Adoption of photovoltaic power supply systems: a study of key determinants in India. Renew. Energ. **31**, 2272–83 (2006)
Reddi, K.R., Moon, Y.B.: System dynamics modeling of engineering change management in a collaborative environment. Int. J. Adv. Manuf. Tech. **55**, 1255–1239 (2011)
Retortillo, P., Mediavilla, M., Miguel, L.J., Castro, C.: An attempt to automate the analysis of complex system dynamics models: an example of WORLD 3. Proceedings of the 26th International Conference of the System Dynamics Society, Athens, Greece (2008)
Rodrigues, L.L.R., Dharmaraj, N.: System dynamics approach for change management in new product development. Manag. Res. News **29**(8), 512–523 (2006)
Roger, F., Naill, R., Belanger, S., Klinger, A., Petersen, E.: An analysis of the cost effectiveness of U.S. energy policies to mitigate global warming. Sys. Dynam. Rev. **8**(2), 111–128 (1990)
Rogers, E.: Diffusion of innovation. Free Press, New York (1995)
Rogers, E.M.: Diffusion of innovations. Free Press, New York (2003)
Saysel, A. K., Barlas, Y.: Model simplification and validation testing. Sys. Dynam. Rev. **22**(3), 241–262 (2006)
Sterman, J.D.: Business dynamics: systems thinking and modeling for a complex world. Irwin/McGraw-Hill, Boston (2004)
Sugeno, M.: Industrial applications of fuzzy control. Elsevier Science Pub. Co. (1985)
Tessem, B., Davidsen, P.I.: Fuzzy system dynamics: an approach to vague and qualitative variables in simulation. Sys. Dynam. Rev. **10**(1), 49–62 (1994)
Trappey, A., Trappey, C., Hsiao, C.T., Ou, J., Chang, C.T.: System dynamics modeling of product carbon footprint life cycles for collaborative green supply chains. Int. J. Comp. Integ. M. **25**(10), 934–945 (2011)
Trappey, A.J.C., Trappey, C., Hsiao, C.T., Ou, J.J.R., Li, S.J., Chen, K.W.P.: An evaluation model for low carbon island policy: the case of Taiwan's green transportation policy. Energ. Policy **45**, 510–515 (2012)
Uemura, Y., Kai, T., Natori, R., Takahashi, T., Hatate, Y., Yoshida, M.: Potential of renewable energy sources and its applications in Yakashima Island. Renew. Energ. **29**, 581–91 (2003)
Van den Bulte, C., Joshi, Y.V.: New product diffusion with influentials and imitators. Market. Sci. **26**(3), 400–421 (2007)
Vicki, G., Tomas, M.K.: Breaking the cycle producer and consumer perspectives on the non-adoption of passive solar housing in the US. Energ. Policy **36**, 551–66 (2008)
Wang, J., Lu, H., Peng, H.: System dynamics model of urban transportation system and its application. J. Transport. Sys. Engg. Info. Tech. **8**(3), 83–89 (2008)
Wong, R., Sheng, S.Y.: Business application of the system dynamics approach: word-of-mouth and its effect in an online environment. Tech. Innov. Manage. Rev. (2012)
Zadeh, L.: A fuzzy sets. Inf. Control **8**, 338–353 (1965)
Zadeh, L.A.: Fuzzy sets as a basis for a theory of possibility. Fuzzy. Set. Syst. **1**, 3–28 (1978)

Resurrecting a Forgotten Model: Updating Mashayekhi's Model of Iranian Economic Development

Saeed P. Langarudi and Michael J. Radzicki

Introduction

In 1978, *Ali Naghi Mashayekhi* developed a system dynamics model to investigate the dependency of the Iranian economy on oil revenue (Mashayekhi 1978). Although this study created a general awareness about Iranian oil-dependency among academics and politicians, the model itself has, by and large, been forgotten. The purpose of this chapter is to revisit and update Mashayekhi's model (the "M-model") and show that it deserves more attention. In particular, it will be demonstrated that the M-model has the potential to become a well-known starting point for future Iranian macroeconomic modeling efforts, especially in the area of energy–economy interactions.

The M-model was created as a part of Mashayekhi's Ph.D. dissertation at the Massachusetts Institute of Technology. Simulations of the M-model in the late-1970s revealed that Iran would face a severe depression during the 1980s if its government pursued a policy of importing intermediate goods purchased with revenue from oil exports.

Figure 1 is a simulation run from the original formulation of the M-model that illustrates this potential crisis. It presents the base run time paths for Iranian oil reserves (curve 1) in terms of billion barrels, oil production (curve 2) in terms of million barrels per year, oil revenues (curve 3) in terms of million rials[1] per year, gross national product (GNP) (curve 4) in terms of million rials per year, and non-oil outputs (curve 5) in terms of million rials per year.

[1] Rial is the Iranian currency unit.

S. P. Langarudi (✉) · M. J. Radzicki
Department of Social Science & Policy Studies,
Worcester Polytechnic Institute, 100 Institute Road,
Worcester MA 01609-2280, USA
e-mail: slangarudi@wpi.edu

M. J. Radzicki
e-mail: radzicki@comcast.net

H. Qudrat-Ullah (ed.), *Energy Policy Modeling in the 21st Century,*
Understanding Complex Systems, DOI 10.1007/978-1-4614-8606-0_11,

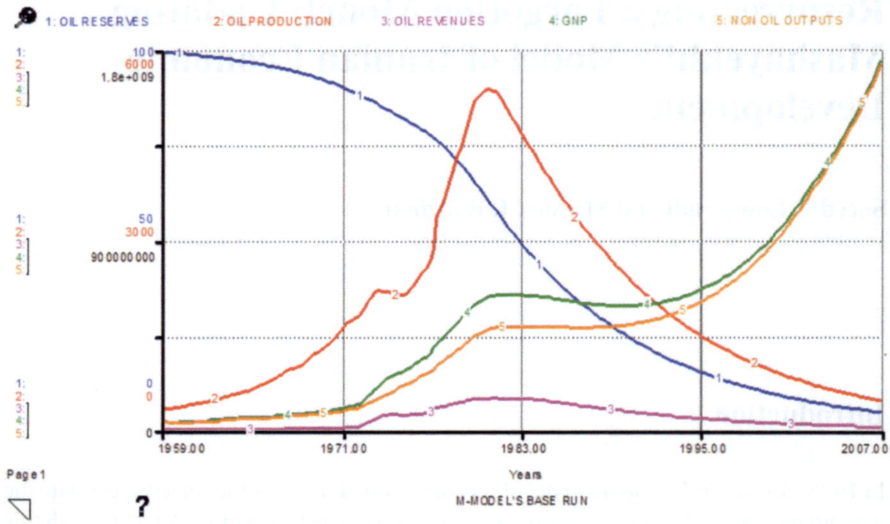

Fig. 1 Base simulation run of the M-model

The dynamics of this base simulation run are as follows. Iranian oil revenue grows from the late 1950s to the early 1980s. In the middle of the 1980s, however, they begin to decline due to the depletion of Iranian oil reserves. Consequently, Iran's stock of foreign exchange begins to shrink and it begins to limit the importation of intermediate goods. The shortage of intermediate goods causes the production capacity of the economy to fall and the growth rate of non-oil output to approach zero, and even briefly turn negative, during the 1980s. The stagnation of both oil and non-oil output leads to a severe depression that lasts until beginning of the 1990s.

In general, the M-model demonstrated that the high dependency of the Iranian economy on imports of intermediate goods, financed with oil revenue, would sooner or later cause Iran to run into serious economic difficulty. Although the specific scenario shown in Fig. 1 never occurred, the potential problems for the Iranian economy suggested by the M-model still exist. For example, Fig. 2 shows that during the period 1965–2008 the ratio of oil revenue to total revenue of the Iranian government ranged from 25 to 86 %, with an average value of 57 % (CBI 2012). This situation was mitigated somewhat by a downward trend in the ratio during the years 1999–2008, although its value is currently hovering around its historical average.

At the same time, Fig. 3 shows that Iranian imports of raw material and intermediate goods have increased dramatically in recent years. From these data it is clear that Iran continues to be dependent on oil revenue and must import raw material and intermediate goods as aggressively as ever. As a consequence, it makes sense to update the M-model and restate its message so that Iranian policy makers can be reminded of the strategic issues it raises.

Although Mashayekhi was a pioneer in identifying the problems associated with the dependency of the Iranian economy on oil revenue, his model and its conclusions have arguably never received the attention they deserve. There are several reasons for this including:

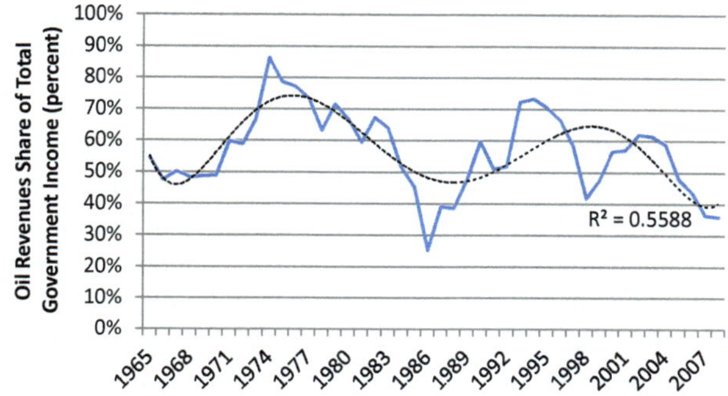

Fig. 2 Oil's share of total Iranian government revenue. (CBI 2012)

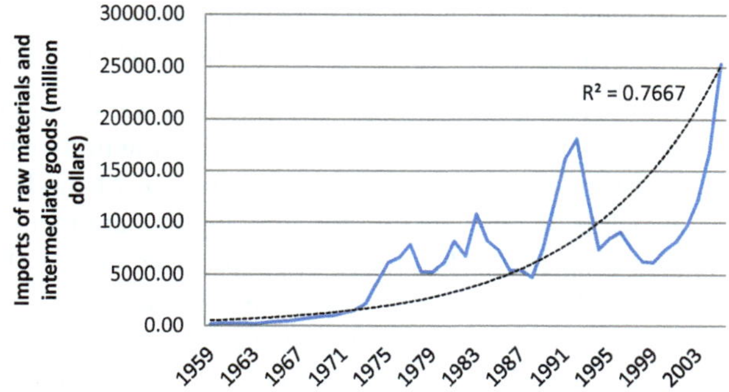

Fig. 3 Iranian imports of raw materials and intermediate goods. (CBI 2012)

- *The Islamic revolution*. The creation of the M-model coincided with the birth of the Islamic revolution. The revolution led to fundamentally different decision making processes within the highest levels of Iranian political and economic institutions. As a consequence, the usefulness of the M-model became ambiguous. Moreover, in 1980, 1 year after the revolution's success, an eight-year war began when Iraq invaded Iran. This national emergency changed the Iranian government's priorities from economic reform to financing the war and stabilizing the political economy of the country (Ahmadi Amouee 2006). Not surprisingly, few policy analysts paid attention to the oil-dependency issue during this period of time.
- *Unfamiliarity with system dynamics*. The intellectual origin of system dynamics is engineering and management, not economics. As a consequence, most of economists in Iran were—and still are—unfamiliar with the system dynamics methodology. In fact, during the late 1970s there was virtually no one in Iran who could fully understand and appreciate the M-model. Even now, there are few economists in Iran who know about system dynamics and it is thus not surprising that the first effort to apply system dynamics to the Iranian economy was largely ignored.
- *Competing obligations*. Mashayekhi himself believes[2] that the main reason his model has failed to make a significant impact on Iranian policy making is that his graduation from MIT and return to Iran coincided with the rise of the Islamic

[2] Telephone interview with Ali Mashayekhi on May 12, 2011.

regime and the new government asking him to help reconstruct the Iranian higher education system. As such, he was left with little time to publish, promote, and extend his model.

Despite these setbacks, the M-Model has the potential to be updated and used for energy–economy analysis in Iran. By reintroducing the M-model, top-level Iranian political and economic decision makers can be reminded that oil dependency can be a potential danger to the long-term economic growth and stability of the country. Moreover the model can provide a foundation and road map for additional system dynamics modeling projects in the Iranian energy–economy space. Finally, this chapter will show how system dynamics models can be updated and expanded, which is a very important, yet often neglected, part of the system dynamics modeling process.

To achieve these goals, this chapter presents an updated and revalidated version of the M-model. The updating and revalidation process involves three main issues:

- *Improvements in software*. The M-model was developed in 1978. Since then there have been significant improvements in system dynamics validation methods and software tools. For example, the original M-model was written in DYNAMO which is an obsolete tool for applying modern methods of model validation. For this chapter, the M-model was reprogrammed in iThink,[3] which offers a wide range of validation and verification options.
- *Structural changes and historical data*. In 1978, Mashayekhi simulated the M-model forward in time to project the implications of various policy choices on the growth and stability of the Iranian economy. In the present day, of course, what was once the future is now the past. As such, it is possible to determine how accurately the M-model predicted the future. Not surprisingly, some inconsistencies between the projections of the M-model and the historical data have been identified. Although, system dynamicists believe that the point-by-point fit of a model to time series data is a weak proof of model validity (Forrester and Senge 1996; Sterman 1984; Saeed 1992; Radzicki 2004); modelers such as Sterman (1984) argue that it is an important consideration because it builds confidence in the eyes of model users. Hence, in order to increase the M-model's potential for acceptance by Iranian policy makers it will be shown that updating exogenous oil export and price data, along with some structural changes and parameter recalibrations, can significantly improve the model's ability to reproduce the historical behavior of the Iranian economy.
- *Model revalidation and publication*. Mashayekhi never published a comprehensive analysis of his model's ability to pass a traditional list of tests necessary to build confidence in a system dynamics model (Peterson 1980). This was probably due to software and/or time limitations, and/or to the level of knowledge of Iranian academics about the system dynamics methodology at that time. As a consequence, revalidating the model and publishing its results will potentially increase its creditability among those economists who insist that valid models require the application of statistical techniques to numerical data.

[3] iThink Analyst v9.1.4, 1985–2010.

In the next section a revalidation of the M-model according to criteria that are standard in the field of system dynamics (Sterman 2001) will be presented.

Revalidating the M-model

In the field of system dynamics, models are never considered to be purely "valid" or "invalid." Instead, they are evaluated according to their ability to generate confidence in their users. A model never can be validated absolutely because all models are wrong. All models are simplified and abstract versions of real systems. So they can never be regarded exactly as corresponding real systems. So, why do we look for validating a model? The answer is that you, as a leader, have to use a model to make your decisions. You may use only your mental models or a mathematical one, etc. Whatever you use, the question is which model you want to use; not whether you can use a model or not (Sterman 1991, 2001, 2002).

Indeed, putting a model through a validation process helps decision makers feel confident that the results they are seeing are legitimate and useful.

Over the years, system dynamicists have assembled a comprehensive list of tests to which a model can be subjected in an effort to build confidence among its users.[4] These tests include:

1. Boundary adequacy tests
2. Structure assessment tests
3. Dimensional consistency
4. Parameter assessment
5. Extreme condition tests
6. Integration error tests
7. Behavior reproduction tests
8. Behavior anomaly tests
9. Family member tests
10. Surprise behavior tests
11. Sensitivity analysis
12. System improvement tests

The application of these tests to the M-model will now be described.

Boundary Adequacy Tests

A model's boundary defines what is included in and excluded from its structure. Boundary adequacy tests evaluate the appropriateness of a model's boundary vis-a-vis the purpose for which it was created.

---

[4] See for example Forrester (1973); Peterson (1975, 1980); Mass and Senge (1978); Forrester and Senge (1996); Sterman (1984); Radzicki (2004); Barlas (1996); Sterman (2001); and Oliva (2003).

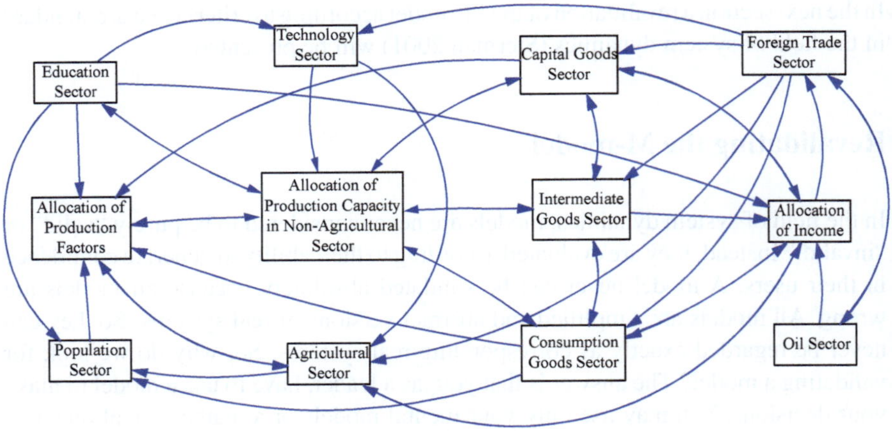

Fig. 4 Sector-level view of the structure of the M-model

Table 1 List of endogenous, exogenous, and excluded variables of M-model

Endogenous	Exogenous	Excluded
GNP	Labor market	Oil imports
Population	Oil exports	Financial market
Education	Oil prices	Exchange market
Capital accumulation		Alternative energies
Energy consumption		
Foreign trade		
Technology		

GNP gross national product

Figure 4 presents a sector-level view of the structure of the M-model. It consists of 325 variables and constants embodied in 12 interacting subsystems of the Iranian socioeconomic system. In addition, Table 1 lists some important macroeconomic variables that are endogenous, exogenous, and excluded from the M-model. The relevant question is whether or not this structure is still adequate for the M-model's purpose.

The endogenous variables can be examined first. Since the purpose of the M-model is to analyze the effect of oil revenue on the Iranian economy, it makes perfect sense to have it calculate a major economic summary index such as GNP. Furthermore, to replicate the dynamics of the aggregate production process in Iran, it is crucial to model population (as generator of the labor force), education (as an important input into the aggregate production function), capital accumulation (as the process that generates capital, which is another important production factor), and technology (again, an important input into the aggregate production function) as endogenous processes. Energy consumption is represented endogenously because it is a process that can limit oil exports and thus Iranian oil revenue. Finally, foreign trade is modeled as an endogenous process in order to capture the dynamics that drive the importation of intermediate goods and to show how foreign exchange is utilized.

In terms of exogenous variables, the dynamics of the labor market in the M-model are represented autonomously. More specifically, the model simply assumes that 56 % of the adult population is employed every year. This assumption is employed in order to avoid the complexity of the Iranian labor market. Since the main goal of the M-model is to reproduce the dynamics of Iranian oil dependency, it appears that this simplification is reasonable. Although including labor market dynamics can enhance the M-model's capacity to analyze a wider range of policies and scenarios, this capability is outside the focus of both the original, and present, studies. If the social consequences of oil dependency were the focus of the M-model, then a more sophisticated representation of the labor market would be required.

Oil exports are also largely determined exogenously. More precisely, they are set to their historical value for the years 1959–1978 and then determined endogenously thereafter. This is arguably a weakness in the original formulation of the M-model as oil exports are a key factor in generating the model's internal dynamics. The good news is that this problem can be eliminated by adding a comprehensive energy sector to the original M-model.[5]

Oil prices are also represented exogenously in the M-model. As with oil exports, the price of oil is set to its historical value for the years 1959–1978. However, unlike oil exports the price of oil is held constant from 1979 to the end of each simulation. Although different assumptions about the price of oil from 1979 forward can be tested, a superior formulation would generate oil prices endogenously because they are a major determinant of oil revenue.

The M-model's endogenous and exogenous variables represent factors that are part of its structure and are thus *inside* of its boundary. On the other hand, there are some important variables that are entirely excluded from the M-model's structure and hence lie *outside* of its boundary. For instance, the M-model assumes that the importation of oil to Iran is not possible. This assumption is both a boundary inadequacy and a structural deficiency. It implies that the Iranian economy has no source for oil other than its domestic supply. Of course, this is not true and when domestic oil resources decline significantly Iran will have to begin importing oil. Unfortunately, this scenario is impossible to be generated in the original formulation of the M-model.[6] Langarudi et al. (2011), however, present a remedy for this deficiency.

Financial and exchange markets are also excluded from the original version of the M-model. These exclusions have reduced the model's ability to fully analyze the impact of oil revenue on the Iranian economy. For example, the dynamics of the so-called "Dutch disease" cannot be explored. An economy afflicted with the Dutch disease experiences a rise in real exchange rates due to an unexpected increase in foreign exchange revenue generated by its natural resource exports. This in turn causes a fall in total output and employment in the nonnatural resource sectors (usually the manufacturing sector) as the stronger domestic currency makes nonnatural resource

[5] This has been done by Langarudi et al. (2011).

[6] Mashayekhi employed this assumption because the simulation period for the original M-model was 50 years and during this period domestic energy resources were sufficient for domestic energy consumption (see Footnote 10).

exports relatively more expensive (Van Wijnbergen 1984). Although this is clearly an issue with the boundary of the M-model, this chapter will demonstrate that it still provides an excellent foundation for a more complete model that can be used to analyze a wide range of Iranian macroeconomic issues.

Finally, another significant deficiency of the M-model's structure is its reliance on a single energy resource—oil. It can be argued, however, that this assumption poses no significant threat to the model's results because it was not designed to analyze the impact of competing energy resources. Nevertheless, adding alternative energy sources to the M-model, in particular natural gas, can certainly improve its usefulness for strategic planning in the energy sector.[7]

In sum, the M-model's boundary is somewhat inadequate for the purpose for which it was built. To better study the effects of oil revenue on the Iranian economy, the M-model's boundary must be expanded to include a financial market, an exchange market, an energy market, and the process of energy production. As these improvements are possible, the M-model is arguably still an appropriate base platform for undertaking Iranian socioeconomic analysis.

Structure Assessment Tests

Structure assessment tests check to see if a model is consistent with knowledge of the real system that is relevant to the purpose for which the model was created. These tests are concerned with the level of aggregation in a model, the fidelity of the model to basic physical facts, and the realism of the decision rules utilized by the agents in the model.

Structure assessment tests were performed in all steps of reviewing, recalibrating, and analyzing the M-model.[8] The result of this assessment is that, although the M-model has no egregious structural deficiencies, it contains two structural imperfections. These imperfections will be addressed after a detailed review of the M-model's general structure.

Resource Allocation Mechanism

The M-model's agricultural and nonagricultural production processes utilize three inputs: capital, labor, and education. Capital is calculated by accumulating investment in both machinery and construction, and then adding-in the flow of imported capital goods. Labor is supplied by the population sector while education is represented by the average number of years an Iranian citizen spends in school. These three production factors are allocated between the M-model's two production sectors: agricultural and nonagricultural (industrial) sectors.

---

[7] Langarudi et al. (2011) have also addressed this issue.

[8] For a comprehensive description of the model's structure see Mashayekhi (1978)

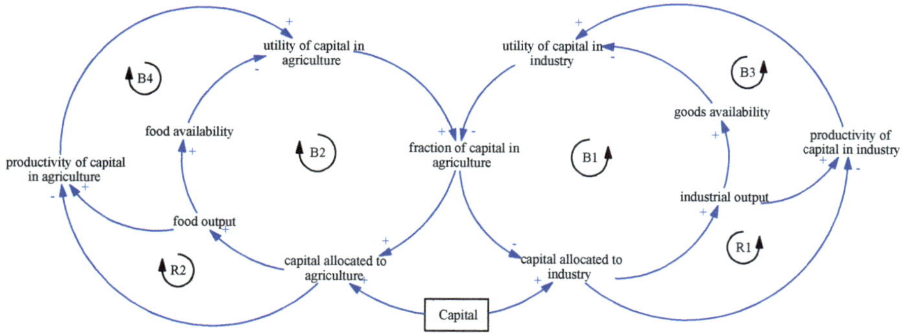

Fig. 5 Causal loop diagram of the resource allocation mechanism in the M-model

The resource allocation mechanism is based on the relative productivity of the three factors of production and the availability of each sector's output. The availability of a sector's output is a measure of demand relative to supply.

To illustrate how this mechanism works in the M-model, Fig. 5 presents a causal loop diagram of the process.[9] The allocation mechanisms for the other factors of production (labor and education) have the same structure.

Agricultural Sector

The model agricultural sector supplies the food demanded by the population. The production function in this sector utilizes the factors of production allocated to it, as well as available farmland and the sector's level of technology. The most important interactions between food production and the rest of the model are shown in Fig. 6.

Allocation of Production Capacity in the Nonagricultural Sector

Similar to the agricultural sector, a unique production function determines the total production capacity of the industrial (i.e., nonagricultural) sector. This production function utilizes the factors of production allocated to it, as well as the sector's level of technology. The total production capacity of the sector is allocated among four competing demands: capital goods production, intermediate goods production, consumption goods production, and educational capacity.

Figure 7 presents a causal loop diagram of the major processes that determine how the M-model allocates its nonagricultural (industrial) production capacity to consumption goods production. From an examination of the figure it is clear that the

---

[9] A causal loop diagram presents only the essential feedback structure of a system dynamics model so that the most important elements of cause and effect can be examined. The actual resource allocation mechanism in the M-model is substantially more sophisticated.

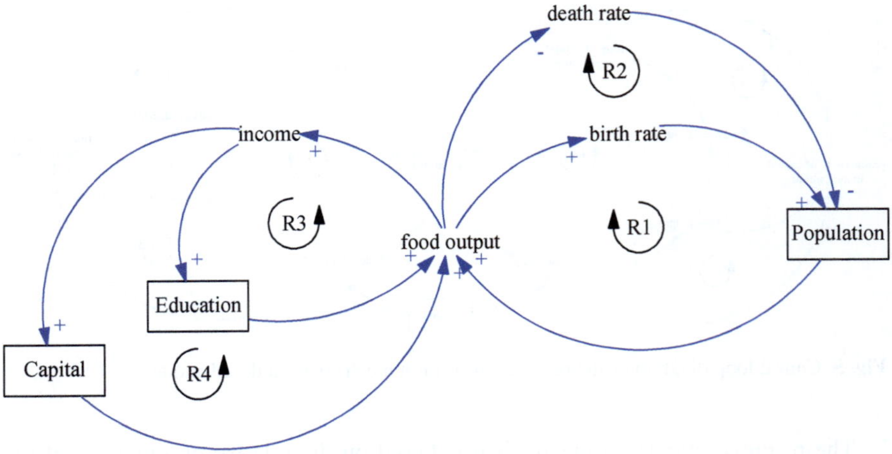

Fig. 6 Causal loop diagram of food production interactions in the M-model

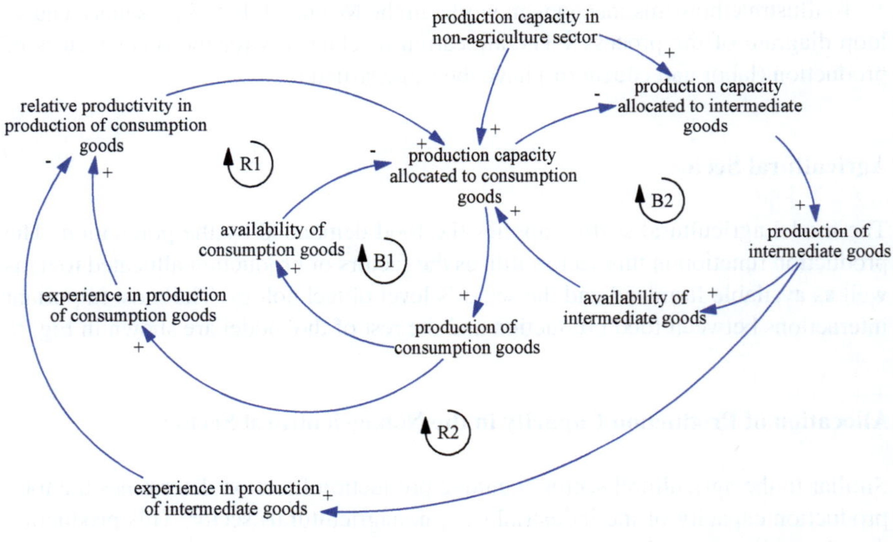

Fig. 7 Causal loop diagram of the resource allocation mechanism for consumption goods in the M-model

production capacity allocated to consumption goods depends on the total production capacity in the nonagricultural sector, the relative productivity of the production factors in consumption goods, the availability of consumption goods, and the availability of intermediate goods. Similar interactions are used to allocate production capacity to the production of capital goods, the production of intermediate goods, and educational capacity.

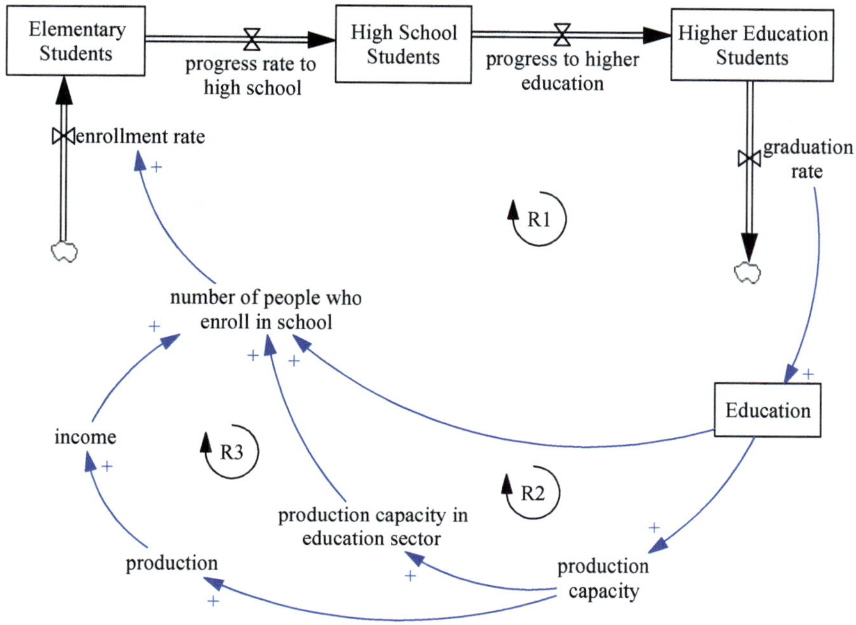

Fig. 8 Aggregate causal relationships in the education sector of the M-model

Education Sector

The output of the education sector is people possessing person-years of schooling. Educational output increases when the M-model's education production capacity and the demand for utilizing this capacity, increase. Educational production capacity depends on the demand for education and government policies. The demand for education is a direct function of both personal income and the educational level of Iranian adults. Figure 8 shows the aggregate causal relationships in this sector.

Technology Sector

The technology sector determines how technical progress diffuses into the Iranian economy. The M-model assumes that technical progress depends on technology transfer from developed countries. The transfer rate is determined by two factors: (1) the availability of required technologies that have not yet been transferred to Iran (i.e., the difference between the technology level in advanced countries and the corresponding technology level in Iran) and (2) Iran's ability to transfer technologies. Iran's ability to transfer technology depends on the education level of its work force and its level of foreign trade with developed countries. Figure 9 illustrates this mechanism for the industrial sector. Technical progress in the agricultural sector has a similar structure.

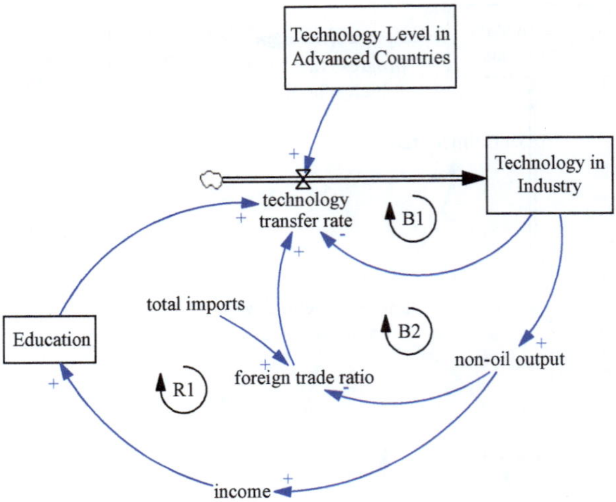

Fig. 9 Technology transfer in the industrial sector of the M-model

Allocation of Income

The allocation of income sector determines the allocation of Iranian national income among five competing demands: expenditures on (1) consumption goods, (2) services, (3) food, (4) saving, and (5) investment. The structure of this sector is based on standard microeconomic theory which specifies that the income elasticity of the demand for food and consumption goods is lower than the income elasticity of investment and saving.

Foreign Trade Sector

Any discrepancy between supply and demand in different sectors of the M-model is addressed through foreign trade. A demand surplus would be imported and a supply surplus would be exported. Imports and exports are also restricted by the availability of foreign exchange and government policies. Figure 10 depicts the feedback structure that determines Iranian foreign trade in consumption goods. The M-model utilizes the same structure to generate the dynamics of Iranian foreign trade in food, capital goods, and intermediate goods.

Oil Sector

In the oil sector, oil is produced and exported to provide Iran with the foreign exchange it needs for its imports. This sector also computes Iran's domestic energy consumption. The feedback structure of the oil sector is shown in Fig. 11.

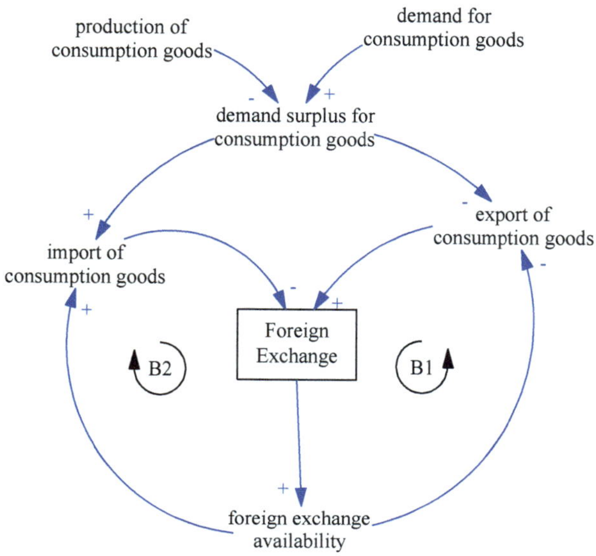

Fig. 10 Feedback structure determining Iranian foreign trade in consumption goods

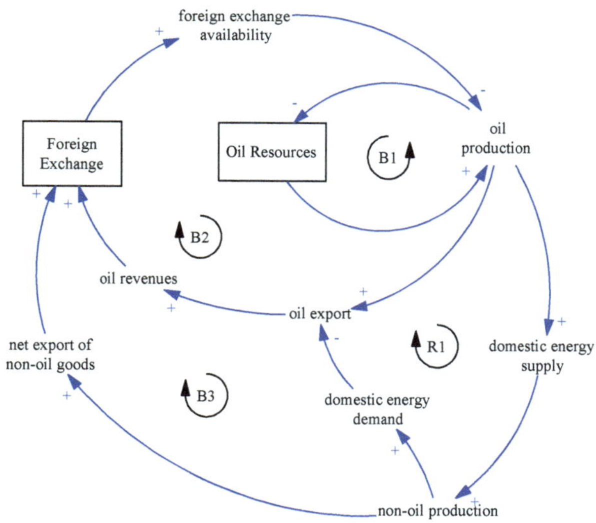

Fig. 11 Feedback structure of the oil sector of the M-model

Population Sector

The population sector of the M-model supplies both the workforce for the economy's production sectors and the consumers of the output from the production sectors. The Iranian birth rate depends on the adult population, available food per capita, the level of Iranian industrialization, and the level of Iranian education. Similarly, the Iranian death rate depends on available food per capita and the level of industrialization. Figures 12 and 13 show the feedback structure of this sector.

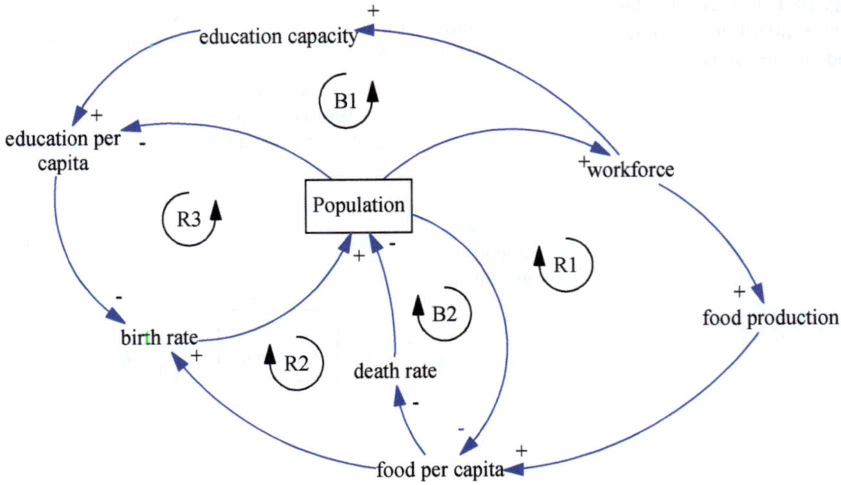

Fig. 12 Feedback structure of the population sector of the M-model (Part 1)

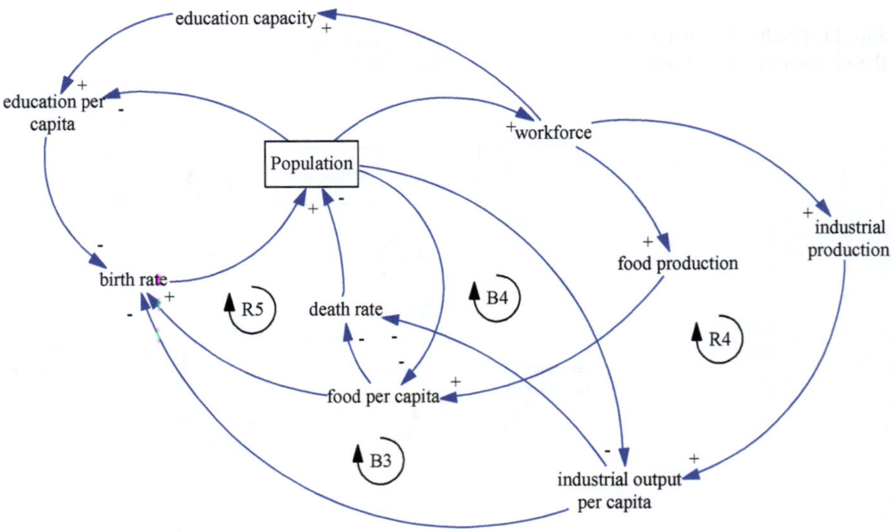

Fig. 13 Feedback structure of the population sector of the M-model (Part 2)

Minor Structural Imperfections in the M-Model

Although the overall structure of the M-model is excellent, there are two areas in which it is deficient. The first involves oil production; more specifically, oil production in the original version of the M-model can be doubled in just 1 year. Although this assumption might have been reasonable for the period before the Islamic revolution of 1979, when the Iranian state was able to attract as much foreign investment as it needed due to its good relationship with the developed world, it is not a valid assumption for the postrevolution era. Indeed, after the 1979 revolution the Iranian government could not persuade major oil companies to invest in its oil and gas industry (Katouzian 2009).

The second area in which the structure of the original M-model is deficient involves energy supply. The original model assumes that there is only one source of Iranian energy—domestic oil production. This assumption implies that the importation of energy is impossible. Even if the domestic supply of energy is sufficient for domestic energy demand, this assumption weakens the robustness of the M-model vis-a-vis extreme conditions. In other words, good system dynamics modeling practice requires that a model behaves correctly under extreme conditions, even if those conditions have never occurred in the actual system and/or will only occur in the model under extreme circumstances.

If a simulation run of the M-model depletes all Iranian energy resources, the economy still survives. Of course, this is extremely unrealistic. Mashayekhi (1978) argues that people will use wood when oil resources are scarce and the M-model implicitly assumes that burning wood is costless—which is simply not true. Despite these criticisms, available energy data shows that the net export of energy for Iran will be positive for at least next eight decades.[10] As a consequence, the structure of the M-model can be said to be adequate in simulation runs shorter than fifty years in duration.

Dimensional Consistency

Good system dynamics modeling practice requires that all of a model's equations be dimensionally consistent. This means that all of a model's equations must produce stocks that are measured in "units" and flows that are measured in "units/time." All of the equations in the M-model were checked and no dimensional inconsistencies were found.

Parameter Assessment

The parameter assessment test determines whether a model's parameter values are consistent with relevant descriptive and numerical knowledge of the actual system, and whether all the model's parameters have real world counterparts.

To answer these questions, all parameters of the model were checked and no inconsistencies were found among them and their real world counterparts. Mashayekhi's dissertation presents a comprehensive documentation of the M-model's parameters and how they were obtained.

---

[10] Simulations by Langarudi et al. (2011) show that Iran's net export of energy won't become negative until 2094. This result is yielded under this assumption that world demand for Iran's oil is infinitive so Iran can export that portion of its produced oil remaining after domestic consumption.

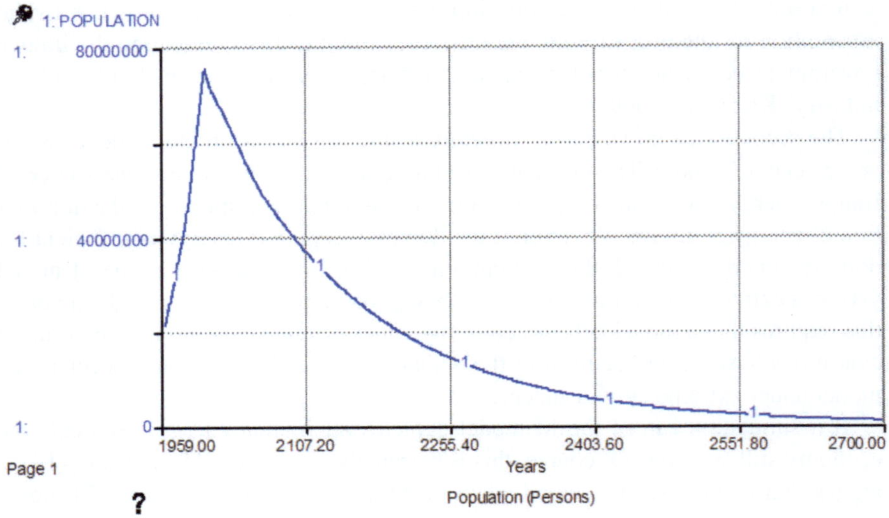

Fig. 14 Iranian population when the birth rate is set to zero after the year 2000

Extreme Condition Tests

Tests of extreme conditions are designed to evaluate whether or not each equation in a model makes sense when its inputs take on extreme values. In other words, they test whether or not a model's equations respond reasonably when subjected to extreme policies, shocks, and parameters.

To test the M-model for extreme conditions, each equation was evaluated, in isolation, for its response to extreme values for each of its inputs, alone and in combination. In addition, the overall M-model was subjected to large shocks and extreme conditions and then inspected for conformance to basic physical laws (e.g., an absence of inventory should mean there will be no shipments; zero labor should mean zero production).

All these tests revealed no serious problems with the M-model. However, some minor defects were detected. For example, the birth rate in the population sector was set to zero for all years after 2000. The result is shown in Fig. 14.

The Iranian population should have reached zero in approximately 150 years. However, Fig. 14 show that the Iranian population is still positive at year 2700. This implies that there are some individuals who can live for more than 700 years! The good news is that this deficiency does not significantly influence the primary results of the M-model and it can be corrected in a future version of the model.

Integration Error Tests

System dynamics models are continuous time models run on discrete machines (digital computers) and are thus solved via numerical integration. As a result, modelers must choose both a numerical integration method, and a time step, to approximate the continuous dynamics of the underlying system. Too large a time step utilized in concert with a particular numerical integration technique may yield too much integration error and thus simulated time paths that are too inaccurate for the problem at hand. Too small a time step utilized in concert with a particular numerical integration technique may yield simulated time paths that are unnecessarily precise for the problem at hand and thus simulation runs that are needlessly computationally intensive (i.e., slow).[11] Good system dynamics modeling practice, therefore, requires picking a time step/numerical integration combination that is no more accurate than is necessary for the problem at hand. This is typically accomplished by selecting an initial time step/numerical integration technique combination, running the model, cutting the time step in half, rerunning the model, and inspecting the pre- and postcut synthetic time paths for significant differences. When no significant differences in dynamic behavior can be detected, the model is deemed to be accurate enough for the problem at hand.[12]

The M-model was systematically tested with different numerical integration methods and time steps.[13] Euler's method (the default simulation method in most system dynamics modeling packages due to its simplicity and computational ease) proved to be fine and a time step reduction to 0.1 year yielded no significant change in model behavior.

Behavior Reproduction Tests

Many system dynamicists believe that historical fit is a weak test for model validity (Forrester 1973; Forrester and Senge 1996; Sterman 1984; Radzicki 2004). As Forrester (2003, p. 5) has written:

> There is no reason that a generic model should reproduce any specific historical time series. Instead, it should generate the kind of dynamic behavior that is observed in the systems that are being represented. If one runs the model with different noise sequences one will get simulations that have the same character, but not the same values at different points in time. Likewise, the time series from an actual economy represent only one of a multitude

[11] In the extreme, the smallness of a model's time step is limited by the precision of the digital computer being used.

[12] Mathematical rules of thumb relating a model's time step to its smallest time constant also exist in system dynamics modeling.

[13] Various numerical integration techniques have well-known strengths and weaknesses that come into play under different circumstances.

of detailed behaviors that might have occurred if the random effects in the real system had been different. In other words, historical data from a real economy should be interpreted as only one of a multitude of possible data histories.

The consensus view in the field of system dynamics is that, although reproducing historical behavior is only one of many tests required to build confidence in a system dynamics model, it can often be essential. Failure to convince a reviewer that a model's historical fit is satisfactory, for example, is often sufficient grounds for him/her to dismiss the model and its conclusions (Sterman 1984).

Sterman (1984) lays out a detailed example of how Theil's inequality statistics (Theil 1966) can be used to analyze the fit of a system dynamics model to historical data. These statistics are used in this chapter to examine the fit of the modified M-model to historical data from Iran. Before applying these statistics, however, the structural deficiencies of the M-model must be addressed and its behavior updated.

The first step in this process is to update the M-model's exogenous variables with the latest available data. Recall that the major exogenous variables in the M-model are oil exports and oil prices.

Utilizing modern data, the initial value of Iranian oil reserves was updated from 100 to 221 billion barrels. Iran's remaining proven reserves at the end of 2011 were estimated to be 154.6 billion barrels (OPEC 2012). The cumulative production of oil in Iran since 1959—which is the starting date for M-model simulations—until the end of 2011 was about 66.5 billion barrels (OPEC 2010). This means that Iran should have had 221.1 billion barrels of total proven reserves in 1959 ($154.6 + 66.5 = 221.1$ billion barrels). Of course, alternative values for initial oil reserves also can easily be tested in the model.

Figure 15 presents a comparison of actual and simulated Iranian non-oil output, from 1959 to 2007, after updating the initial value of Iranian oil reserves in the M-model as described earlier.[14] Other key variables from the M-model were monitored during this recalibration process but are not presented here due to space limitations.[15]

Non-oil output was chosen as a more plausible index of general economic output than GNP because GNP includes oil revenue. Since oil revenue is exogenously determined by a historical time series from 1959 to 2007, a large portion of the M-model's ability to reproduce Iranian GNP would be attributable to an exogenous input. Focusing on non-oil output, on the other hand, can better illustrate the M-model's ability to *endogenously* replicate the real system's behavior.

Figure 15 shows that the updated M-model's *qualitative* behavior, i.e., exponential growth followed by a peak, a decline, and the resumption of exponential growth, is very close to the real system's behavior. The point-by-point fitness of the M-model, however, is clearly not acceptable. Therefore, it is necessary to review the original assumptions of the M-model and if possible, modify them in order to improve the model's ability to replicate Iranian economic history.

---

[14] All real world data presented in this chapter comes from three main sources: (1) the electronic database of the Central Bank of Iran (CBI 2012), (2) OPEC Annual Statistical Bulletin (OPEC 2010, 2012), and (3) the BP Statistical Review of World Energy (BP 2012)

[15] A summary of the M-model's ability to replicate the dynamics of the key variables in the Iranian economy is presented at the end of this section in Table 2.

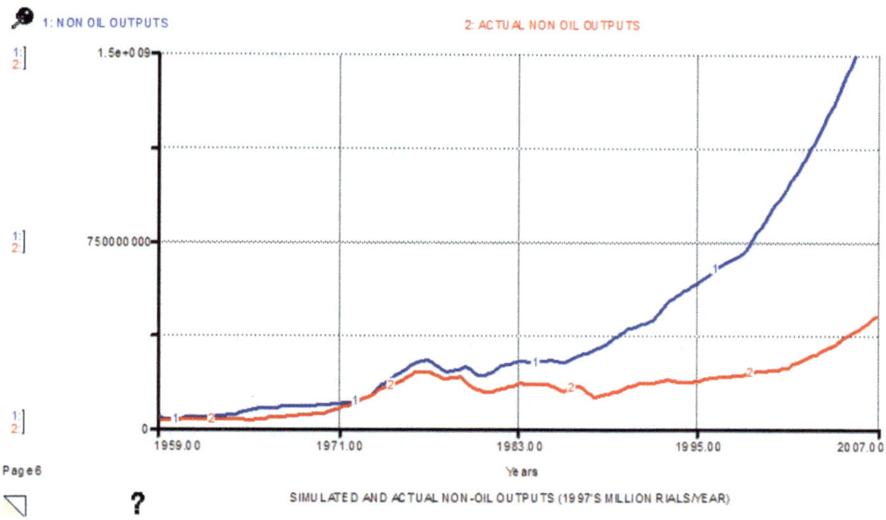

Fig. 15 Actual and simulated Iranian non-oil output from 1959 to 2007 after updating the initial value of Iranian oil reserves

As Fig. 15 demonstrates, the discrepancy between the behavior of the updated M-model and Iranian historical data starts and expands after 1979 when the Islamic revolution took place. As was previously mentioned, this revolution was followed by an eight-year war with Iraq. The most likely cause of the divergence between the actual and synthetic data presented in Fig. 15, therefore, is these events and the changes they caused in Iranian political economy. To test this hypothesis, structural changes representing the revolution and war must be introduced into the M-model. Katouzian (2003) argues that Iranian society had to endure the following impacts from the revolution and the war beginning in 1979:

1. A reduction in oil exports.
2. A reduction in the utilization of production capacity in the economy.
3. A high rate of capital flight.
4. A high rate of brain drain.
5. A reduction in the rate of investment.
6. A deep enmity between Iran and Western countries.

Here is an explanation of how these impacts are introduced into the M-model:

1. *Reduction in oil exports*. Since oil exports are treated as an exogenous input into the M-model until 2007 no further action is required.
2. *Reduction in the utilization of production capacity in the economy*. To introduce this effect a new variable called the "economic security indicator" (ESI) is introduced into the M-model. This variable is positively influenced by the growth rate of Iranian GNP, but only after a significant asymmetrical delay. More precisely, when the Iranian GNP growth rate increases the ESI increases, and when it decreases the ESI decreases. However, the delay between a change in Iranian GNP

and a change in the ESI is *longer* when GNP is rising compared to when GNP is declining. Changes in the ESI then influence the utilization of production capacity in the economic sectors of the M-model. The rationality behind this assumption is that after the 1979 revolution many business owners left or had to leave the country because they were suspected to be in contact with the dethroned Shah or his family (Katouzian 2009). In addition, many factories were underutilized due to an economic recession which was the natural result of the 1979 political turmoil and the war with Iraq (Pesaran 2000).

3. *High rate of capital flight.* To introduce this effect into the M-model, the ESI is also modeled to affect foreign exchange reserves.
4. *High rate of brain drain.* The M-model assumes that a fixed percent of highly educated people emigrate every year from Iran after the 1979 revolution. The percent rate of emigration can be changed by the model user.
5. *Reduction in the rate of investment.* Since investment rates in the M-model are determined by desired investment rates, and desired investment rates are based on the current utilization of a sector's production capacity, the effect of the ESI on the utilization of production capacity automatically adjusts the M-model's investment rates in response to the overall condition of the Iranian economy.
6. *The enmity between Iran and Western countries.* After the 1979 revolution some actions by radical Iranian revolutionists turned the governments of many western nations against the new Iranian state. The response of these governments was to implement political and economic sanctions against Iran. This forced Iran to pay higher prices for imported goods. Another result of this hostility was an increase in the difficulty Iran faced in transferring-in technology from developed countries. These facts are introduced into the M-model by defining a "hostility effect multiplier." This multiplier is an autonomous number between zero to four. When it is zero, it means that there is no hostility effect while "four" represents the highest tension in Iranian foreign relationships. Then, this multiplier affects two variables in the model: it has a negative impact on "technology transfer rate" and a positive impact on the value of "imports" (it increases the import expenses). Users of the M-model can manually change this multiplier to see how it influences the system's dynamics.

Figure 16 presents a comparison of actual and simulated Iranian non-oil output, from 1959 to 2007, after the next set of modifications (described earlier) are introduced into the M-model. A quick visual inspection of the figure reveals that the changes have significantly improved the ability of the M-model to reproduce Iranian historical data.

A more rigorous analysis of the ability of the M-model to reproduce historical data from the Iranian economy involves Theil's Inequality Statistics. Table 2 presents the Theil Statistics for four key variables from the M-model. In this table, r represents the correlation coefficient between simulated and actual data; U represents the inequality coefficient and U^M, U^S, and U^C reflect the fraction of the mean square error (MSE) attributable to bias, unequal variance and unequal covariance, respectively.

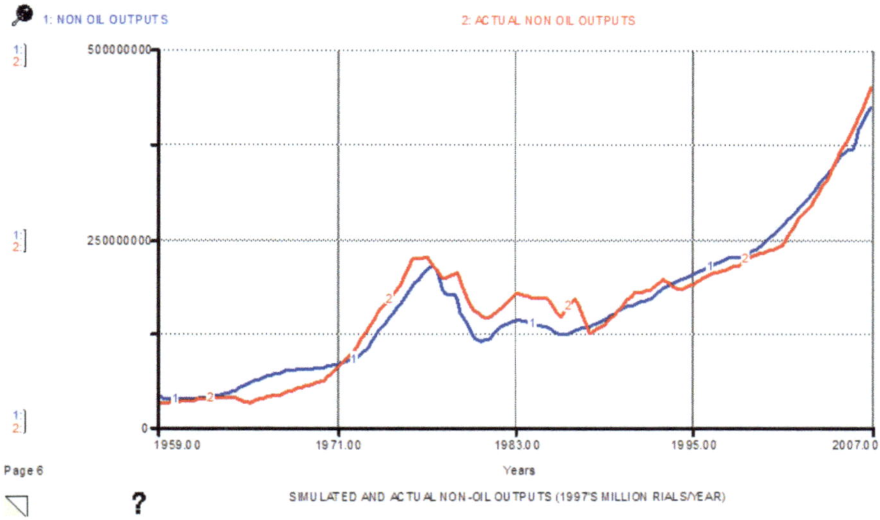

Fig. 16 Actual and simulated Iranian non-oil output from 1959 to 2007 after introducing the second set of changes

Table 2 Theil statistics for four variables from the M-model after introducing the second set of changes

Variables	r	U	U^M	U^S	U^C
GNP	0.984697	0.057733	0.192186	0.100753	0.707062
Non-oil outputs	0.978745	0.062130	0.208761	0.161118	0.630121
Oil production	0.929713	0.090369	0.174969	0.012924	0.812107
Population	0.996077	0.017460	0.045054	0.019337	0.935609

GNP gross national product

Inspection of Table 2 reveals that the correlation coefficient for all four variables is quite high and the inequality coefficient is reasonably low. Moreover, the majority of the MSE for all four variables is concentrated in unequal covariation (U^C). *Sterman* (1984, p. 220) interprets this situation as follows:

> If the majority of the error is concentrated in unequal co-variation U^C, while U^M and U^S are small, it indicates that the point-by-point values of the simulated and actual series do not match even though the model captures the average value and dominant trends in the actual data well. Such a case might indicate a fairly constant phase shift or translation in time of a cyclical mode otherwise reproduced well. More likely, a large U^C indicates one the variables has a large random component or contains cyclical modes not present in the other series. In particular, a large U^C may be due to noise or cyclical modes in the historical data not captured by the model. A large U^C indicates the majority of the error is unsystematic with respect to the purpose of the model, and the model should not be faulted for failing to match the random component of the data.

Therefore, it is reasonable to conclude that the second set of revisions to the M-model enable it to reproduce the actual system's behavior reasonably well. Not only are all the inequality coefficients from the Theil statistics small but most of the errors are unrelated to bias or unequal variance between the simulated and actual data.

Behavior Anomaly Tests

Data limitations often lead to difficulty in establishing the statistical significance of many relationships in a model. The importance of these relationships can, nevertheless, be examined by "behavior anomaly tests." These tests involve determining whether or not anomalous model behavior arises when a relationship is deleted or modified. Anomalous behavior generated due to the elimination of a relationship would be a sign of the importance of the relationship. Most of the relationships in the M-model were tested and no unnecessary or useless structure was found.

Family Member Tests

The family member test examines a model's ability to generate the behavior of other cases within the same class as the system being modeled. The greater the number of cases a model can mimic, the more general the theory the model represents.

The M-model was developed to address Iranian macroeconomic issues. Iran is an oil-exporting country and is highly dependent on its oil revenue. It also has a relatively large population and a notable agricultural sector with about 16 % average share of total GNP value (during 1959–2007) (CBI 2012).

In the real world it is possible to identify clusters of nations that are somewhat similar to Iran. Karl (1997, 1999), for example, argues that Iraq, Nigeria, Algeria, Indonesia, Venezuela, Ecuador, and Mexico possess many common characteristics and refers to them as "capital-deficient" countries. To pass a family member test, the M-model must be able to generate the macroeconomic behavior of at least some of these countries after a reasonable amount of modification to reflect each nation's unique features. Although this sort of effort is beyond the scope of this chapter, it can be argued that the M-model possesses a generic structure that can be applied to all "capital-deficient" countries.

Surprise Behavior Tests

Inconsistency between a model's behavior and its expected behavior reveals that there are some deficiencies in the formal model, the modeler's "mental model," or both. According to Sterman (2001, p. 882):

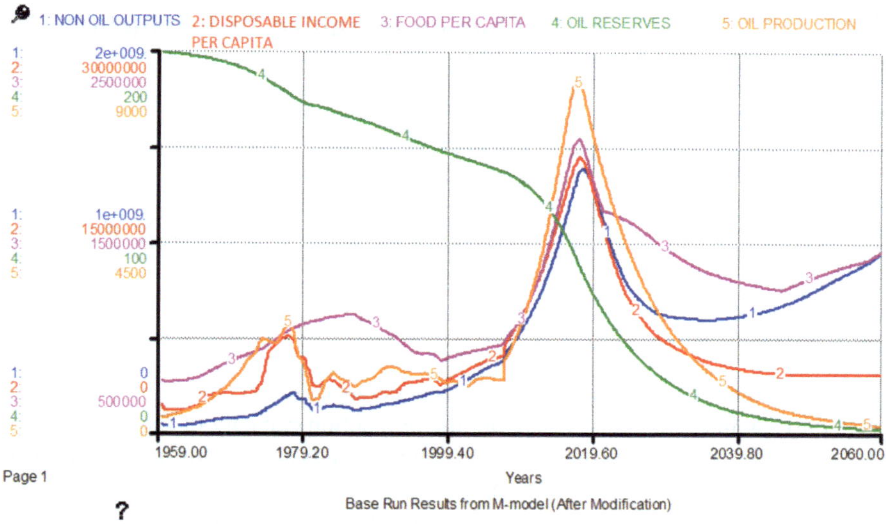

Fig. 17 Base run of the M-model from 1959 to 2060 after the second set of modifications have been introduced

> Often, of course, discrepancies between model output and our understanding of the system's dynamics indicate flaws in the formal model. Occasionally, however, it is our mental model and our understanding of the data that require revision.

The surprise behavior test is passed when a model generates a certain behavior, previously unrecognized, and it does indeed occur in the real system.

To test the M-model for surprise behavior it must be run under a variety of scenarios and its results carefully examined. Figure 17 presents a base run for the M-model from 1959 to 2060 after the second set of modifications has been introduced. The synthetic variables presented include: non-oil outputs (million rials per year), disposable income per capita (rials per person per year), food output per capita (rials per person per year), oil reserves (billion barrels), and oil production (million barrels per year).

After the second set of modifications the M-model produces the following behavior from the year 2007 forward. Oil production and oil exports grow exponentially causing Iranian oil reserves to deplete rapidly. The increase in oil exports provides the Iranian government with an influx of foreign exchange. This huge windfall of oil revenue makes it possible for Iran to import the goods it needs. It also improves the people's purchasing power, so the demand for consumption goods rises. The urgent need for an increased supply of consumption goods shifts the economy's production factors from its other sectors to the consumption goods sector. As a result, the intermediate goods sector weakens.

Of course, the economy needs intermediate goods to keep the production capacity of the consumption goods sector fully utilized. A quick response to this pressure is to import intermediate goods. This would normally address the problem in the short-term. However, over time Iran's oil reserves deplete at a rapid rate causing oil revenue

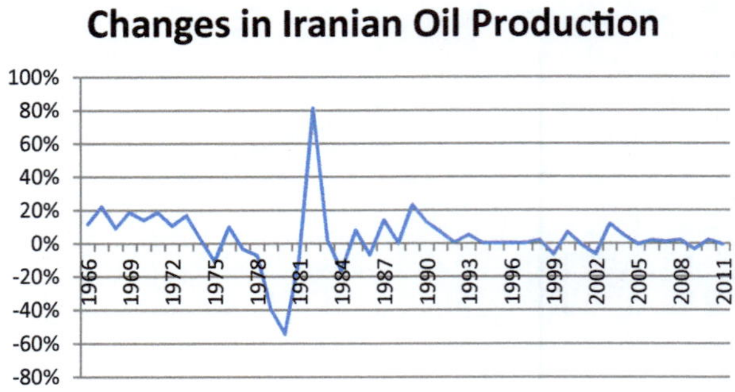

Fig. 18 Historical data for changes in Iranian oil production 1965–2011. (BP 2012)

to decrease and the importation of intermediate goods to become limited. Since the economy cannot seamlessly substitute the production capacity of the consumption sector for the production capacity of the capital goods sector, this leads to a severe depression. Stated differently, the depletion of Iranian oil reserves shown in Fig. 17 occurs so quickly that the economy cannot react to it in a timely fashion.

Even though the modified M-model generates an internally consistent story, some aspects of its behavior are surprising. For example, there is no way that Iranian oil production could expand as quickly as is shown in Fig. 17. The sharp increase in oil production is related to an unrealistic assumption embedded in the model. More specifically, the M-model assumes that Iran is able to double its oil production in a period as short as 1 year. An examination of the historical data, however, shows that this cannot possibly be true, particularly in the postrevolution era.

Figure 18 shows the history of changes in Iranian oil production from 1965 to 2011. From an inspection of the figure it is clear that Iranian oil production has not been able to rise dramatically in any given year since the 1980s. Indeed, the last extraordinary high growth rate (23 %) occurred in 1989. Moreover, significant increases in oil production appear to occur in years following a deep *fall* in oil production. This implies that high rates of growth are due to the reutilization of an underutilized *existing* production capacity, rather than from an increase in overall production capacity.

After the 1979 revolution the expansion of Iran's oil production capacity became more difficult because of a dramatic change in the new state's foreign policies that made foreign investment problematic. Energy economists believe that the main contemporary challenge in the Iranian energy sector is the lack of funding and investment (Barkeshli 2006). It is perhaps reasonable to assume that the prerevolutionary Iran could double its oil production in 1 year, but it is not a realistic assumption for the postrevolutionary Iran.

Clearly, the surprise behavior of the M-model presented in Fig. 17 reveals a need to further modification of the M-model's structure. The next version of the model assumes that the maximum growth rate of Iranian oil production is 14 %[16]. The model is run again and the results are shown in Fig. 19.

[16] Except for 1981, the highest growth rate after the revolution is 10.98 % in 1988 (BP 2012)

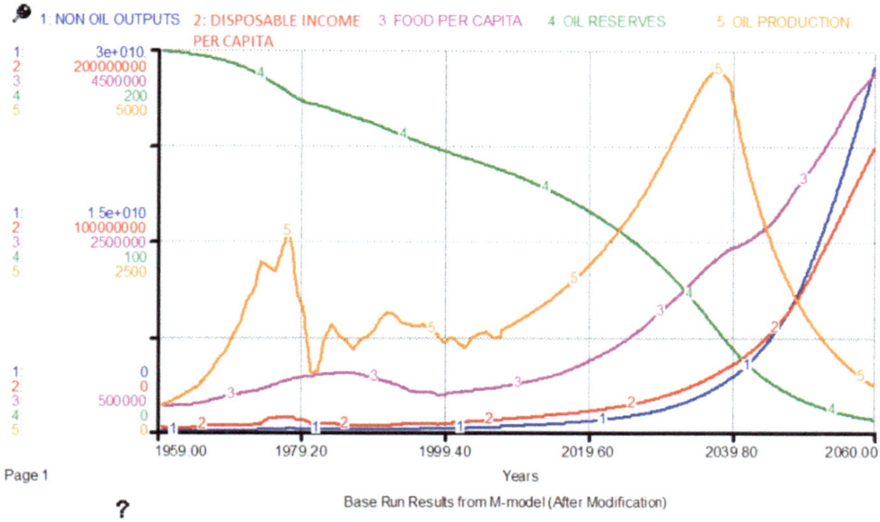

Fig. 19 Base run of the M-model from 1959 to 2060 after introducing the final modification

Figure 19 reveals that there is no severe depression in the economy from 2007 to 2060, despite the egregious depletion of Iranian Oil reserves. The reason is that Iranian oil production grows more slowly than in previous versions of the M-model (compare to Fig. 17) and the model economy, therefore, has enough time to wean itself from oil revenue. In other words, Iran's inability to absorb enough investment to develop its oil industry leads to less dependency on oil revenue. In fact, this was a policy originally proposed by *Mashayekhi* (1978) to alleviate the economic recession he had predicted for the 1990s. He had suggested that the Iranian government could slow down the production of oil *as a policy choice* so that the economy could adapt to the difficulties that would arise from the reduction in oil revenue he predicted for the 1990s. In reality, the 1979 revolution and war with Iraq *forced* Iran to slow down the production of oil. But, regardless of whether the slowdown was voluntary or mandatory its results are consistent with Mashayekhi's prediction.

Sensitivity Analysis

Sensitivity analysis reveals the robustness of a model's results with respect to changes in the values of its parameters over a reasonable range of uncertainty. There are three types of model sensitivity: "numerical," "behavior mode," and "policy." A model is numerically sensitive when a change in the values of its parameters changes the numerical values associated with its behavior. Of course, no mathematical model can be perfectly numerically insensitive. A model is behaviorally sensitive when the patterns of behavior it generates change with a change in the values of its parameters.

For instance, a model would demonstrate behavior mode sensitivity if reasonable alternative parameter values changed its behavior from, say, overshoot and collapse to s-shaped growth. Policy sensitivity exists when the impact or suitability of a suggested policy change is significantly altered by a change in the values of a model's parameters.

Since the purpose of the M-model is to determine whether or not Iran will experience economic growth during its transition from an oil-rich to an oil-poor nation, the focus of this section will be on behavior mode and policy sensitivity.

Behavior Mode Sensitivity

The behavioral sensitivity of the modified M-model will be reported in this section. To conduct the test, the values of important parameters in the M-model were systematically varied over a range of uncertainty and an examination of how the M-model's behavior changed in response was conducted. "Disposable income per capita" is chosen as the proxy for the model's overall behavior and as an example of the test.[17] Each of the parameter values was randomly altered twenty times over a range of $\pm 50\,\%$ of their base case values, using a uniform probability distribution.

The overall results of the behavior mode sensitivity test showed that, generally speaking, the modified M-model's behavior is *not* sensitive to changes in its parameters. However, a few parameters did prove to be more influential than others. The sensitivity test for five of the model's parameters is shown later.

Consider the modified M-model's two Cobb–Douglas production functions—one for the industrial sector and the other for the agricultural sector. Sensitivity testing revealed that the model is numerically *very* sensitive to the elasticity parameters for the inputs to the two functions. For example, Fig. 20 presents the sensitivity of the modified M-model to changes in "exponent of labor in agricultural sector production function" (ELA). The base value of this parameter is 0.45 and the range for the sensitivity test was 0.225–0.675. Although the modified M-model is *numerically* very sensitive to the value of *ELA*, its *behavior mode* does not change significantly. Sensitivity analysis for the parameters of the other production function in the modified M-model yields similar results.

The next parameter examined is "fraction of investment in capital equipment" (FICE). This parameter determines the portion of domestic demand that is allocated to capital goods production and the portion that is allocated to construction. The base value for FICE is set to 0.317 and the range for the sensitivity test was 0.158–0.475. Figure 21 presents the results. From a visual inspection of the figure, it is obvious that the M-model's behavior is not sensitive to the changes in *FICE*.

Another parameter which was selected for examination is "normal reserve coverage time" (NRC). This parameter determines how quickly the government depletes Iranian oil reserves. The base value for this parameter is set to 15 years. This means

[17] The behavior of all of the M-model's key variables was examined during the behavior mode sensitivity test but space limitations prevent their presentation in this chapter.

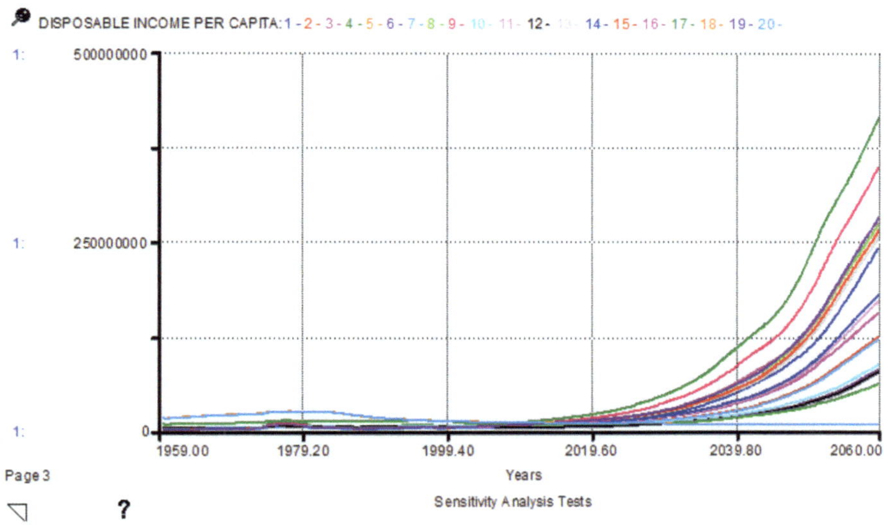

Fig. 20 Behavioral sensitivity of disposable income per capita in the modified M-model to changes in ELA

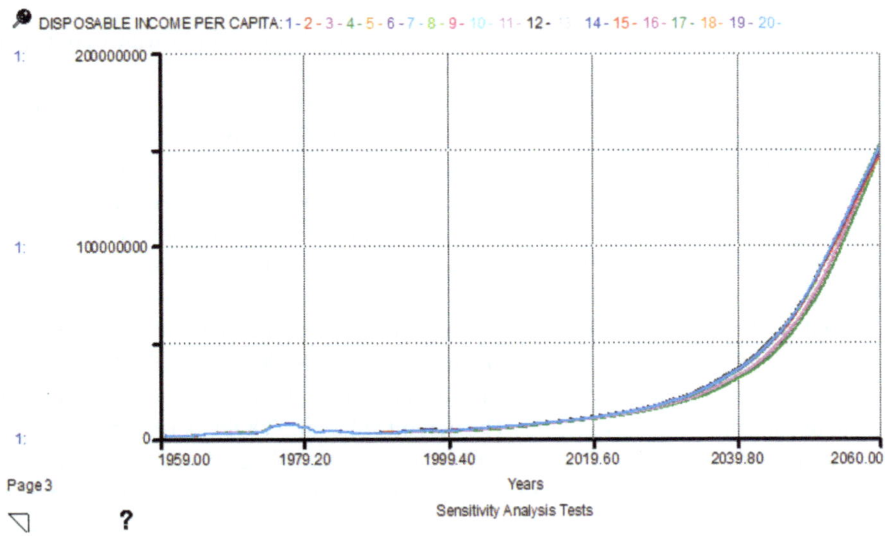

Fig. 21 Behavioral sensitivity of disposable income per capita in the modified M-model to changes in fraction of investment in capital equipment (FICE)

that the Iranian government adjusts its oil production rate such that existing oil reserves will last 15 more years. The range of values for the sensitivity test was chosen to be between 8 and 22 years. The results of the test are shown in Fig. 22. Again, from a visual inspection of the figure, it is obvious that the modified M-model's behavior is insensitive to changes in NRC.

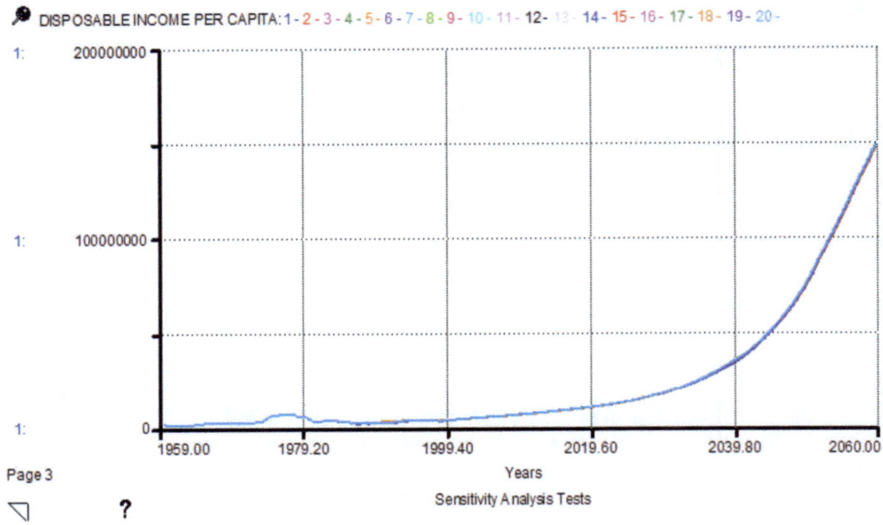

Fig. 22 Behavioral sensitivity of disposable income per capita in the modified M-model to changes in normal reserve coverage time (NRC)

The next parameter selected for presentation is "normal industrial output per capita" (NIPC). In the population sector, the birth and death rates depend on the level of industrialization in the country. NIPC provides a base value against which industrial output per capita generated by the model can be compared. This comparison provides an indication of the rate of Iranian industrialization. The base value of NIPC is 1,015,272 rials and the range of its value during the sensitivity test is 507,636–1,522,908 rials. Figure 23 presents the results of the test. Once again there is no evidence that the modified M-model is behaviorally sensitive to changes in NIPC.

The next test of model sensitivity involves the parameter "non-oil output per capita normal" (NOOPCN), which provides a base value against which non-oil output per capita (NOOPC) can be compared. If the ratio of NOOPC to NOOPCN is greater than 1 Iranian education capacity expands. Alternatively, if the ratio is less than 1 Iranian education capacity shrinks. The default value of NOOPCN is 3,555,739 rials per year per person (RPYPP) and the range of values explored during the sensitivity test is 1,777,870–5,333,608 RPYPP. The results of the sensitivity test are shown in Fig. 24.

Clearly, the behavior of the modified M-model is insensitive to this parameter. For the last example presented in this chapter, all five of the parameters discussed earlier are varied simultaneously. Figure 25 shows that the modified M-model is *numerically* sensitive, but *behaviorally* insensitive, to the combined set of changes. Indeed, all but one of the simulation runs generates exponential growth in disposable income per capita.

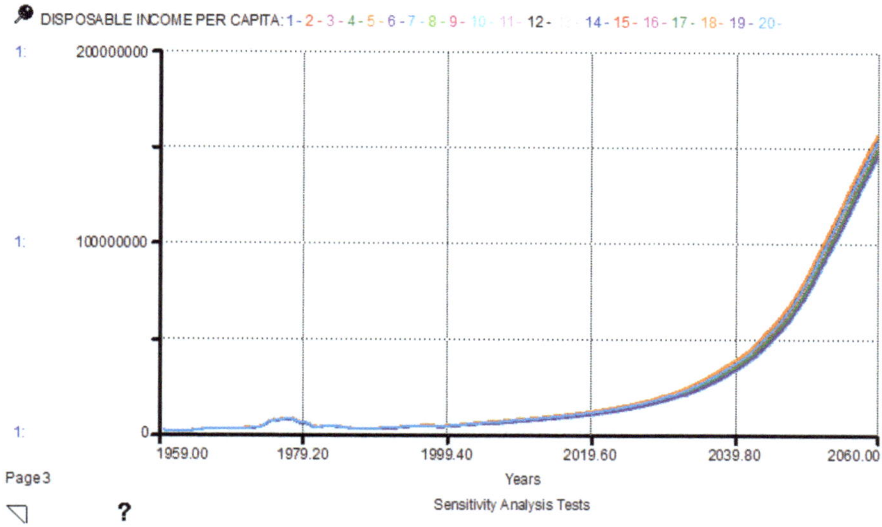

Fig. 23 Behavioral sensitivity of disposable income per capita in the modified M-model to changes in normal industrial output per capita (NIPC)

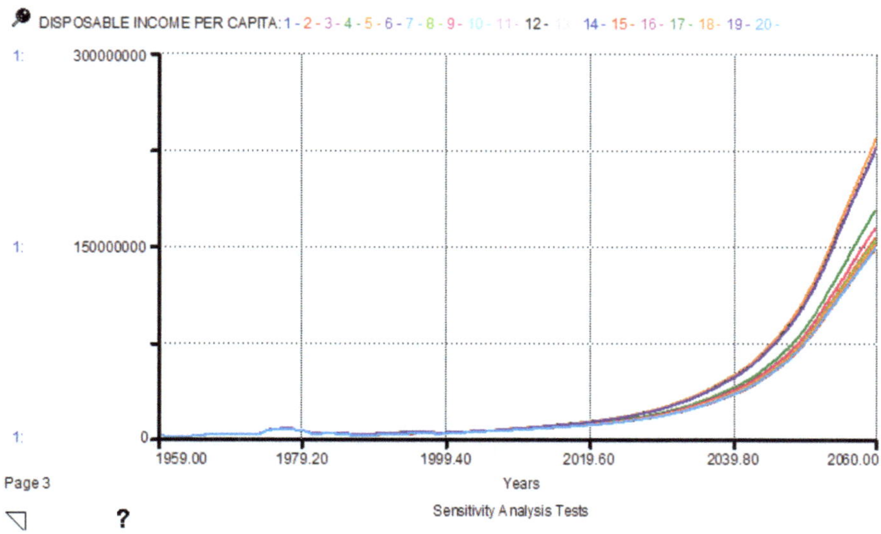

Fig. 24 Behavioral sensitivity of disposable income per capita in the modified M-model to changes in non-oil output per capita normal (NOOPCN)

Policy Sensitivity

If decision makers are to have confidence that the policy prescriptions generated by a system dynamics model are likely to yield the same results in the real system as they do in the virtual world, the policy prescriptions have to be robust. That is, the

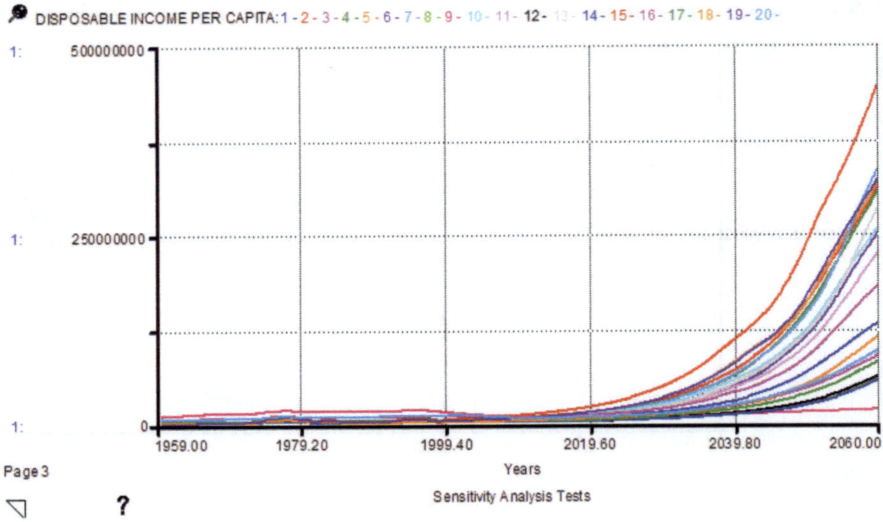

Fig. 25 Behavioral sensitivity of disposable income per capita in the modified M-model to simultaneous changes in all five parameters

policy prescriptions should not change when a model's parameters are varied over a reasonable range of values.

This section will present the results of policy sensitivity tests run on the modified M-model. As a prerequisite, however, some policy conclusions need to be drawn from the model.

Recall that, although the M-model has the potential to be a foundational platform for Iranian macroeconomic modeling, its current structure is quite limited. More specifically, its current structure is appropriate for examining issues related to the oil dependency of the Iranian economy, but not for answering broader macroeconomic questions in the areas of fiscal, monetary, or income redistribution policies.

Recall also that one of the counterintuitive conclusions drawn from simulations of the M-model is that *slowing down* or *limiting* investment in Iranian energy production will yield long-term benefits for the economy. This conclusion is in sharp contrast to the viewpoint held by many energy experts who believe that the Iranian government should attract *more* investment to speed-up Iranian oil and gas production (Barkeshli 2006).

The "more investment now" viewpoint is principally based on two perceptions. First, most of Iran's proven oil reserves are in the second half of their life cycles (MOE 2008). As a result, if secondary or tertiary oil recovery methods are not brought on-line, it will become increasingly difficult to exploit these reserves in the future (Ahmed 2006). To bring these methods on-line, however, Iran will have to invest more in its energy sector. Second, Iran shares some of its oil and natural gas fields with its neighbors (e.g., Qatar). If these jointly-owned reserves cannot be exploited in a timely fashion, they will impose some opportunity costs on the Iranian economy. Investing more in its energy sector now, rather than later, will increase the probability that the jointly-owned reserves can be utilized by Iran.

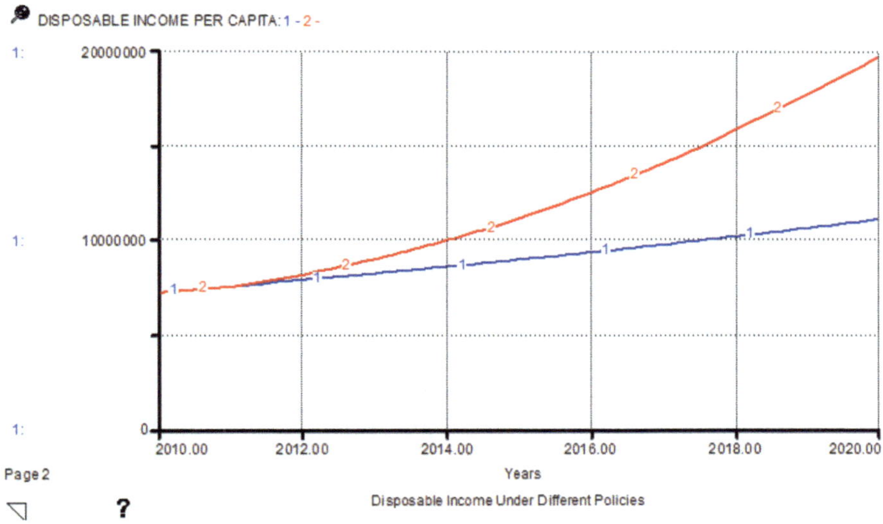

Fig. 26 Time paths for disposable income per capita from 2010 to 2020 under the "slow investment down" and "more investment now" strategies

The implications of the "more investment now" versus the "slow investment down" strategies can be tested with the modified M-model by simulating a change in Iranian foreign policy. To test the "more investment now" strategy, the assumption is made that the Iranian government improves its relationships with the developed countries and, as a result, is able to accelerate its production of oil. More precisely, it is assumed that this about-face in foreign affairs will allow Iran to double its oil production in as little as 1 year.[18] The simulation run embodying this assumption can be compared with an earlier one in which the maximum oil production growth rate was assumed to be 14 %. This is the "slow investment down" strategy. The short-term and long-term implications of these strategies can be compared separately. Figure 26 shows the time paths for "disposable income per capita" in the short-term (2010–2020).

Curve 1 represents the "slow investment down" strategy and curve 2 represents the "more investment now" strategy. Over this narrow period of time it is clear that increasing Iranian oil production capacity at a faster rate can yield higher economic welfare.

Figure 27, on the other hand, shows that the story is different in the long run. Although superior in the short-term, the "more investment now" strategy leads to an economic depression in the long-term because the economy cannot adjust to a lack of oil revenue caused by depletion.

The robustness of this policy conclusion can now be examined. To conduct this sensitivity test the protocol from section "Behavior Mode Sensitivity" will again be

[18] Recall that this was an assumption in the original M-model.

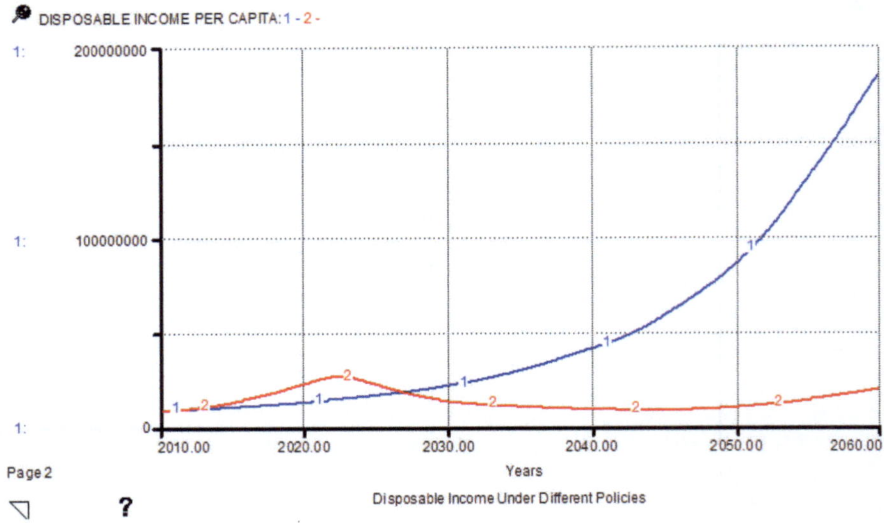

Fig. 27 Time paths for disposable income per capita from 2010 to 2060 under the "slow investment down" and "more investment now" strategies

utilized. Further, the same parameters that were varied in section "Behavior Mode Sensitivity" will be changed with the only difference being that, for each sensitivity run, each parameter is given only one new value. Similar to the results presented in Fig. 25, all of the parameters will be altered simultaneously.

Figures 28 and 29 show representative results from the policy sensitivity test in both the short and long runs, respectively. In both cases, curves 2 and 4 repeat the same results shown in Figs. 26 and 27 while, curves 1 and 3 show the time paths generated by the modified M-model with new, randomly chosen, parameter sets. In both, the short and long runs, the *behavior modes* are insensitive to the parameter changes. The overall conclusion of the test is that the modified M-model's policy recommendations are robust—that is insensitive to changes in model parameters.

System Improvement Tests

Solving a problem in an actual system is the ultimate goal of a system dynamics modeling project. A system improvement test is designed to determine whether or not the modeling process led to the achievement of this goal.

There are three parts to a system improvement test: (a) the model must generate policies which can improve the behavior of the system; (b) those policies must be applied in reality; and (c) the policy changes should enhance the performance of the real system in the ways suggested by the model. However, evaluating the impact of a model in practice is almost impossible. As Sterman (2001, pp. 887–888) explains:

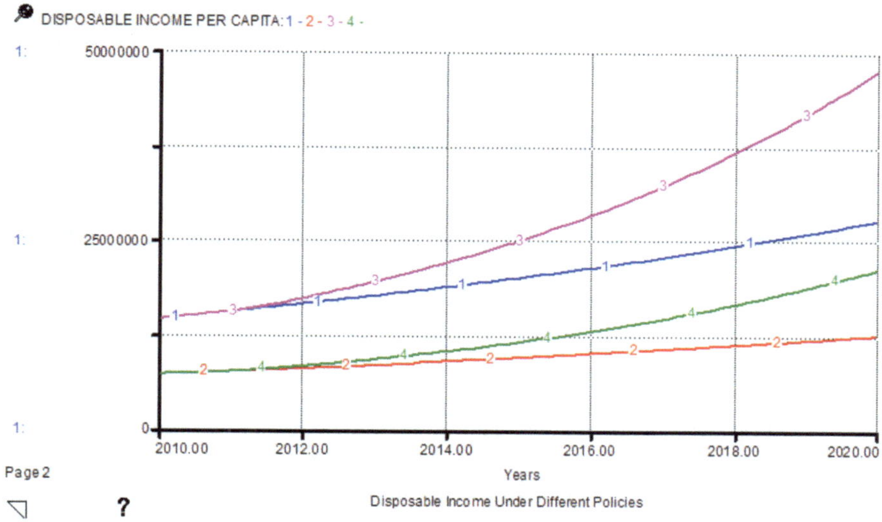

Fig. 28 Sensitivity of time paths for disposable income per capita from 2010 to 2020 under the "slow investment down" and "more investment now" strategies

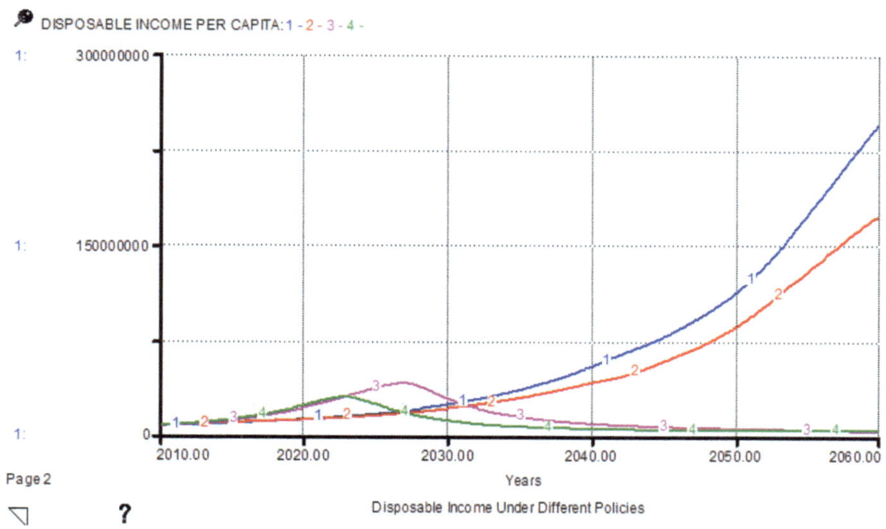

Fig. 29 Sensitivity of time paths for disposable income per capita from 2010 to 2060 under the "slow investment down" and "more investment now" strategies

It is hard to assess the extent to which the modeling process changed people's mental models and beliefs. It is rare that clients adopt the recommendations of any model promptly or without modification. When new policies are implemented, it takes a long time for their effects to manifest. Many other variables and conditions change at the same time new policies are implemented, confounding attempts to attribute any results to the policies. Performance

> improvement following a study does not mean the model-based policies were responsible; the system may have improved for reasons unrelated to the modeling process. Likewise, deteriorating performance after policy implementation does not mean the model failed since the outcome could have been even worse without the new policies.

In the present case, the ability of the M-model to pass a system improvement test is even more difficult to assess because it was never used by Iranian decision makers. Nevertheless, a case can be made that the M-model is able to pass this test. First, as was argued earlier in this chapter, the political conflict between Iran and Western countries after 1979 led to the reduction of Iranian oil production, which was one of the policies that Mashayekhi had suggested in his study. In other words, one of Mashayekhi's policy recommendations stemming directly from the M-model was implemented in the actual system, albeit accidentally rather than deliberately.

Second, after the publication of M-model's conclusions in Iran (Mashayekhi 1984) the topic of oil dependency came to the forefront in the academic arena. It also turned out to be an appealing subject in Iranian presidential elections. Indeed, all presidential candidates during and after the 1990s have emphasized the need for Iran to become independent of oil revenues and have promised, if elected, to implement policies that will lead this outcome (Katouzian 2009). Since other publications devoted to the topic of Iranian oil dependency appeared during the same period of time,[19] it is difficult to say which study had the greatest impact on Iranian society, but Mashayekhi's study had two advantages: (1) a version of it was published in Farsi, and (2) it was very easy to read and understand. It can be argued therefore that, although the M-model was unable to change the government's behavior regarding the oil dependency issue, it may have contributed to changing the mental models of the Iranian people.

Conclusions

In this chapter, a classic system dynamics model developed by Ali Naghi Mashayekhi in 1978 was resurrected, updated and revalidated. The goal of the model is to investigate the issue of Iranian oil dependency. The original model had predicted that Iran would face a harsh economic recession during the 1980s due to a steep fall in oil revenue caused by natural resource depletion. Thirty-five years later, however, Iran's oil reserves remain intact and the country has not encountered the sort of severe depression that was predicted.

An examination of the original M-model showed that it did not contain the structure necessary to capture the dynamics of the Islamic revolution or the war with Iraq that occurred during the 1980s. Updating the M-model's exogenous variables, modifying some of its assumptions, and recalibrating some of its parameters significantly improved its ability to reproduce Iranian economic history.

Revalidation of the M-model has shown that it is fairly robust and generally reliable. Although it is an excellent tool for analyzing questions directly related to the issue of Iranian oil dependency, however, due to its relatively narrow boundary it

[19] For example see Mahdavy (1970); Vakil (1977); Katouzian (1978); and Amuzegar (1983).

is an inadequate platform for analyzing many contemporary Iranian macroeconomic policies. Broadening the boundary of the M-model by adding sectors such as a financial market, a foreign exchange market, a labor market, and an energy market would greatly enhance its versatility. As such, it can be argued that the M-model can serve as a foundational platform for future Iranian macroeconomic modeling efforts.

Finally, this chapter can serve as a starting point and archetype for those who wish to develop a system dynamics macroeconomic model of a resource-dependent developing nation. Future research involving the use of the M-model for this purpose should, therefore, address the following issues:

1. As previously mentioned, the boundary of the M-model should be broadened to include a financial market, foreign exchange market, labor market, and an energy market.
2. The energy sector of the M-model should be revised to address energy–economy interactions. For example, the original M-model and its current modified version contain only one source of energy—oil. The boundary of the energy sector needs to be broadened to include alternative sources of energy and the economics of their substitutability.
3. The importation of energy is impossible in both the original M-model and its current modified version. This is not acceptable, particularly when the purpose of the model is to analyze energy–economy interactions.
4. The production functions in both the original M-model and its current modified version are very sensitive to their elasticity parameters. The formulation of these functions should be modified to eliminate this fragility.
5. The modified M-model should be recalibrated to see if it can reproduce the behavior of other "capital-deficient" oil exporting nations such as Nigeria, Algeria, Indonesia, Venezuela, Ecuador, or Mexico that have large populations and significant agricultural sectors.

References

Ahmadi Amouee, B.: Eqtesad Siasi Jomhouri Islami (Political Economy of Islamic Republic), 3rd ed. Gaame Now, Tehran (2006) (ISBN: 964-7387-22-9)

Ahmed, T.: Reservoir engineering handbook. Gulf Professional Publishing (2006) (ISBN: 0080480683, 9780080480688)

Amuzegar, J.: Oil exporters' economic development in an interdependent world. International Monetary Fund (1983) (ISBN: 1557750785, 9781557750785)

Barkeshli, F.: Daghdaghehaye Bakhshe Energy (Concerns of the energy sector). Energy Econ. Surv. **6**, 7–15 (2006)

Barlas, Y.: Formal aspects of model validity and validation in system dynamics. Syst. Dyn. Rev. **12**:183–210 (1996)

BP: BP Statistical review of world energy 2012. British Petroleum, London (2012)

CBI: Economic time series database (2012) Economic Time Series Database, Economic Research and Policy Department, Central Bank of the Islamic Republic of Iran (CBI). http://tsd.cbi.ir/DisplayEn/Content.aspx

Forrester, J.W.: Confidence in models of social behavior with emphasis on system dynamics models. Unpublished paper no. D-1967. Cambridge, MIT (1973)
Forrester, J.W.: Economic theory for the new millennium, in: 21st International Conference of the System Dynamics Society, New York City, The System Dynamics Society, 20–24 July 2003
Forrester, J.W., Senge, P.M.: Tests for building confidence in system dynamics models. Modelling for management: simulation in support of systems thinking **2**, 414 -434 (1996)
Karl, T.L.: The paradox of plenty: oil booms and petro-states. University of California Press (1997) (ISBN: 052091869X, 9780520918696)
Karl, T.L.: The perils of the petro-state: reflections on the paradox of plenty. J. Int. Affairs-Columbia Univ. **53**, 31–52 (1999)
Katouzian, H.: Oil versus agriculture a case of dual resource depletion in Iran. J. Peasant Stud. **5**, 347–369 (1978)
Katouzian, H.: Iranian history and politics: the dialectic of state and society. Routledge (2003) (ISBN: 1134430957, 9781134430956)
Katouzian, H.: The Persians: ancient, medieval, and modern Iran. Yale University Press (2009) (ISBN: 0300169329, 9780300169324)
Langarudi, S., Boushehri, N.S., Shirani, B.A., Mashayekhi, A.N.: Yek Model System Dynamics baraye Tajzie Tahlil System Eqtesad-Energy Iran (A System Dynamics Model for Analyzing Iran's Energy-Economy System) 26–1, 71–87 (2011)
Mahdavy, H.: The patterns and problems of economic development in rentier states: the case of Iran. Life **1000**, 1 (1970)
Mashayekhi, A.N.: Strategy of economic development in Iran: a case of development on exhaustible resources (Ph.D. Dissertation). MIT, Cambridge, MA (1978)
Mashayekhi, A.N.: Durnamaye Ayandeye Toeseye Eqtesadie Iran bar Payeye Naft (Strategy of Economic Development in Iran: A Case of Development on Exhaustible Resources). Markaze Nashre Daneshgahi, Tehran: Iran (1984)
Mass, N.J., Senge, P.M.: Alternative tests for the selection of model variables. Systems, Man and Cybernetics, IEEE Transactions on 8, 450–460 (1978)
MOE: Iran and world energy facts and figures. Deputy for Power & Energy Affairs: Ministry of Energy, Tehran (2008)
Oliva, R.: Model calibration as a testing strategy for system dynamics models. Eur. J. Oper. Res. **151**, 552–568 (2003)
OPEC: Annual statistical bulletin 2009. Organization of the Petroleum Exporting Countries, Vienna. http://www.opec.org/opec_web/static_files_project/media/downloads/publications/ASB2009.pdf (2010)
OPEC: Annual statistical bulletin 2012. Organization of the Petroleum Exporting Countries, Vienna (2012)
Pesaran, M.H.: Economic trends and macroeconomic policies in post-revolutionary Iran. In: Alizadeh, P. (ed.) The economy of Iran: the dilemmas of an Islamic state, pp. 63–100. Tauris, London (2000)
Peterson, D.W.: Hypothesis, estimation, and validation of dynamic social models: energy demand modeling. Massachusetts Institute of Technology. http://18.7.29.232/bitstream/handle/1721.1/27424/02200997.pdf?sequence=1 (1975)
Peterson, D.W.: Statistical tools for system dynamics. In: Randers, J. (ed.) Elements of the system dynamics method, pp. 143–161. Productivity Press, Cambridge (1980)
Radzicki, M.J.: Expectation formation and parameter estimation in uncertain dynamical systems: the system dynamics approach to post Keynesian-institutional economics. In: Proceedings of the Int System Dynamics Conference, Oxford, UK (2004)
Saeed, K.: Slicing a complex problem for system dynamics modeling. Syst. Dyn. Rev. **8**, 251–261 (1992)
Sterman, J.D.: Appropriate summary statistics for evaluating the historical fit of system dynamics models. Dynamica **10**, 51–66 (1984)

Sterman, J.D.: A skeptic's guide to computer models. In: Barney, G.O., et al. (ed.) Managing a nation: The microcomputer software catalog, pp. 209–229. Westview Press, Boulder (1991)
Sterman, J.D.: Business dynamics: systems thinking and modeling for a complex world. McGraw-Hill Education (2001) (ISBN: 0071068120, 9780071068123)
Sterman, J.D.: All models are wrong: reflections on becoming a systems scientist. Syst. Dyn. Rev. **18**, 501–531 (2002)
Theil, H.: Applied economic forecasting. North-Holland Pub. Co. (1966)
Vakil, F.: Iran's basic macroeconomic problems: a twenty-year horizon. Econ. Dev. Cult. Change **25**, 713–729 (1977)
Van Wijnbergen, S.: The 'Dutch disease': a disease after all? The Econ. J. **94**(373), 41–55 (1984)

Making Progress Towards Emissions Mitigation: Modeling Low-Carbon Power Generation Policy

Isaac Dyner, Carlos J. Franco and Laura M. Cardenas

Introduction

Climate change has become a major global concern in recent decades. The magnitude and scope of the effects of climate alteration are increasing worldwide. Droughts, melting glaciers, and the devastating impact of hurricanes and floods are just some of the manifestations that have been observed in different parts of the world (UNFCCC 2005).

These effects have promoted international agreements such as the Kyoto Protocol that aim at reducing greenhouse gases (GHGs) through the use of clean technologies and increasing energy efficiency. To achieve the goals that have been set, different mechanisms have been established, such as the Clean Development Mechanism (CDM) and Emissions Trading (UNFCCC 2002).

CDM promotes projects that can certify emission reductions in developing countries (Olsen 2007). Emissions trading involve negotiating carbon credits in exchanges such as the European Union Emissions Trading System (EU ETS), which accounts for 76 % of the global carbon market (Kossoy and Ambrosi 2010).

These mechanisms intend to contribute to the diffusion of clean or low carbon technologies in all countries. Note that the largest emitting sectors include electricity generation and urban transport; and that electricity generation—essential in modern

I. Dyner (✉)
CIIEN, CEIBA, Facultad de Minas, Universidad Nacional de Colombia,
sede Medellín, Colombia
e-mail: idyner@unal.edu.co

C. J. Franco
CIIEN, CEIBA, Facultad de Minas, Universidad Nacional de Colombia,
Carrera 80 No 65-223 Medellin, Colombia
e-mail: cjfranco@unal.edu.co

L. M. Cardenas
Facultad de Minas, Universidad Nacional de Colombia,
Carrera 80 No 65-223 Medellin, Colombia
e-mail: lmcardenasa@unal.edu.co

H. Qudrat-Ullah (ed.), *Energy Policy Modeling in the 21st Century*,
Understanding Complex Systems, DOI 10.1007/978-1-4614-8606-0_12,

societies for comfort and wealth creation, and the focus of this paper—may efficiently contribute to the diffusion of low carbon (Collantes 2007; Negro et al. 2010).

The adoption of clean energy technologies aims at breaking the vicious cycle that involves economic competitiveness and the technologies currently used for power generation, many of which are highly polluting (Bartz and Kelly 2007). Perhaps the predominant strategy to this end is the development and adoption of renewable energy technologies (RETs), which transform natural resources into useful forms of energy. Thus, the diffusion of these technologies is essential for achieving sustainable development goals (Negro et al. 2010; Rao and Kishore 2010). However, under the most optimistic scenarios, we are still in the early stages of this process, especially in emerging economies (Negro et al. 2010; Schwarz and Ernst 2009).

This chapter assesses strategies—through simulations—that may promote a low carbon electricity sector, illustrating this for the Colombian case. In this sense, we structure the discussion as follows: Section *The Future of Global Agreements and their Effect on Electricity* establishes the framework of international agreements on GHGs. Section *The Colombian Power Sector: Regulation of Non-Conventional Technologies* briefly presents the current setup of the Colombian electricity sector, including how unconventional energy sources are being promoted. This is followed, in the section *The Model*, by the description of an SD model that has been developed to assess alternative low-carbon electricity policy. Section *Model Components* discusses the model components, prior to the section *Policy Analysis*, which shows the simulations of policy effects for the Colombian case in terms of three scenarios, and finally, conclusions and recommendations are discussed in section *Conclusions and Recommendations*.

The Future of Global Agreements and their Effect on Electricity

The first commitment period of the Kyoto Protocol finished in 2012, and the world has not yet come to an agreement that could guarantee a second commitment period. Nonetheless, global concerns may lightly extend these policies, given that:

- The EU ETS will remain in place until 2020, regardless of the Kyoto Protocol (CEC 2009).
- The EU ETS will take into account CDM certificates, even after 2013, under the condition that they had been acquired before 2012 (CEC 2009).
- The establishment of the Kyoto Protocol has changed the global trend in the electricity sector, inducing this sector to head towards low carbon generation (Zachmann and Hirschhausen 2008), which has driven the development of cleaner technologies, making their adoption easier and cheaper.
- Efficient and low-carbon technologies have significantly penetrated the industrial, commercial, and residential sectors worldwide; and R&D in this field continues to produce results that have led to significant improvements.

Table 1 Colombian installed capacity, 2012 (XM 2012)

Resource		MW	%
Hydro		9,185	63.7
Thermal		4,545	31.5
	Natural gas	3,053	
	Coal	991	
	Fuel oil	501	
Small		635	4.7
	Hydro	533	
	Thermal	83	
	Wind	18	
Co-generators		55	0.4
Total SIN		14,420	100

In spite of the earlier mentioned stimulus, the general lack of progress following the Kyoto Protocol has nevertheless resulted in a state of inertia due to the long delays present in climate systems (Sterman and Sweeney 2002), as: (a) the events occurring today are manifestations of interventions undertaken some time ago, and (b) the degradation of the global environment has continued to worsen in recent years. As a result of these two factors, it is expected that during the years to come the electricity sector and others—i.e., transport, industry, and residential sectors—will continue to intensify R&D and the use of efficient and low carbon emission technologies globally.

However, the incorporation of these technological innovations might—most likely will—induce major changes in society, markets, and organizations, as well as in global economies. The implementation of these technological changes and their effects are still unknown in their entirety, and so they deserve to be analyzed in some depth.

The Colombian Power Sector: Regulation of Nonconventional Technologies

In Colombia, the electricity sector is characterized by a market structure established by Laws 142 and 143, 1994. The regulatory framework was initially established in 1995, Resolution No. 24 (Comisión de Regulación de Energía y Gas 1995).

In this market, energy utilities and the demand sector (or their agents) meet to make agreements on quantities and prices without government intervention. In Colombia, the Regulatory Commission for Energy and Gas (CREG) is responsible for overseeing the system.

By the end of 2012 the system installed capacity was 14,420 MW, distributed as follows (Table 1):

The system demand was 57,150 GWh during 2011, and is expected to grow 3.0–4.5 % during the coming years (UPME 2012).

Electricity prices are established according to two main elements: (a) market conditions (short- and long-term supply and demand trading), and (b) a security of

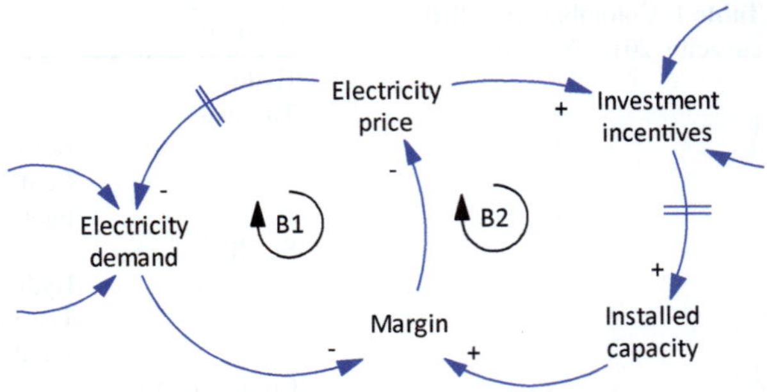

Fig. 1 Dynamics of electricity markets

supply mechanism that establishes a floor to price aimed at guaranteeing system reliability (reliability charge). This regulation is disadvantageous for wind and solar because of the intermittency embedded in these technologies. This paper examines alternative policy, using system dynamics (SD) modeling, to assess its effect on the development of nonconventional technologies such as wind and solar. Due to space constrains, system reliability is not studied here.

The Model

The internalization of emission costs has been largely accepted to reduce carbon emissions globally (Stern 2007). This policy is implemented through GHG certificates and their incorporation in the production costs. In the case of electricity, this means increasing the costs of fossil-based technologies—i.e., those intensive in GHGs—which has an effect on electricity prices. The SD model that we present in this paper was built using PowerSim Studio 7.

The system dynamics literature on environment-related issues in electricity markets has been extensive. Fiddaman (1997) assesses policy mitigation of GHGs in relation to electricity by way of levies: carbon taxes, energy taxes, and depletion taxes; Ford (1990, 2008, 2010) studies the electricity market to evaluate efficiency standards, and cap and trade mechanisms in the United States. Qudrat-Ullah and Davidsen (2001) present a model for evaluating policy incentives to the private sector, taking into account the evolution of CO_2 emissions in Pakistan. Dyner et al. (1995), and Dyner and Franco (2004) assess efficiency issues in Latin America. Sterman et al. (2012) present an energy version of a model that assesses carbon prices and subsidies globally. Preliminary versions of the model presented in this chapter have been discussed in Dyner et al. (2011, 2012, 2013).

In a significant number of countries, electricity systems have moved away from central planning structures to market-based mechanisms (Newbery 1999). As in any other market, prices are determined depending on abundance or scarcity. In this context, as shown in Fig. 1, electricity price increases as reserve margin contracts, which provides signals for both demand and supply. On the one hand, as price increases demand decreases in the intermediate future, given the demand–price elasticity; on

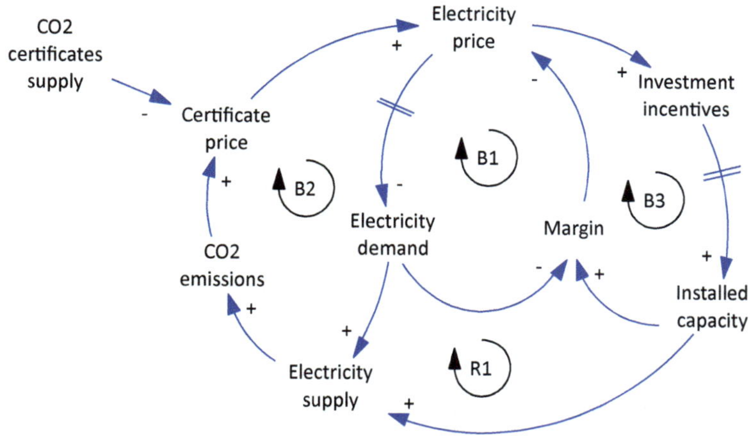

Fig. 2 Dynamic hypothesis of a policy of internalizing costs of greenhouse gas (GHG) emissions in the electricity sector

the other hand, sustained higher prices become a determining signal, along with other variables such as suitable regulation and an appropriate business environment, for capacity investment in the long term. Both higher demand and larger capacity contribute, in opposite ways, to the reserve margin, closing both control cycles. In this case, delays usually contribute to oscillations, which have been reported in actual electricity systems (Bunn and Larsen 1994).

Recent shifts in electricity markets policy that incorporate environmental trends seek to discourage fossil fuel-based generation. Figure 2 presents a dynamic hypothesis of a policy that internalizes the costs of GHG emissions in the electricity sector as a strategy to address the challenge of moving towards low-carbon economies. Here, we are discarding second-order effects coming from other variables that influence the industry.

This hypothesis comprises four cycles. The first one is the balance loop (B1), which indicates that as electricity demand increases, system margin decreases, inducing an opposite effect on electricity price, which in turn negatively influences demand. Loop B2 indicates that as demand increases, along with a higher installed capacity, greater electricity supply will be in place, which will induce higher emissions, promoting fossil-fuelled technologies. Thus, more carbon certificates will be needed, resulting in additional costs for electricity generation, increasing price, which has an opposite effect on demand.

Balance loop B3 establishes how installed capacity is influenced by the new generating capacity obtained from expected higher revenues, as a consequence of price hikes. This additional capacity will increase the system margin, with a negative effect on electricity price. Finally, rises in capacity and demand will induce surges in electricity supply, boosting the market for carbon certificates.

The institutional framework (North 1990) is reflected in the model through its influence on costs via alternative policies (Fig. 3). This paper assesses these two

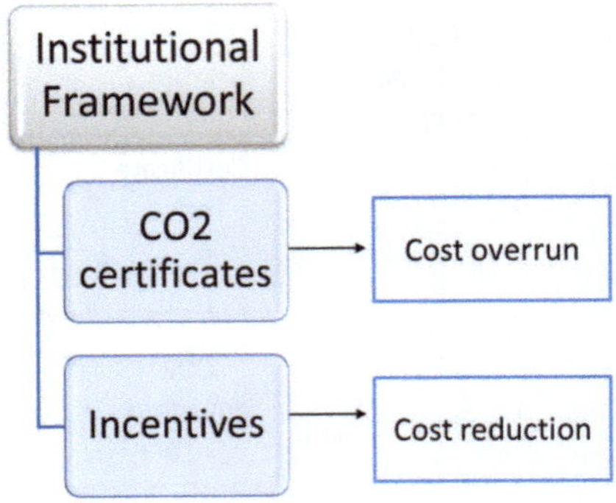

Fig. 3 Institutional framework for low carbon policies

alternative policies: (a) one that considers CO_2 certificates, which establishes extra cost for fossil-fuelled technologies, and (b) an alternative one that promotes clean technologies through operational subsidies.

The certificate policy is represented in Figs. 2 and 3 (loop R1). This establishes price incentives for cleaner technologies as they are not exposed to the cost of these certificates. Electricity supply from fossil fuels (or clean energy), depending on the technology that is used, increases (or reduces) the amount of CO_2 emissions, affecting the price of CO_2 certificates and therefore electricity price, which provides a signal for investment in new capacity. Balance loops B4 and B5 (Fig. 4) represent the direct incentive policy for clean technologies.

The incentive policy is introduced in the dynamic hypothesis to account for CO_2 emissions as emissions stimulate investments in clean technologies.

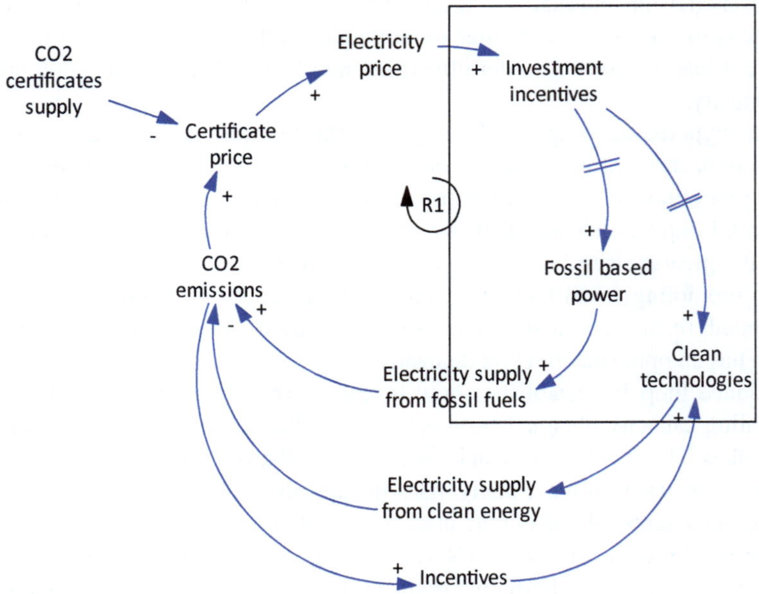

Fig. 4 Dynamic hypothesis considering low-carbon policy

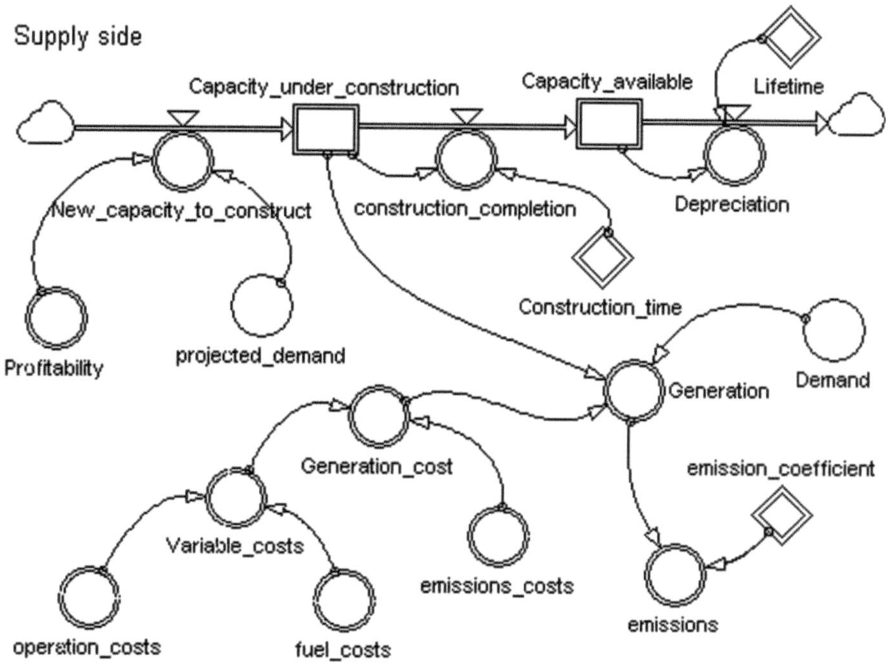

Fig. 5 Model structure for the supply side

Model Components

The model built, contains four components: supply, demand, price formation, and low-carbon policies. Some details are provided next.

The Supply Component

The electricity supply component includes the alternative technologies available, the installed capacity in place, and their main characteristics: generation price and emission coefficient, according to kWh generated. Figure 5 presents the model structure of the supply side.

Capacity under construction depends on the new capacity to construct and construction completion (Eq. 1). New capacity to construct depends on the profitability of each technology and projected demand; and this is added to the capacity available according to the construction time taken to build the corresponding plant (Eqs. 2–4).

$$CUC_i(t) = CUC_i(0) + \int_0^t (NCC_i(t) - CC_i(t))dt \tag{1}$$

$$CA_i(t) = CA_i(0) + \int_0^t (CC_i(t) - D_i(t))dt \tag{2}$$

$$NCC_i(t) = PF_i(t), PD(t) \tag{3}$$

$$CC_i(t) = \frac{CUC_i(t)}{TC_i(t)} \tag{4}$$

The generation costs for each technology depend on their variable costs (which consist of operating costs and fuel costs) and emissions costs.

The Demand Component

Demand depends on the growth rate inertia (determined by population and economic growth). This is, however, affected by the demand elasticity to price, which is estimated by a parameter (normally between − 0.2 and − 0.5), expressed in Eqs. 5–7, as can be seen in Fig. 6.

$$Dem(t) = Dem(0) + \int_0^t DemG(t)^{-\alpha(t)}dt \tag{5}$$

$$\alpha(t) = Price\_elasticity\_on\_demand \tag{6}$$

$$DemG(t) = (P(t), GDP(t)) \tag{7}$$

Price Formation

Price formation in the model is given by the balance between demand and supply. Each of the technologies offers quantities and price. These offers are organized according to merit, from the lowest to the highest price, thus forming the supply curve. In the spot market, demand establishes the pool price, as shown in Fig. 7.

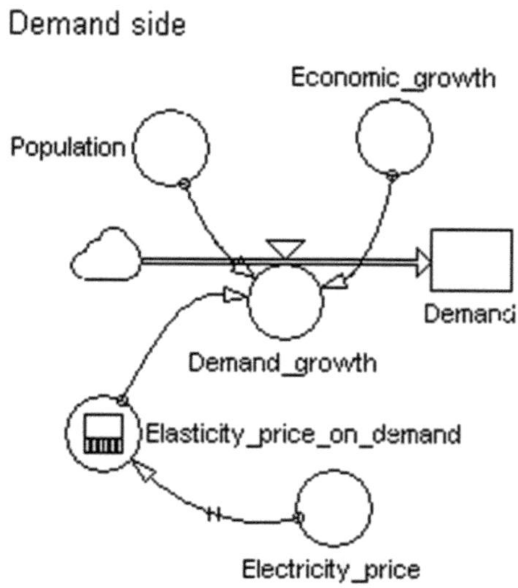

Fig. 6 Model structure for the demand side

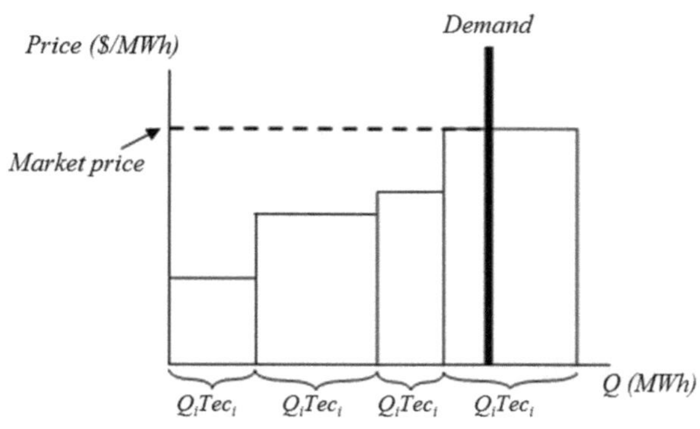

Fig. 7 Price formation in the model

Low Carbon Policies

The model considers two alternative policies: (1) CO_2 certificates, which establishes extra cost on fossil-fuelled technologies, and (2) subsidies to promote clean technologies through an established tariff. Figure 8 shows how the inclusion of a carbon price generates emission costs, which affect generation costs of fossil fuel-based technologies.

The structure of the second policy (incentives for clean technology) is presented in Fig. 9, which sets out a direct subsidy for renewable technologies that increase their profits from the sale of electricity.

We validated the model structurally and behaviorally, including dimensional consistency and extreme conditions tests, as well as sensitivity analyses. Results of these assessment exercises are not presented in this chapter as they go beyond the scope of the chapter.

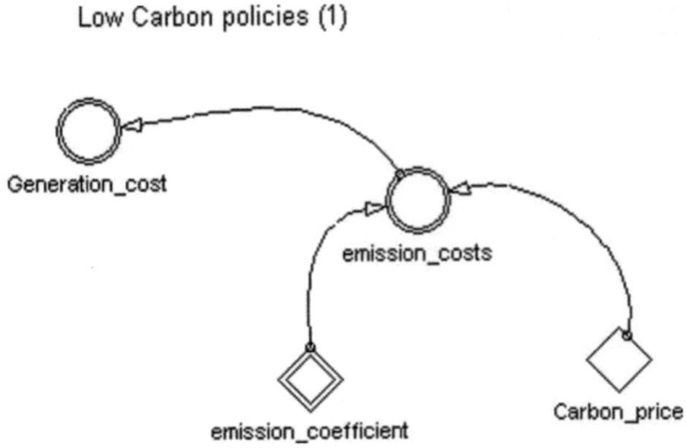

Fig. 8 Model structure for CO_2 certificates policy

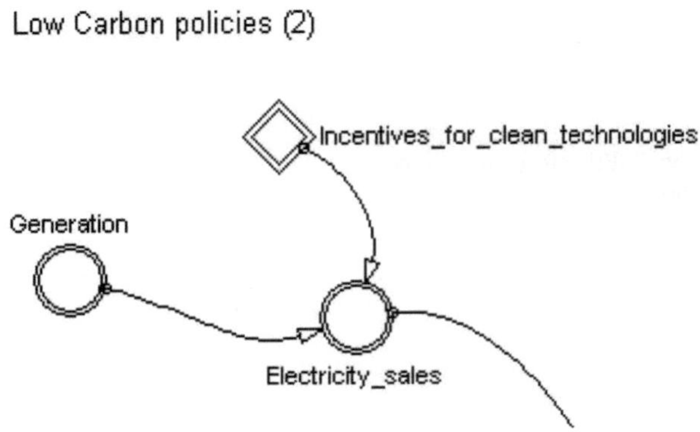

Fig. 9 Model structure for operational subsidies for clean technologies

Policy Analysis

The built model simulates the period 2013–2035. The model considers each of the Colombian electricity technologies: natural gas, fuel oil, coal, wind, hydro, and solar. We next describe three scenarios to provide the grounds for policy analysis (Fig. 10):

- Business as usual: There is neither presence of carbon certificates for electricity generators that use fossil fuels nor a policy to stimulate clean technologies.
- Emission certificates: This scenario establishes that generation utilities that pollute have access to carbon emission certificates—the polluter pays.
- Incentives for clean energy: This scenario establishes that government provides incentives for generators that use clean technologies (i.e., wind and solar). Incentives are provided via financial resources to generators that use clean technologies. The extra charges for the system are at the expense of higher electricity prices.

Next we discuss simulation results for each of the scenarios described in this section. The oscillations correspond to the cycles that occur as a result of the investment decisions that have been modeled.

Fig. 10 Scenarios for the policy analysis

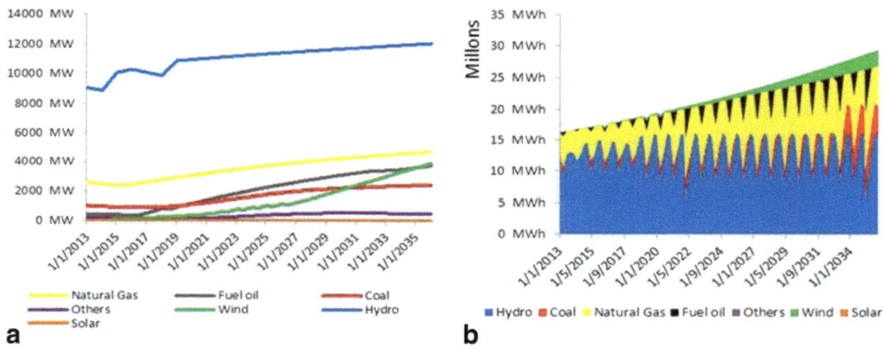

Fig. 11 **a** Installed capacity, **b** generation in business as usual scenario

Business as Usual Scenario

Figure 11 shows the simulated evolution of the generation capacity and the monthly generation of each of the modeled technologies.

This scenario, as expected, shows significant increases in hydroelectricity, given its low operational cost. Thermoelectricity shows oscillations with average growth-trends. Renewable-based technologies show high increases in the case of wind technology (yet insignificant as a percentage of total), but low rises in the case of solar, given its high investment cost. Spikes in generation are the result of the dispatch mechanism in place, which is a consequence of the seasonal variability of the system.

Emission Certificates Scenario

Figure 12 shows the evolution of installed capacity and generation for each of the technologies modeled in the Emission certificates market scenario.

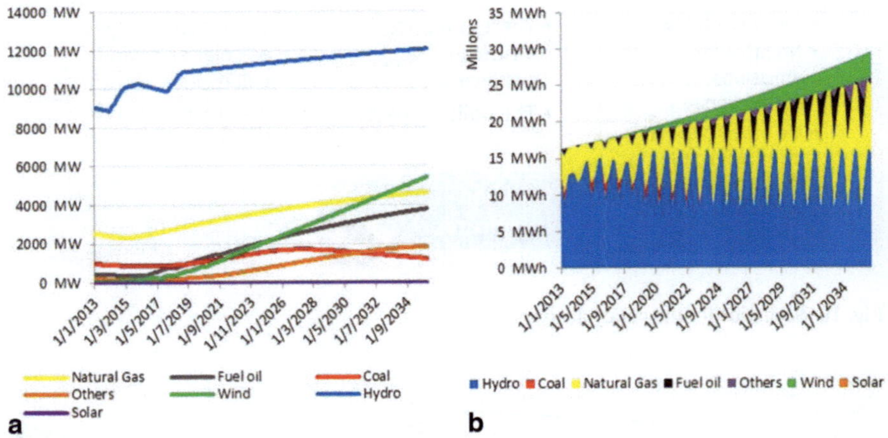

Fig. 12 **a** Installed capacity, **b** generation in emissions certificates scenario

The difference as compared with the Business as usual scenario is that all technologies that use fossil fuels tend to decline significantly, because of their emission costs, which make them less profitable. In this scenario there are greater increases in clean technologies (wind) than in the Business as usual scenario, but the increases are insignificant with respect to the total; solar remains insignificant in terms of market participation.

Incentives for Clean Technologies Scenario

Figure 13 shows results in terms of capacity and generation for the scenario: incentives for clean technologies.

This scenario highlights growing trends in clean technologies (wind and solar) due to the presence of clear and focused incentives. This is the result of positive returns-on-investment for these technologies, which clearly incentivize investment. This comes at the expense of higher electricity prices. It is important to point out decreases in fossil fuel technologies, though not as sharp as in the previous scenario.

Figure 14 shows simulations of CO_2 emissions under the previously discussed scenarios. As expected, the worst scenario is business as usual. As time passes, the system increases investments on clean technologies and the difference with respect to the base-case scenario becomes significant. Policy that promotes clean technologies manages to attain reductions of as much as 60 % of CO_2 emissions.

We now turn to our final section, on conclusions and recommendations.

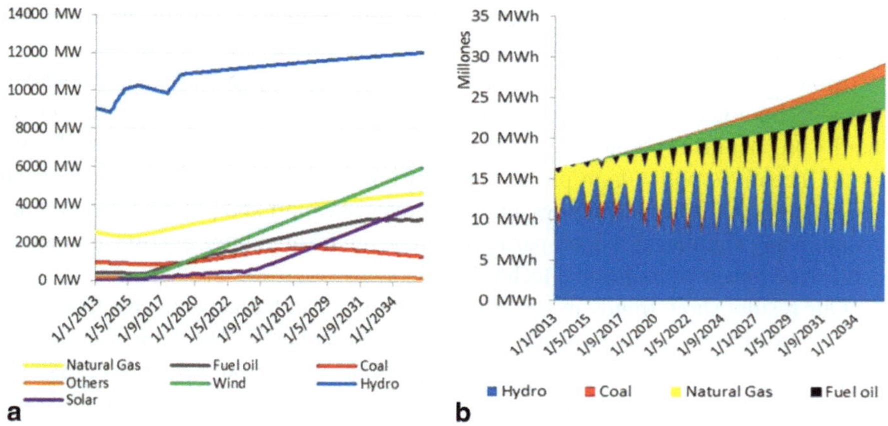

Fig. 13 **a** Installed capacity, **b** generation in incentives scenario

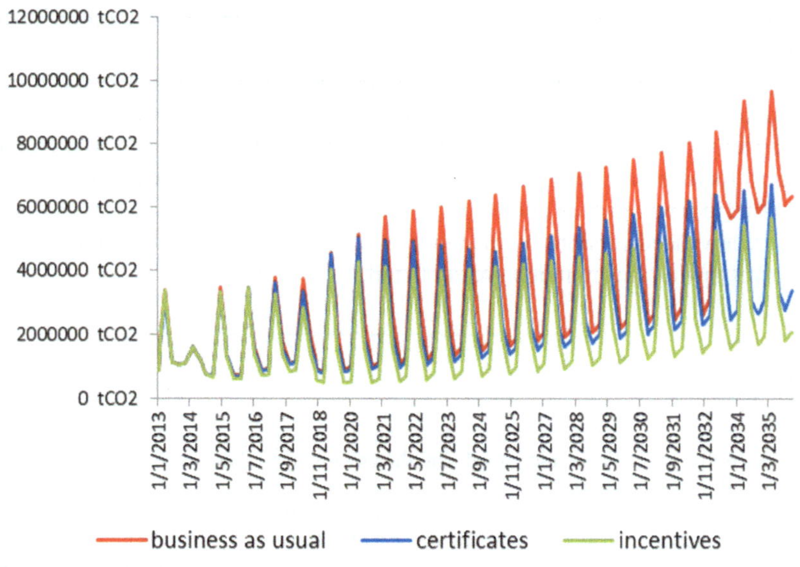

Fig. 14 CO_2 emissions counts for the three scenarios

Conclusions and Recommendations

This research has been fruitful on at least two counts: It signals the benefits of the modeling approach to assessing a low CO_2-emission policy and, second, it shows the modeling capability of the proposed approach for assessing the penetration of renewable-based technologies.

The modeling approach that has been undertaken facilitates policy assessment regarding CO_2 emissions. The literature supports the general idea proposed by the authors of this paper but there is novelty regarding how to apply this to the power supply sector.

Modeling provides capabilities to assess and analyze the penetration of renewable-based technologies. It becomes clear that the penetration of wind technologies would hardly prosper in the absence of a CO_2-emission policy. However, when a Certificates or an Incentives policy is in place, the diffusion of clean technologies prospers. It is also clear that solar technology requires greater support and policy focus in order to attain any significant market participation, given its high present costs, compared with wind power or traditional technologies.

Our results show trends in the composition of installed capacity for Colombia when considering policies towards a low carbon economy. The policy, under the premise that "the polluter pays", has a significant impact on reducing emissions, but a direct and more focused policy—Incentives—has a greater effect, stimulating expensive technologies, such as solar.

Although a Certificates scenario reduces the increase of polluting technologies, this is not sufficient to promote clean technologies as much as the one that focuses on Incentives. This has been shown in practice in Europe and in some Latin American countries.

The results of this paper indicate that if the aim of the policy in place is to reduce emissions, then the Certificates scenario presents interesting options, as this affects the capacity of fossil fuel technologies, with high GHG emissions. The Incentives scenario focuses on expanding the capacity of clean technologies without affecting fossil fuel technologies as much.

More advanced versions of this model will incorporate the demand sector as well as market institutions in much greater detail.

References

Bartz, S., Kelly, D.L.: Economic growth and the environment: theory and facts. Resour. Energy Econ. **30**, 115–149 (2007)

Bunn, D., Larsen, E.: Assessment of the uncertainty in future UK electricity investment using an industry simulation model. Util. Policy. **4**, 229–236 (1994)

CEC (Commission of the European Communities: Directive 2009/29/EC of the European Parliament and of the Council of 23 April 2009 amending Directive 2003/87/EC so as to improve and extend the greenhouse gas emission allowance trading scheme of the Community (Text with EEA relevance), OJ L 140, 5.6.2009, pp. 63–87. Brussels (2009)

Collantes, G.O.: Incorporating stakeholders' perspectives into models of new technology diffusion: the case of fuel-cell vehicles. Technol. Forecast. Soc. Chang. **74**, 267–280 (2007). doi:10.1016/j.techfore.2006.02.001

Comisión de Regulación de Energía y Gas.: Resolución No 24 de 1995. Por la cual se reglamentan los aspectos comerciales del mercado mayorista de energía en el sistema interconectado nacional, que hacen parte del Reglamento de Operación. Colombia. http://www.creg.gov.co/html/i_portals (1995). Accessed 18 Oct 2011

Dyner, I., Franco, C.J.: Consumers' bounded rationality: the case of competitive energy markets. Syst. Res. Behav. Sci. **389**, 373–389 (2004)

Dyner, I., Smith, R.A.R., Peña, G.E.: System dynamics modelling for residential energy efficiency analysis and management. J. Oper. Res. Soc. **46**, 1163–1173 (1995)

Dyner, I., Cardenas, L., Franco, C.J.: A system dynamics model for assessing the UK carbon market. 29th International Conference of the System Dynamics Society. p. 15 (2011)
Dyner, I., Franco, C., Cárdenas, L.: Making progress towards defossilisation: modelling low-carbon policy for power generation. 30th International Conference of the System Dynamics Society, p. 11. St. Gallen, Switzerland (2012)
Dyner, I., Franco, C.J., Cárdenas, L.: Assessing policy for mitigating greenhouse gases in electricity. The 31st International Conference of the System Dynamics Society, p. 20. Cambridge (2013)
Fiddaman TS: Feedback Complexity in integrated climate-economy models. Ph.D. Thesis, MIT Sloan School of Management, Cambridge (1997)
Ford, A.: Estimating the impact of efficiency standards on the uncertainty of the northwest electric System. Oper. Res. **38**, 580–597 (1990)
Ford, A.: Simulation scenarios for rapid reduction in carbon dioxide emissions in the western electricity system. Energy Policy. **36**, 443–455 (2008). doi:10.1016/j.enpol.2007.09.023
Ford, A.: Greening the economy with new markets: System Dynamics Simulations of Energy and Environmental Markets. In: System Dynamics Society (ed.) The 28th international conference of the System Dynamics Society, pp. 1–26. Seoul, Korea (2010)
Kossoy A, Ambrosi P: State and trends of the carbon market 2010, p. 89. World Bank, Washington DC (2010)
Negro, S., Hekkert, M., Alkemade, F.: Seven Typical system failures that hamper the diffusion of sustainable energy technologies. ERSCP-EMSU Conference, pp. 1–21. Delft, Netherlands (2010)
Newbery D: Privatization, restructuring, and regulation of network utilities. MIT Press, Cambridge (1999)
North, D.: Institutions, institutional change, and economic performance. Cambridge Universitiy Press (1990)
Olsen, K.H.: The clean development mechanism's contribution to sustainable development: a review of the literature. Clim. Chang. **84**, 59–73 (2007). doi:10.1007/s10584-007-9267-y
Qudrat-Ullah, H., Davidsen, P.I.: Understanding the dynamics of electricity supply, resources and pollution. Pak. Case Energy. **26**, 595–606 (2001). doi:10.1016/S0360-5442(01)00019-6
Rao, K.U., Kishore, V.V.N.: A review of technology diffusion models with special reference to renewable energy technologies. Renew. Sustain. Energy Rev. **14**, 1070–1078 (2010). doi:10.1016/j.rser.2009.11.007
Schwarz, N., Ernst, A.: Agent-based modeling of the diffusion of environmental innovations—An empirical approach. Technol. Forecast. Soc. Chang. **76**, 497–511 (2009). doi:10.1016/j.techfore.2008.03.024
Sterman, J.D., Sweeney, L.B.: Cloudy skies: assessing public understanding of global warming. Sist. Dyn. Rev. **18**, 207–240 (2002)
Sterman, J., Fiddaman, T., Franck, T., et al.: Climate interactive: the C-ROADS climate policy model. Syst. Dyn. Rev. **28**, 295–305 (2012). doi:10.1002/sdr
Stern N: The economics of climate change: the stern rReview, p. 712. Cambridge University Press, Cambridge (2007)
UNFCCC: Para comprender el cambio climático: guía elemental de la convención marco de la Naciones Unidas y el Protocolo de Kyoto, p. 21. Bonn. http://unfccc.int/resource/docs/publications/beginner_sp.pdf (2002) Accessed 7 June 2010
UNFCCC: Cuidar el clima, Guía de la Convención Marco sobre el Cambio Climático y el Protocolo de Kyoto. 50 (2005). doi:92-9219-022-9
UPME: Plan de Expansión de Referencia Generación—Transmisión 2012–2025. Unidad de Planeación Minero Energética—UPME. Bogotá, Colombia. http://www.upme.gov.co/Docs/Plan_Expansion/2012/PLAN_COMPLETO_2012–2025.pdf (2012). Accessed 5 March 2013
XM: Descripción del sistema eléctrico colombiano. In: Información operativa y comercial. http://www.xm.com.co/Pages/DescripciondelSistemaElectricoColombiano.aspx (2012). Accessed 1 Oct 2012
Zachmann G, Von Hirschhausen C: First evidence of asymmetric cost pass-through of EU emissions allowances: examining wholesale electricity prices in Germany. Econ. Lett. 99(3), 465–469 (2008)

Exploring Energy and Economic Futures Using Agent-Based Modeling and Scenario Discovery

P. Wang, M. D. Gerst and M. E. Borsuk

Introduction

Providing quantitative support for climate change policy is a challenging problem because doing so involves representing linked social and technological systems over long time spans. Such systems, which are complex and adaptive, are difficult to model with reasonable scientific accuracy because they contain both irreducible (also known as aleatoric or statistical) and reducible (also known as epistemic or knowledge) uncertainties. For example, the likelihood that research and development (R&D) programs will reduce renewable energy costs to be competitive with energy produced from fossil fuels is considerably uncertain and fundamentally unknowable. Past results of R&D can be used to provide a guide of what is possible, but ultimately the uncertainty surrounding cost reductions is irreducible. Other uncertainties, such as how households or firms make decisions, are, in theory, reducible, but the state of our knowledge often still requires considering multiple hypotheses of real-world behavior.

Historically, construction of scenarios has proven valuable as a means for organizing and communicating the many uncertainties associated with climate policy support. A scenario can be thought of as a 'coherent, internally consistent, and plausible description of a possible future state of the world' (McCarthy et al. 2001). By illuminating the span of possible futures, consideration of diverse scenarios has the potential to highlight the interaction of complex uncertainties that would otherwise be difficult to analyze (Groves and Lempert 2007).

Climate policy scenarios have mostly been produced by a sequential, piecewise process. Subject-matter experts are convened to create storylines that qualitatively describe plausible, internally consistent outcomes for irreducibly uncertain processes,

M. E. Borsuk (✉) · P. Wang · M. D. Gerst
Dartmouth College, Thayer School of Engineering, 14 Engineering Drive, Hanover, NH 03755, USA
e-mail: mark.borsuk@dartmouth.edu

P. Wang
e-mail: peng.wang@dartmouth.edu

H. Qudrat-Ullah (ed.), *Energy Policy Modeling in the 21st Century*, Understanding Complex Systems, DOI 10.1007/978-1-4614-8606-0_13, such as future population change, economic growth, and technological progress. These storylines are then translated into quantitative projections that are thought to be representative of the storyline themes. Finally, the exogenous projections are used as inputs to formal models that produce key outputs such as energy technology market shares, greenhouse gas emissions, and atmospheric CO_2 concentration.

The most well-known application of the sequential scenario process to climate policy has been the Special Report on Emissions Scenarios (SRES; Nakicenovic and Swart 2000). It adopted the scenario axis method adopted by Schwartz (1991), which uses quadrants of a two-dimensional space to define four scenarios. In SRES, the axes are defined by degree of globalization and degree of sustainable development. Following the sequential process, the quadrants were used to sketch four storylines and quantify four sets of projected exogenous variables, which were used as model inputs for many climate policy studies. However, after more than a decade of utilization, the modeling community began to indicate that the scenario axis and sequential methods often hindered effective use of scenarios (Moss et al. 2010; Parson et al. 2007). Because storylines were drafted separately from model construction, it was often difficult for the models to completely engage with scenario themes. Furthermore, how to interpret the scenarios in a decision-making context was often unclear, as disagreement among modelers and practitioners surrounded the issue of assigning probabilities to scenario outcomes.

A recent effort to overcome these issues has been the Representative Concentration Pathway framework (RCP; Moss et al. 2010). In contrast to SRES, RCP scenarios are first defined by outcomes instead of driving forces: four radiative forcing stabilization pathways, ranging from ambitious climate stabilization at 2.6 W/m^2 forcing to a more baseline scenario of 8.5 W/m^2 forcing, which correspond, respectively, to atmospheric greenhouse gas concentrations of about 430 and 1,230 ppm CO_2-eq. in the year 2100. Then, pathways are used in one of the two ways: (i) as forcing inputs into complex climate system models or (ii) as targets for climate policy models.

Beginning scenario planning with policy targets defined by physical variables introduces new challenges and opportunities. On the positive side, modeling teams have more freedom to define social, economic, and technological scenario attributes. However, this new flexibility adds an additional layer of uncertainty to the comparison of model results because storyline and model assumptions are now likely to be different. As a result, the scientific community has begun the task of defining a set of Shared Socioeconomic Pathways (SSPs) to serve as a baseline for comparison (Kriegler et al. 2012). The first step in that direction has been to compare existing scenarios, looking for consistent patterns of socioeconomic drivers across differing emissions scenarios. Using scenarios from EMF-22 (energy modeling forum; Clarke et al. 2009), AR4 (fourth assessment report; Fisher et al. 2007; Nakicenovic et al. 2006), and the RCPs (Moss et al. 2010), van Vuuren et al. (2012) found that much overlap existed in the range of socioeconomic drivers for any given emission trajectory. This indicates that RCPs, or emissions trajectories, alone may not sufficiently identify individual socioeconomic scenarios. Resultantly, van Vurren et al. (2012) have proposed a matrix framework whereby RCP forcing targets define four matrix rows, and SSP drivers, such as mitigative and adaptive capacity, define columns. How to fill in the matrix elements remains an open question. Among the many issues are how to ensure consistency among rows and columns and how to address co-variance among SSP drivers.

In an initial attempt at addressing these questions, Rozenberg et al. (Rozenberg et al. 2012) use 286 simulations of the IMACLIM-R model (Rozenberg et al. 2010) and Bryant and Lempert's (2010) scenario discovery method to generate self-consistent scenarios to populate the matrix. Scenario discovery operates in the opposite direction of sequential approach. First, probabilistic simulations from a quantitative model are generated. Then, using nonparametric statistical methods, model outputs are grouped according to chosen metrics and determinant driving forces for each group are identified. As discussed in Gerst et al. (2013a), Bryant and Lempert's method, while clearly a step forward, requires selecting a priori performance thresholds in order to group model outputs. This introduces the possibility that interesting dynamics might be overlooked, as it is difficult to determine whether selected thresholds appropriately delineate multidimensional model output.

Our previous work (Gerst et al. 2013a) demonstrated a more generalized version of scenario discovery that allows for multiple performance dimensions without the need for a priori threshold selection. In the current contribution, we further demonstrate the utility of this approach by using an enhanced version of the agent-based ENGAGE model (Gerst et al. 2013b) to identify socioeconomic pathways for the 4.5 W/m^2 RCP. While ENGAGE remains a relatively simple model, we believe the results demonstrate how the combination of agent-based modeling and scenario discovery might be used to 'fill in' the matrix framework relating to RCPs and SSPs.

Method

ENGAGE is an agent-based, energy–economy model that is patterned after the family of evolutionary economic models recently developed by Dosi et al. (2010), (Fig. 1). This model consists of four types of agents—households, consumption goods firms, capital goods firms, and government—and one resource, labor. It is particularly well suited as a starting point for investigating the technological and economic aspects of climate policy because technological change is modeled as the driving force of economic growth and is represented as being both stochastic and endogenous. In our previous work (Gerst et al. 2013b), we expanded the original model to include energy as a resource, which involved adding firms that produce energy technologies and a single form of energy used by households and firms. In this section, we provide a brief description of the model and detail a new functionality added for the current study, including: (i) probabilistic population growth, (ii) probabilistic fuel costs, and (iii) endogenous climate policy. We encourage readers to refer to Gerst et al. (2013b) for details on the structure, parameterization, and motivation for ENGAGE.

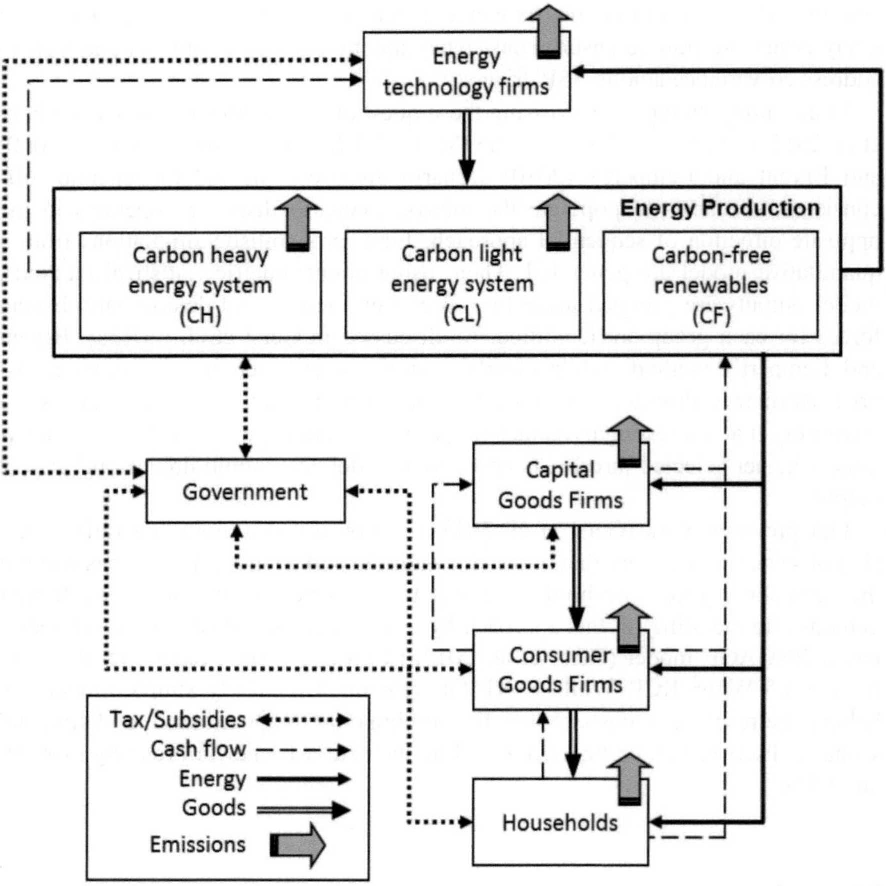

Fig. 1 Model schematic with boxes showing the various classes of agents and arrows indicating their interactions

Households

In our simplified economy, households supply labor to firms and spend all earned income on purchasing new generic consumption goods, which we call 'thneeds,' and energy to use existing thneeds. We do not explicitly model the labor market: wages (w) earned by households track closely with economy-wide changes in labor productivity and unemployed households receive an income subsidy provided by the government.

In the model version described by Gerst et al. (2013b), which represented the US economy, it was assumed that the number of households remained constant over time—a major simplifying assumption. In the current study, we relax this assumption

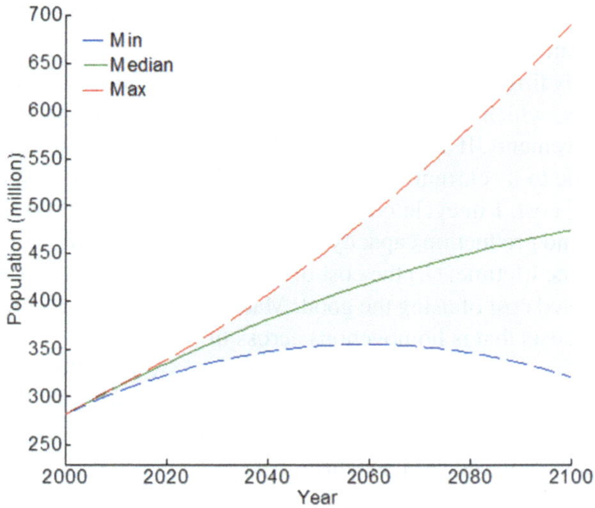

Fig. 2 Summary of probabilistic population projections used in the model. *Lines* indicate minimum (*blue*), median (*green*), and maximum (*red*) trajectories of 500 simulations

through a representation of the US population change that is fit to the probabilistic projections of Raftery et al. (2012). Specifically, population (P) is represented by a quadratic function

$$P[t] = a_1(t - 2000)^2 + a_2(t - 2000) + P_{2000}, \tag{1}$$

where population at $t = 2000$ (P_{2000}) is 282.5 million and the coefficients a_1 and a_2 are linked to the uncertain population in 2100 (P_{2100}) by

$$a_1 = 1.10 \cdot 10^{-7}\left(P_{2100}^2\right) - 1.21 \cdot 10^{-5}(P_{2100}) - 2.82 \cdot 10^{-2}, \tag{2}$$

$$a_2 = -1.10 \cdot 10^{-5}\left(P_{2100}^2\right) + 1.12 \cdot 10^{-2}\,(P_{2100}) - 3.88 \cdot 10^{-3}. \tag{3}$$

We represent P_{2100} by a log-normally distributed variable with arithmetic mean = 481.2 million and s.d. = 56.8 million (Fig. 2).

Capital Goods Sector

In our model, innovation activity is centered in the capital goods sector. Capital goods firms hire labor and use energy to produce machines, which are purchased by consumer goods firms for the purpose of producing consumer goods. Machines have five properties related to labor and energy use: (i) thneed production labor productivity (thneeds produced per worker), (ii) thneed production energy intensity (MWh per good), (iii) machine production labor productivity (machines produced per worker), (iv) machine production energy intensity (MWh per machine), and (v) thneed use energy intensity (kWh per good).

Capital goods firms reinvest a fraction of their past sales in innovation and imitation activities, which have uncertain outcomes with regard to labor and energy intensity improvements. If a firm successfully innovates or imitates, it then compares the new machine to its currently produced machine and chooses the one having the lowest lifecycle cost. Lifecycle cost is composed of the sum of three terms: (i) the machine price and production capacity annualized by the annual interest on debt and expected machine lifetime; (ii) the cost of using the machine to produce goods; and (iii) the discounted cost of using the good. Machines are priced according a mark-up over operating costs that is homogenous across firms.

Importantly, the market for machines is defined by imperfect information. We model this by limiting the number of consumer goods firms to which capital goods firms may advertise. If a capital goods firm cannot find customers, then we assume it is subsequently replaced by a new firm.

Consumer Goods Sector

Consumer goods firms use their stock of purchased machines to produce thneeds. Each firm plans its desired level of thneed production according to expected demand, desired inventory, and the actual inventory. To meet increasing demand or to replace end-of-life machines, firms use lifecycle cost to compare the desirability of advertised machines. Firms may also replace machines before their end-of-life, but must consider the sunk cost of replacing a machine with a remaining useful life.

Like capital goods firms, consumer goods firms set prices using a mark-up over operating costs. However, the mark-up varies from firm to firm and is dependent on the firm's market share. Market shares evolve as a function of firm competitiveness relative to average sector competitiveness weighted by market share, where individual firm competitiveness is a function of price and cost of use.

Machine purchases and thneed production may be funded internally or through borrowing. Firms, however, have a limit to their debt to sales ratio. A consumer goods firm with a near-zero market share and negative liquid assets or an unfilled demand ceases operations and is replaced by a new firm.

Energy Sector

In our model, energy is represented by a generic form and is produced by a single-energy production firm. The firm meets overall energy demand by maintaining a stock of three 'stylized' energy technologies: carbon-heavy, carbon-light, and carbon-free.

New additions are made to the stock to replace end-of-life technologies or to meet increasing demand. The choice of which technology to purchase is made by a levelized, cost-decision rule:

$$
\begin{aligned}
c_{E,dis}[k,t] = {} & \frac{p_T[k,t]}{8760 \cdot u_T[k]} + \frac{w[t]}{AET[k,t] \cdot 10^6} \cdot \sum_{t'=1}^{\eta_T[k]} (1 + r_E)^{-t'} \\
& + \frac{EFP[k,t]}{10^6} \cdot \sum_{t'=1}^{\eta_T[k]} c_F^*[t+t'] \cdot (1 + r_E)^{-t'} \\
& + \frac{\sigma[k]}{10^3} \cdot \sum_{t'=1}^{\eta_T[k]} tax_C[t] \cdot (1 + r_E)^{-t'}.
\end{aligned} \tag{4}
$$

The levelized cost comprises the sum of four terms: (i) the price of the energy technology (p_T) accounting for the capacity factor (u_T); (ii) discounted labor costs calculated from the prevailing annual wage (*w*, \$ per worker), labor productivity of energy production (*AET*, GWh per worker); (iii) discounted fuel costs calculated from the heat rate (*EFP*, BTU per kWh) and forecasted fuel cost (c^*_F; \$ per 10^6 BTU), and (iv) discounted carbon emissions costs calculated from an emission factor (σ, tonnes CO_2 per MWh) and carbon tax rate (tax_C, \$ per tonne CO_2). All discounting calculations are based on an annual discount rate (r_E), Energy technologies are manufactured by three separate firms. We assume that carbon-heavy is a mature technology, and thus its costs remain constant. Carbon-light and carbon-free technologies undergo uncertain learning-by-searching and learning-by-doing, which act to reduce technology capital costs. Learning-by-searching is a function of cumulative research and development effort and learning-by-doing is dependent on cumulative built capacity.

We repeat the simplifying assumption in Gerst et al. (2013b) that the carbon-heavy and carbon-light technologies use the same global fossil-fuel resource stock with a cost–supply curve based on the aggregation of coal, oil, and natural gas resources. Here, however, we adopt probabilistic cost–supply curves based on the method of Mercure and Salas (2012). In this setup, the supply of a particular energy resource available at a given cost is represented by the cumulative distribution function

$$
N(c) = A \cdot e^{\left(-\frac{B}{C-C_0}\right)}, \tag{5}
$$

where *A* represents the total energy supply potential for that resource, *B* represents the scaling of costs (e.g., due to inflation), and C_0 represents fuel extraction cost changes (e.g., due to learning-by-doing).

Parameters *B* and C_0 can be calculated using values for *A* and any two points on the cost–supply curve (C_1, Q_1) and (C_2, Q_2) from the following expressions:

$$
C_0 = \frac{C_2 ln \frac{Q_2}{A} - C_1 ln \frac{Q_1}{A}}{ln \frac{Q_2}{A} - ln \frac{Q_1}{A}}, \tag{6}
$$

Table 1 Fuel cost–supply curve parameters

Resource	Total technical potential, A (1000 EJ)	Cost at 1 % of technical potential, $C_{0.01}$ (USD/GJ)	Cost at 95 % of technical potential, $C_{0.95}$ (USD/GJ)
Crude oil	Tri(7, 11, 11)	1.7	Tri(5.5, 6.8, 8.2)
Oil shale	Tri(0, 27, 56)	6.8	Tri(6.8, 8.5, 10.2)
Oil sands	Tri(1, 29, 31)	8.5	Tri(13.7, 17.1, 20.5)
Conventional gas	Tri(7, 12, 16)	0.5	Tri(4.56, 5.7, 6.8)
Shale gas	Tri(0, 29, 47)	3.8	Tri(6.9, 8.6, 10.3)
Tight gas	Tri(0, 6, 12)	2.6	Tri(6.1, 7.6, 9.1)
Methane gas	Tri(0, 32, 32)	4.4	Tri(6.9, 8.6, 10.3)
Hard coal	Tri(24, 220, 419)	1.7	Tri(2.7, 3.3, 4)
Soft coal	Tri(5, 37, 75)	2.7	Tri(5.3, 6.7, 8)

$$B = -(C_1 - C_0)\ln\frac{Q_1}{A}. \tag{7}$$

We adopt the values for A and costs at the 1st and 95th percentiles of A ($C_{0.01}$, $Q_{0.01}$) and ($C_{0.95}$, $Q_{0.95}$) provided by Mercure and Salas (2012). To represent uncertainty, for each model simulation we draw a value for A from a triangle distribution with mode, lower value, and upper value equal to the most probable, lower bound, and upper bound values on technical potential given by Mercure and Salas (2012). We also draw a value for $C_{0.95}$ representing the uncertainty in cost reduction due to technological innovation in extraction from a triangular distribution with mode at the value given by Mercure and Salas, lower bound at 80 % of the modal value, and upper bound at 120 % of the modal value. These randomly selected values are then used together with the given values of $C_{0.01}$, $Q_{0.01}$, and $Q_{0.95}$ to calculate values for C_0 and B from Eqs. (6) and (7) and all parameters are held constant over time for each simulation.

Distributions for the nine primary fossil-based energy resources are summarized in Table 1. For any given simulation, the nine cost–supply curves are assumed to be independent and are therefore aggregated by summing across all resources for each cost value (Fig. 3).

Government Agent

In the original DFR model, the government has the ability to collect a tax on other agents and use the revenue for a variety of purposes (e.g., to subsidize R&D by firms). Gerst et al. (2013b) use this modeling capability to assess the impact of a carbon tax on energy technology, energy use, carbon emissions, and economic growth. They use an exogenously-specified, increasing carbon tax and compare the impacts of three different revenue recycling schemes: (a) returning revenues to households in the form of a tax rebate, (b) using revenues to subsidize innovation by capital goods firms, and (c) investing revenues in renewable technology R&D.

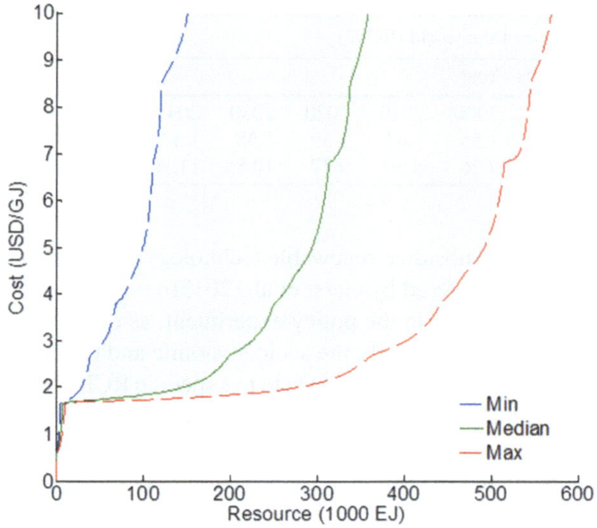

Fig. 3 Summary of probabilistic fossil-based energy cost–supply curves used in the model. *Lines* indicate minimum (*blue*), median (*green*), and maximum (*red*) cost curves based on 500 simulations

Gerst et al. (2013b) found that, on its own, the carbon tax does not provide enough of a price signal to markedly alter the energy technology mix in the model: the carbon-light energy technology achieves significant market share only about 5 years earlier in schemes (a) and (b) than in a no-tax reference specification. Only when the carbon tax revenue is used to subsidize renewable energy technology R&D (scheme c) does the energy system transition away from carbon-emitting technologies within the next century. As mentioned earlier, however, the model of Gerst et al. (2013b) assumes a stable population, fixed fuel cost curve, and exogenous carbon tax schedule. All of these limitations can be expected to have a significant effect on results, both in terms of most likely outcomes and estimates of uncertainty.

Endogenous Policy Experiment

To simulate endogenous policy formation, we assume that in the year 2000, nations agree to emissions pathways that will lead to a stabilization of climate forcing of 4.5 W/m^2 by 2100. The necessary annual emissions commitments are given in Table 2 and are consistent with RCP4.5, as calculated by GCAM (global change assessment model; Thomson et al. 2011).

We assume that to meet its commitments the US government adopts a carbon tax with initial value of US$ 25 per tonne CO_2, increasing at a nominal rate of 5 % per year. The effectiveness of the tax is monitored every 10 years by comparing actual cumulative carbon emissions against cumulative emissions commitments resulting from Table 2. If actual cumulative emissions are at or below the target, then the carbon tax growth rate remains the same. If cumulative emissions are above the target, then the annual carbon tax growth rate is adjusted upward by 0.5 %. All carbon tax revenue

Table 2 Annual emissions commitments in PgC (petagrams of carbon) per year for the USA and the rest of the world (ROW)

Regions	Year										
	2000	2010	2020	2030	2040	2050	2060	2070	2080	2090	2100
USA	1.55	1.62	1.59	1.58	1.47	1.26	1.06	0.81	0.50	0.47	0.44
ROW	7.26	8.50	9.77	10.88	11.30	11.03	9.40	7.12	4.18	4.19	4.20

is used to subsidize renewable technology R&D, consistent with the most effective policy considered by Gerst et al. (2013b).

Our interest in the policy experiment, as described, is to determine the extent to which we can identify the socioeconomic and technological factors (the columns of the van Vuuren matrix) that lead to a specific RCP (the rows of the matrix framework). We accomplish this by generating a large number of stochastic model simulations to which we apply the multidimensional scenario discovery method described by Gerst et al. (2013a).

Model Simulation

Our model was calibrated to the U.S. historical rates of growth for GDP (gross domestic product) per household (1.7 % per year) and residential energy use per household (0.7 % per year) by adjusting the distributions representing stochasticity of labor productivity and energy efficiency improvement by the capital goods sector. For the purposes of calibration, we simulated the period 1820–2000, assuming a historical energy price increase of 1.0 % per year and constant average economy-wide labor and energy unit costs. This assumption was necessary to ensure that modeled technological improvements kept pace with increases in wages and energy price. Other model parameters mostly adopted the values of Dosi et al. (2010), as reported by Gerst et al. (2013b).

For our policy simulation, starting conditions were specified by selecting the simulated year 2000 state from the final calibration that most closely matched the actual investment fraction of GDP and household fraction of total energy use observed in 2000. Wages, energy price, and other parameters were then scaled to match the observed year 2000 values. This procedure preserved the agent heterogeneity generated in the calibration exercise, while allowing initial conditions to accord with overall macro variables observed for the year 2000.

For computational tractability, our model of the US economy is scaled to be represented by 50 capital goods firms, 200 consumer goods firms, and 250,000 households in the year 2000. The number of households then scales proportionally with population change, as represented by Eqs. (1–3). In the current version of the model, the number of firms remains constant, although production and labor demands can change with population.

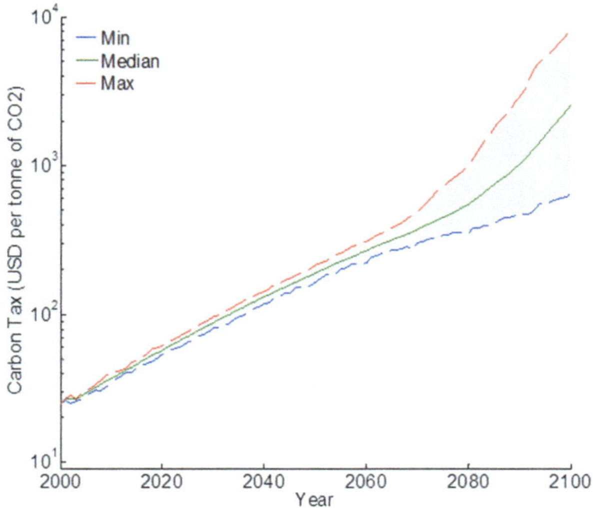

Fig. 4 Modeled inflation-adjusted carbon tax. *Lines* indicate minimum (*blue*), median (*green*), and maximum (*red*) values of 500 simulations

To represent the range of possible model outcomes, 500 simulations were used to generate all figures and statistics. These simulations represent stochastic realizations of the model's dynamics emerging from the same set of initial conditions and model parameter values. Stochasticity arises from the uncertain technological development process. Random draws are taken each year from the distributions characterizing innovation and imitation success of firms seeking to improve labor productivity and energy efficiency. Similarly, energy technology firms reduce the cost of manufacturing low-carbon and carbon-free energy technologies through a two-factor learning curve characterized by stochastic rates of learning-by-searching and learning-by-doing effects.

Results

The price signal introduced by a growing carbon tax (Fig. 4) potentially acts through two channels to influence technological change and carbon emissions: (i) the machine purchasing decisions of consumer good firms (and therefore the incentive structure of capital goods firms) and (ii) the capital budgeting decisions of energy producers. As already revealed by Gerst et al. (2013b), the carbon tax on its own does not lead to substantial improvements in energy efficiency of produced machines or consumer goods beyond what would otherwise be achieved under a no-carbon tax scenario. Thus, even with an inflation-adjusted tax level of over US$ 100 per tonne CO_2, model results indicate that the USA is unlikely to achieve the annual emissions commitments of Table 2 by mid-century (Fig. 5).

On the energy supply side, however, the effects of the carbon tax can be substantial—not necessarily because of the price signal, which is small compared

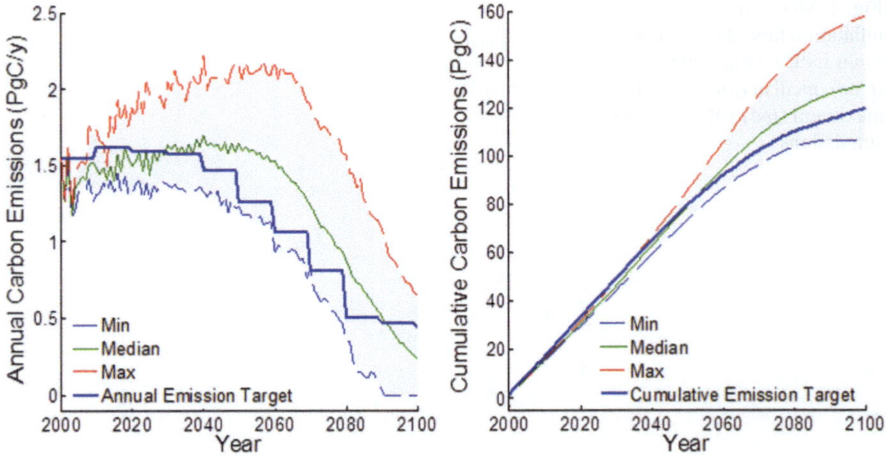

Fig. 5 Predicted annual and cumulative carbon emission. *Lines* indicate minimum (*blue*), median (*green*), and maximum (*red*) predicted trajectory from 500 simulations. Bold lines represent target emissions

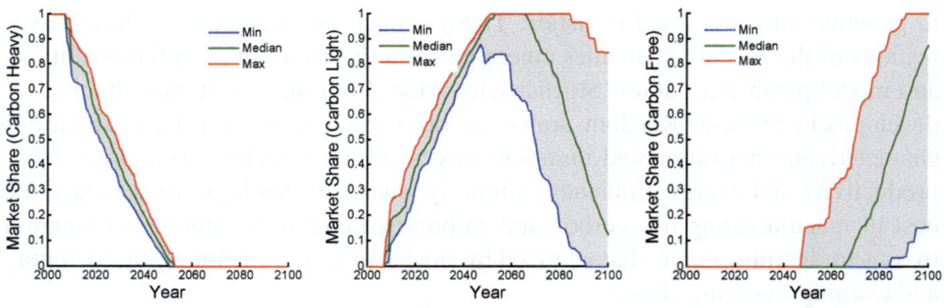

Fig. 6 Predicted market share for each energy technology. *Lines* indicate minimum (*blue*), median (*green*), and maximum (*red*) of 500 simulations

to the possible rise in future fuel costs (Fig. 3), but because of the dramatic influence of the subsidization of energy technology R&D that a carbon tax enables. By mid-century, carbon-free renewable energy begins to achieve significant market penetration in most simulations (Fig. 6). There is substantial uncertainty in the breakthrough year, due to the inherent stochasticity of technology improvement, giving rise to large uncertainty in predicted emissions in mid-century (see Fig. 5). This uncertainty is exacerbated by uncertainty in the fuel cost and population growth curves. However, once carbon-free sources take hold as a major contributor to the national energy mix, the economy becomes essentially uncoupled from fuel costs, resulting in the potential for dramatic economic growth by the end of the twenty-first century (Fig. 7a).

Due to technological improvement of energy technology, capital goods, and consumer goods, the economy-wide energy use per dollar GDP is predicted to decrease substantially over time (Fig. 7). Our model predicted decrease in energy intensity,

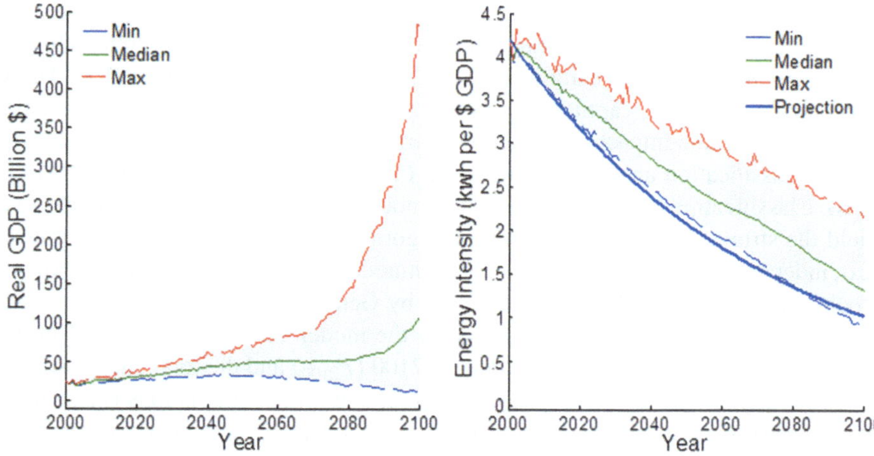

Fig. 7 Predicted trajectories for real GDP and energy intensity. *Thin lines* indicate minimum (*blue*), median (*green*), and maximum (*red*) trajectories of 500 simulations. *Bold line* indicates the projected change in energy intensity at the historical average rate of 1.39 % per year

however, is less than the average historical annual decline of 1.39 % from 1949–2009 (projected as the bold line in Fig. 7). The high historical decline in energy intensity is known to be due, at least in part, to broad structural changes that have occurred over the past 60 years, such as shifts from a manufacturing to a service-oriented economy, and changes in the international trade balance (Sue Wing 2008). These trends may or may not continue over the next century, but in any case, they are not currently represented in the model.

Scenario Discovery

Description of Method

We employ the method for multidimensional scenario discovery described by Gerst et al. (2013a). Each simulation is first represented by the values of two or more selected outcome variables. The full set of simulations is then subject to a hierarchical clustering algorithm to identify statistically similar groups according to these selected outcomes. Finally, these clusters, or 'candidate scenarios,' are subject to a classification analysis to identify the stochastic model inputs that serve as key scenario drivers. The results of this classification are then taken to represent the final 'discovered' scenarios. The notion of multidimensional similarity is what distinguishes our cluster-based technique from threshold-based methods (Bryant and Lempert 2010) or full-factorial "quadrant-based" scenario definitions.

We implemented our hierarchical cluster analysis using available functions in MATLAB. Distances between points were calculated using Euclidean distance and clustering-employed Ward's method. For our application, we chose to cluster according to two dimensions: average GDP per capita growth rate (excluding climate damages) and cumulative carbon emissions, both for the period 2000–2100. These two outcome variables capture the key tradeoff of the climate policy: weighing the potential economic impacts of abatement versus the potential for climate impacts.

For classification analysis, we used the *ClassificationTree.fit* function of MATLAB. Classification trees represent dichotomous splits of independent variables that yield the strongest associations with a categorical dependent variable. In our context, independent predictors consisted of the nine constructed variables characterizing stochastic technological development used by Gerst et al. (2013b), as well as two additional probabilistic parameters used in the model extensions described in the present contribution: the US population in 2100 (P_{2100}) and the total energy supply potential across all fuel types (A_{tot}). The groups of model simulations (i.e., candidate scenarios) identified by the cluster analysis served as the dependent variable. The independent variables that best predict candidate scenario membership are then interpreted as the key driving forces, and the combination of conditions on these variables is then taken to define the final 'discovered' scenarios. To maintain an easily interpretable tree, we set the minimum number of simulations for splitting each node to 160 and the minimum number of simulations for each final branch to 45.

Scenario Results

As already shown in Figs. 5 and 7, there is large variation in carbon emissions and GDP growth under our simulated policy setting. This makes the results especially conducive to scenario discovery. Hierarchical cluster results (not shown) indicate that the model simulations, as represented by the two selected outcome variables, naturally divide into four clusters. A bivariate scatterplot of the cumulative carbon emissions and average GDP per capita growth rate for the four clusters (Fig. 8) indicates that this number represents a range of reasonably distinct groupings. The fact that these groupings do not conform neatly to quadrants of the two-dimensional space suggests that the use of empirical cluster analysis holds some value over threshold-based methods.

As the next step to scenario discovery, the classification tree (Fig. 9) indicates that the four clusters defined in the two-dimensional space of carbon emissions and GDP growth can also be reasonably distinguished by four partitions over three stochastic model variables. The three variables selected empirically as strong predictors are: (i) the population size in 2100 (P_{2100}), (ii) the relative efficacy of R&D with respect to labor productivity to produce consumer goods (EFF_A), and (iii) the relative efficacy of carbon free energy technology experience ($EXPER_{cf}$). The other eight variables in the candidate set of predictors appear not to be strong drivers of policy performance.

We take the partitioning defined by the classification tree in Fig. 9 to be our final set of four 'discovered' scenarios. Although defined with respect to only three variables, these scenarios represent a complete partitioning of the 500 model simulations in the

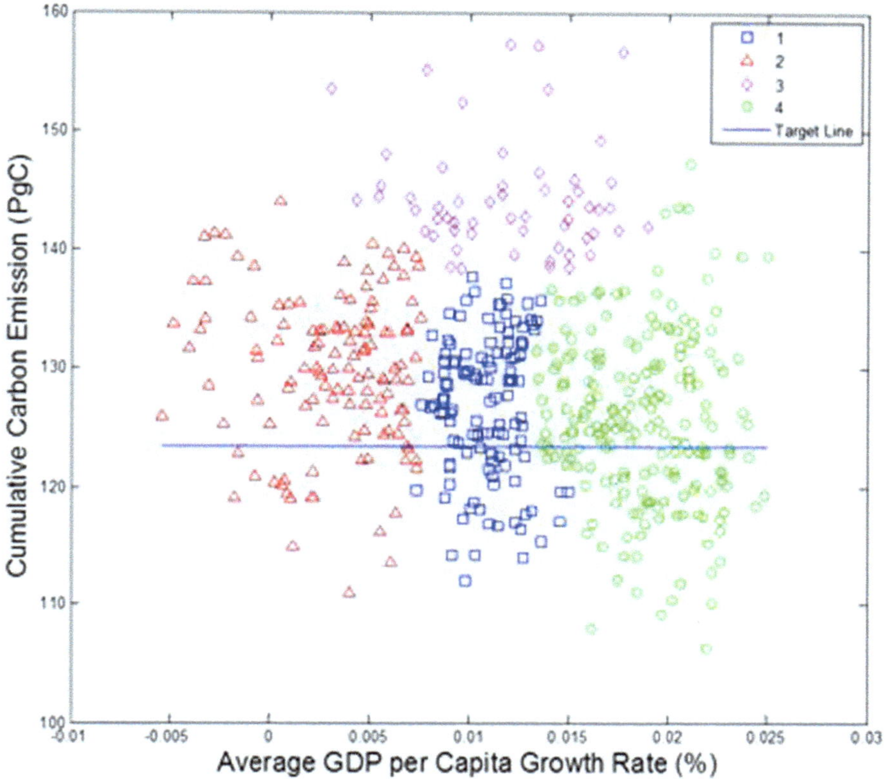

Fig. 8 Scatterplots of cumulative carbon emissions and average GDP per capita growth rate. *Points* represent the 500 stochastic simulation results. *Symbols* represent groupings identified by the cluster analysis and serve as candidate scenarios. The *horizontal line* indicates the cumulative emissions target for 2100

multidimensional space of all stochastic model variables and outcomes. The defining characteristics of these scenarios can be best viewed as a set of boxplots comparing the range of conditions experienced under each scenario (Fig. 10).

Scenario 1 is characterized by low levels of carbon emissions and moderate GDP per capita growth, associated with low to moderate levels of population growth and labor productivity improvement, but high efficiency in converting experience with carbon-free technology into emissions reductions (i.e., learning-by-doing). Scenario 2 on the other hand, has moderate emissions and very low GDP growth, associated with poor efficiency of learning-by-doing. Scenario 3 has the highest emissions levels and moderate-to-high levels of GDP per capita growth, associated primarily with very high population growth (greater than about 546 million by 2100). Finally, scenario 4 might be considered the most successful overall for achieving the lowest emissions and highest GDP growth. These results from low population growth are combined with high efficiency in converting R&D funding into improvements in labor productivity.

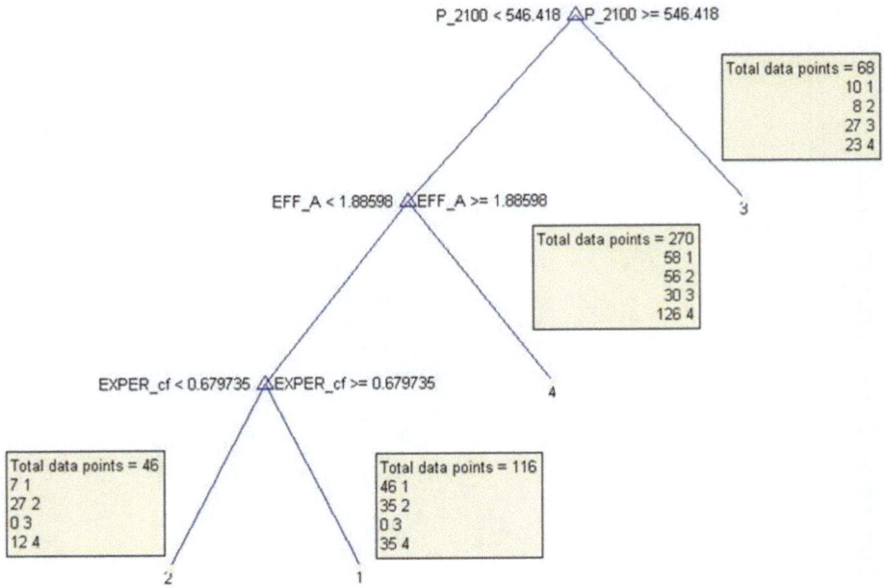

Fig. 9 Classification tree indicating the optimal partitioning of stochastic model variables for predicting candidate scenarios resulting from the cluster analysis. At each split, observations less than the indicated value proceed to the *left branch* and observations greater proceed to the *right*. Each split is conditional on the result of the splits above it in the tree. The bottom branches are labeled with the predicted cluster membership. *Boxes* indicate the total number of simulations that meet all the specified conditions leading to the corresponding branch, as well as the actual categorical membership frequencies among these simulations

Discussion

We demonstrate how the process of scenario discovery as applied to results of ENGAGE, a stochastic, dynamic agent-based model, might be used to generate socioeconomic scenarios relevant to a given emissions target, or RCP. For a carbon tax policy designed to meet the 4.5 W/m^2 RCP, population growth, improvement in labor productivity, and efficiency of learning-by-doing regarding carbon-free energy technology are revealed to be the key factors driving policy success. In particular, a low population growth and a high ability to convert experience in carbon-free energy technology into further cost reductions seem to be jointly, a key to meeting emissions targets with minimal negative economic impact. This implies that these features should form the key elements of the storylines underlying socioeconomic scenarios associated with the 4.5 W/m^2 RCP if they are to provide a meaningful exploration of policy efficacy. Such scenarios, which pair varying levels of population and economic growth with differing degrees of innovation in the energy sector, are consistent with those generated using more conceptual methods in the climate scenario literature (Moss et al. 2010; Parson et al. 2007; van Vuuren et al. 2012). However, by being derived from the results of a quantitative model, our specification is intrinsically consistent with practicable modeling assumptions and parameterizations.

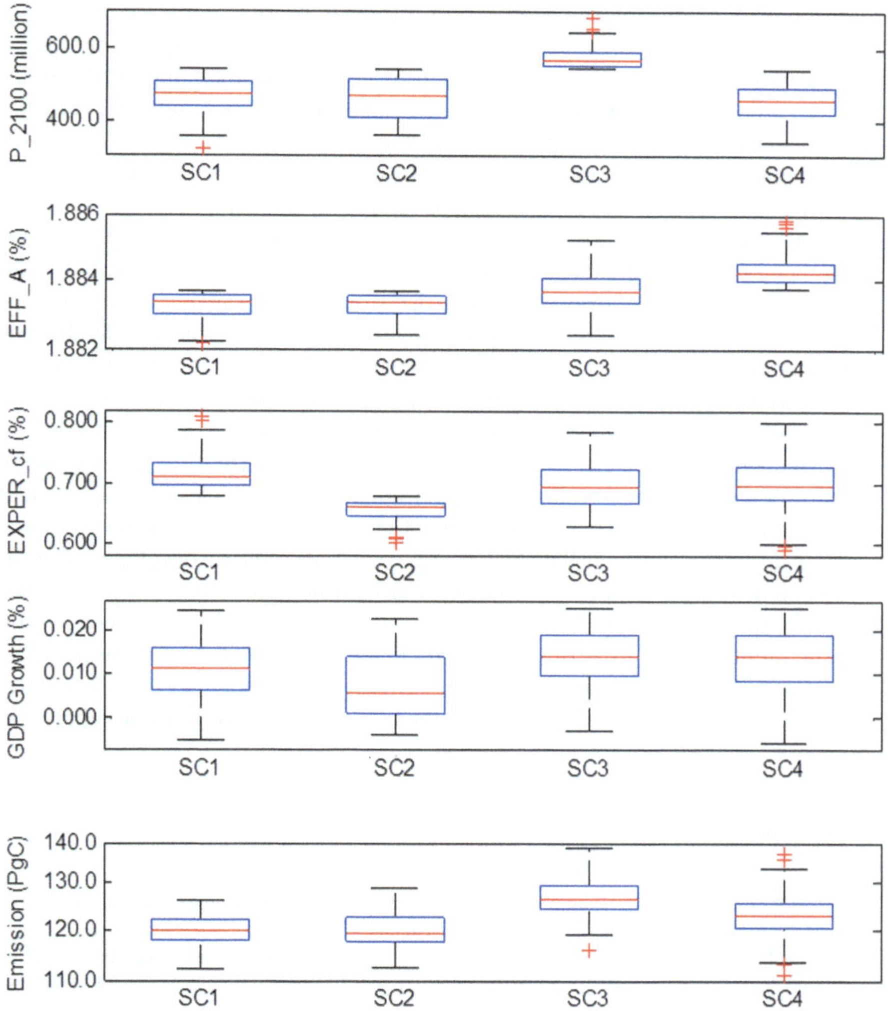

Fig. 10 Boxplots summarizing conditions associated with the four final discovered scenarios. *Boxes* indicate the middle 50 % of the simulation values (interquartile range, IQR) for each scenario, *central lines* indicate median values, *vertical whiskers* extend out to the furthest simulation value within 1.5*IQR of the boxes, and *crosses* indicate further outlying values. Scenario numbering corresponds to Figs. 8 and 9 and variables are defined as described in text. All variables are reported on an annual basis, except for *P*2100 which is the population in the year 2100 and emissions which is cumulative from 2000 to 2100

While in the current contribution, we have overcome some of the key limitations of earlier versions of ENGAGE by allowing for a growing population and uncertain fuel price, there are still a number of simplifying assumptions that we believe are too great to allow direct application of our current results to real-world policy questions. For example, the current simplicity of the energy sector may overlook opportunities for technology innovation and adoption. In particular, we only represent one energy production firm, and it is assumed to utilize the full lifetime of its energy technologies. Thus, it will not prematurely scrap any of its existing stock when improved carbon-light or carbon-free technology becomes available. Also, cost is currently the only factor in the model determining new technology adoption, precluding early adoption to meet moral obligation or public relations objectives. These factors add a significant lag to the achievement of carbon emissions reductions in the model.

Finally, the decision rules of households and firms in our current model are currently homogenous and simplified. For example, firms cannot focus R&D effort toward specific machine attributes or make decisions to hedge against anticipated energy price increases. Similarly, households have homogenous preferences for thneeds that do not represent the true diversity of personal values and beliefs. We are currently working to alleviate these limitations by defining a suite of decision rules that households use to select goods that meet both their individual and social needs.

We recognize that further progress is necessary for ENGAGE to provide useful support for climate policy evaluation and formulation. Nevertheless, we believe that our proposed combination of stochastic, agent-based modeling and multidimensional scenario discovery can contribute to the ongoing climate scenario development effort by complementing traditional approaches. Furthermore, multidimensional scenario discovery may be used with any model that has the capability to generate probabilistic output. Other areas of energy and climate policy that exhibit considerable uncertainty and disagreement over metrics such as impacts, adaptation, vulnerability assessments, and regional infrastructure planning, could benefit from this approach.

References

Bryant, B.P., Lempert, R.J.: Thinking inside the box: a participatory, computer-assisted approach to scenario discovery. Technol. Forecast. Soc. Chang. **77**(1), 34–49 (2010)

Clarke, L., Edmonds, J., Krey, V., Richels, R., Rose, S., Tavoni, M.: International climate policy architectures: overview of the EMF 22 International Scenarios. Energ. Econ. **31**, S64–S81 (2009)

Dosi, G., Fagiolo, G., Roventini, A.: Schumpeter meeting Keynes: a policy-friendly model of endogenous growth and business cycles. J. Econ. Dyn. Control. **34**(9), 1748–1767 (2010)

Fisher, B., Nakicenovic, N., Alfsen, K., Corfee Morlot, J., de la Chesnaye, F., Hourcade, J.-C., Jiang, K., Kainuma, M., La Rovere, E., Matysek, A., Rana, A., Riahi, K., Richels, R., Rose, S., van Vuuren, D.P., Warren, R.: Issues related to mitigation in the long-term context. In: Metz, B., Davidson, O., Bosch, P., Dave, R., Meyer, L. (eds.) Climate Change 2007—Mitigation. Cambridge University Press, Cambridge (2007)

Gerst, M.D., Wang, P., Borsuk, M.E.: Discovering plausible energy and economic futures under global change using multidimensional scenario discovery. Environ. Modell. Softw. **44**, 76–86 (2013a)

Gerst, M.D., Wang, P., Roventini, A., Fagiolo, G., Dosi, G., Howarth, R.B., Borsuk, M.E.: Agent-based modeling of climate policy: an introduction to the ENGAGE multi-level model framework. Environ. Modell. Softw. **44**, 62–75 (2013b)

Groves, D.G., Lempert, R.J.: A new analytic method for finding policy-relevant scenarios. Global. Environ. Chang. **17**(1), 73–85 (2007)

Kriegler, E., O'Neill, B.C., Hallegatte, S., Kram, T., Lempert, R.J., Moss, R.H., Wilbanks, T.: The need for and use of socio-economic scenarios for climate change analysis: a new approach based on shared socio-economic pathways. Global. Environ. Chang. **22**(4), 807–822 (2012)

McCarthy, J., Canziani, O., Leary, N., Dokken, D., White, K. (eds.): Climate Change 2001: Impacts, Adaptations, and Vulnerability. Cambridge University Press, Cambridge (2001)

Mercure, J.-F., Salas, P.: An assessment of global energy resource economic potentials. Energy. **46**(1), 322–336 (2012)

Moss, R.H., Edmonds, J.A., Hibbard, K.A., Manning, M.R., Rose, S.K., van Vuuren, D.P., Carter, T.R., Emori, S., Kainuma, M., Kram, T.: The next generation of scenarios for climate change research and assessment. Nature. **463**(7282), 747–756 (2010)

Nakicenovic, N., Kolp, P., Riahi, K., Kainuma, M., Hanaoka, T.: Assessment of emissions scenarios revisited. Environ. Econ. Policy. Stud. **7**(3), 137–173 (2006)

Nakicenovic, N., Swart, R. (eds.): Special Report on Emissions Scenarios. Cambridge University Press, Cambridge (2000)

Parson, E., Burkett, V., Fisher-Vanden, K., Keith, D., Mearns, L., Pitcher, H., Rosenzweig, C., Webster, M.: Global Change Scenarios: Their Development and Use. U.S. Climate Change Science Program and the Subcommittee on Global Change Research. Department of Energy, Office of Biological & Environmental Research, Washington, D.C. (2007)

Raftery, A.E., Li, N., Sevcikova, H., Gerland, P., Heilig, G.K.: Bayesian probabilistic population projections for all countries. P. Natl. Acad. Sci. **109**(35), 13915–13921 (2012)

Rozenberg, J., Hallegatte, S., Vogt-Schilb, A., Sassi, O., Guivarch, C., Waisman, H., Hourcade, J.-C.: Climate policies as a hedge against the uncertainty on future oil supply. Clim. chang. **101**(3), 663–668 (2010)

Rozenberg, J., Guivarch, C., Lempert, R.J., Hallegatte, S.: Building SSPs for climate policy analysis: a scenario elicitation methodology to map the space of possible future challenges to mitigation and adaptation. FEEM Working Paper No 522012 (2012)

Schwartz, P.: The Art of the Long View: Planning for the Future in an Uncertain World. Currency Doubleday, New York (1991)

Sue Wing, I.: Explaining the declining energy intensity of the US economy. Resour. Energ. Econ. **30**(1), 21–49 (2008)

Thomson, A.M., Calvin, K.V., Smith, S.J., Kylem G.P., Volke, A., Patel, P., Delgado-Arias, S., Bond-Lamberty, B., Wise, M.A., Clarke, L.E.: RCP4. 5: a pathway for stabilization of radiative forcing by 2100. Clim. chang. **109**(1), 77–94 (2011)

van Vuuren, D.P., Riahi, K., Moss, R.H., Edmonds, J.A., Thomson, A., Nakicenovic, N., Kram, T., Berkhout, R., Swart, R., Jenetos, A., Rose, S.K., Arnell, N.: A proposal for a new scenario framework to support research and assessment in different climate research communities. Global. Environ. Chang. **22**, 21–35 (2012)

Index

H. Qudrat-Ullah (ed.), *Energy Policy Modeling in the 21st Century*, Understanding Complex Systems, DOI 10.1007/978-1-4614-8606-0,

MIX
Papier aus verantwortungsvollen Quellen
Paper from responsible sources
FSC® C105338

If you have any concerns about our products,
you can contact us on
ProductSafety@springernature.com

In case Publisher is established outside the EU,
the EU authorized representative is:
Springer Nature Customer Service Center GmbH
Europaplatz 3, 69115 Heidelberg, Germany

Printed by Libri Plureos GmbH
in Hamburg, Germany